Natural Stone and Architectural Heritage

Natural Stone and Architectural Heritage

Special Issue Editors

Lola Pereira
Lidia Catarino
Giovanna Antonella Dino

MDPI • Basel • Beijing • Wuhan • Barcelona • Belgrade

Special Issue Editors

Lola Pereira
Department of Geology,
University of Salamanca
Spain

Lidia Catarino
Department of Earth Sciences,
University of Coimbra
Portugal

Giovanna Antonella Dino
Department of Earth Sciences,
University of Torino
Italy

Editorial Office
MDPI
St. Alban-Anlage 66
4052 Basel, Switzerland

This is a reprint of articles from the Special Issue published online in the open access journal *Sustainability* (ISSN 2071-1050) in 2019 (available at: https://www.mdpi.com/journal/sustainability/special_issues/nsah)

For citation purposes, cite each article independently as indicated on the article page online and as indicated below:

LastName, A.A.; LastName, B.B.; LastName, C.C. Article Title. *Journal Name* **Year**, *Article Number, Page Range*.

ISBN 978-3-03921-550-8 (Pbk)
ISBN 978-3-03921-551-5 (PDF)

Contents

About the Special Issue Editors

Lola Pereira is professor of Geology and Engineering Geology at the University of Salamanca. She is Secretary General of the IUGS Heritage Stones Subcommission and the Chair of the IUGS Publications Committee. She is also the main leader of the UNESCO IGCP-637, dealing with natural stones and heritage built on stone. Her research interests include Petrology and Geochemistry of Igneous and Metamorphic Rocks, as well as natural stones and their use in buildings and restoration of architectural heritage. She is author of more than seventy scientific papers and leader of many national and international research projects.

Lidia Catarino teaches Geological Engineering at the Earth Sciences Department, Coimbra University (Portugal). She has more than twenty five years of experience working with recycling stone wastes and reuse of geological materials. Since 2002 it is involved in interdisciplinary studies related to archaeological materials, but also the use and degradation of stone in heritage buildings, including in university buildings that belong to the Unesco Heritage List. From 2008 to 2012 was responsible for the master of Conservation and Restoration in the same university. These works allow to having interaction and publishing with other areas of knowledge as biology, anthropology, architect and civil engineering.

Giovanna Antonella Dino, graduated in Mining Engineering (Politecnico di Torino) and qualified as mining engineer since 2001. PhD course in Environmental Geoengineering (DIGET—Politecnico di Torino) ended in 2004: the research was about environmental problems connected to extractive industry and to potential recovery of extractive waste. At present, she is working at University of Torino (Earth Sciences Department) as research assistant, interested in quarrying and mining activities (exploitation, application of stones in historical contexts and cultural heritage, working and dressing activities, extractive waste management and potential recovery, environmental impacts associated to mining activities). Responsible of Mineral Dressing and Sampling Laboratory at Earth Sciences Department. Involved from 2005 in several studies concerning exploitation of natural and dimension stones and in particular: Bargiolina (quartzite), Pietra di Luserna (gneiss), Graniti dei Laghi (granites from Verbano Cusio Ossola VCO quarry basin), serizzi and beole (gneiss from VCO), Crevola dolomia and Ornavasso and Candoglia marbles (VCO). Furthermore, from 2001 she has been involved in projects which deals with the impacts, the management and the potential reuse of mining and quarrying waste. From 2001 she published more than 100 scientific papers and abstracts, presented during National and International Conferences or published in scientific journals. Involved as guest editor in scientific journals such as: Resources Policy, Science of the Total Environment, Sustainability and co-chair of scientific sessions about heritage stone, circular economy and landfill mining, resource efficiency and sustainable mining.

Preface to "Natural Stone and Architectural Heritage"

The working group on heritage stones has been growing in the last ten years, and it has become obvious that a Special Issue presenting new stones that have been used in the construction of architectural heritage is demanded. Heritage stones are those stones that have special significance in human culture. Examples include some very important stones that have been either neglected because they are no longer extracted, or stones that have great significance in commercial terms but knowledge of their national and/or international heritage has not been well documented. With this book, we have gone one step forward in the recognition of natural stone as the most important material in architecture.

Lola Pereira, Lidia Catarino, Giovanna Antonella Dino
Special Issue Editors

Article

Seismic Performance of Ancient Masonry Structures in Korea Rediscovered in 2016 M 5.8 Gyeongju Earthquake

Heon-Joon Park [1,*] , Jeong-Gon Ha [2], Se-Hyun Kim [3] and Sang-Sun Jo [3,*]

[1] Ulsan National Institute of Science and Technology (UNIST), Ulsan 44919, Korea
[2] Korea Atomic Energy Research Institute (KAERI), Daejeon 34057, Korea; jgha@kaeri.re.kr
[3] National Research Institute of Cultural Heritage (NRICH), Daejeon 34122, Korea; ksehyun@korea.kr
*   Correspondence: heonjoon@unist.ac.kr (H.-J.P.); ssjo@korea.kr (S.-S.J.); Tel.: +82-52-217-2850 (H.-J.P.); +82-42-860-9216 (S.-S.J.)

Received: 30 January 2019; Accepted: 6 March 2019; Published: 14 March 2019

Abstract: The Gyeongju Historic Areas, which include the millennium-old capital of the Silla Kingdom, are located in the region most frequently affected by seismic events in the Korean peninsula. Despite the numerous earthquakes documented, most of the stone architectural heritage has retained their original forms. This study systematically reviews and categorises studies dealing with the seismic risk assessment of the architectural heritage of the historic areas. It applies research methodologies, such as the evaluation of the engineering characteristics of subsoil in architectural heritage sites, site-specific analysis of the ground motions in response to earthquake scenarios, geographic information system (GIS)-based seismic microzonation according to the geotechnical engineering parameters, reliability assessment of dynamic centrifuge model testing for stone masonry structures and evaluation of seismic behaviour of architectural heritage. The M 5.8 earthquake that hit Gyeongju on 12 September 2016 is analysed from an engineering point of view and the resulting damage to the stone architectural heritage is reported. The study focuses on Cheomseongdae, an astronomical observatory in Gyeongju, whose structural engineering received considerable attention since its seismic resistance was reported after the last earthquake. Dynamic centrifuge model tests applying the Gyeongju Earthquake motions are performed to prove that it is not a coincidence that Cheomseongdae, a masonry structure composed of nearly 400 stone members, survived numerous seismic events for over 1300 years. The structural characteristics of Cheomseongdae, such as the well-compacted filler materials in its lower part, rough inside wall in contrast to the smooth exterior, intersecting stone beams and interlocking headstones are proven to contribute to its overall seismic performance, demonstrating outstanding seismic design technology.

Keywords: Gyeongju Historic Areas; 2016 Gyeongju Earthquake; site characterisation; site-specific ground response analysis; stone architectural heritage; dynamic centrifuge test; Cheomseongdae; ancient seismic design technique

---

1. Introduction

Architectural heritage is a precious cultural asset that provides insight into a nation's historical background. As such, the current generation has the duty to preserve them well for future generations. Since architectural heritage is exposed to the external environment, measures should be taken to protect them from both natural and manmade disasters. Systematic approaches based on scientific and engineering applications should be adopted to protect them from natural disasters such as earthquakes. Unlike seismic performance and design assessments of general buildings, the main focus of seismic risk assessments for architectural heritage should be on the maintenance of their unique values, authenticity and engineering performance.

There are many interesting ways for geotechnical engineers to contribute to conservation issues for historic sites and monuments [1]. Geotechnical engineering also has a very important role to play because there are quite a number of historic sites and monuments affected by geotechnical risks of different types [2]. As all architectural heritages are standing on ground, the seismic risk assessments for architectural heritage can be initiated from geological and geotechnical approaches to understand the composition and the material property distribution of the beneath and surrounding ground, local site effect, and seismic zonation [3–6]. For the structural and architectural engineering perspective, the seismic behaviour of historic monument and architectural heritage composed of stone block or masonry structure have been evaluated analytically, numerically [7–11] and experimentally [12–15].

In Korea, the National Research Institute of Cultural Heritage [16] conducted a multidisciplinary research planning project on disaster prevention for architectural heritages. Detailed mid- and long-term research tasks were derived from this project, and various studies were conducted to develop an overarching cultural heritage disaster-prevention system. Among them, studies on seismic risk assessments were conducted on architectural heritage in historic cities from 2009 to 2012. In addition, a seismic risk assessment methodology for historic areas was proposed based on aspects of geotechnical earthquake engineering.

A seismic risk assessment for architectural heritage at a historical site begins with an understanding of the engineering characteristics of the subsoil and architectural structure. First, an investigation of the subsoil and seismic response analysis are performed at an architectural heritage site to predict the level of ground shaking for each earthquake scenario. Furthermore, an information system based on a geographic information system (GIS) for determining the geotechnical earthquake engineering parameters can be established to estimate the damage potential for each historic area. Architectural heritage has different structural forms, systems and load transfer characteristics. Hence, the seismic behaviour of each architectural heritage must be analysed using an empirical or analytical approach through experimental or numerical modelling. To improve the efficiency of such research, major architectural heritage, each representative of their specific epochs and styles, should be selected and their seismic behaviour characteristics should be evaluated.

In this research project, seismic risk assessments were carried out for the subsoils of historical sites in Gyeongju, Buyeo and Seoul in 2009, 2010 and 2011, respectively. A geotechnical earthquake engineering database of over 70 architectural heritage sites in these three historical cities was established by means of in situ tests [17]. This database served as the basis for setting up a GIS-based geotechnical earthquake engineering information system for Gyeongju and Buyeo [18]. Stone architectural heritage exhibits dynamic seismic behaviour due to the friction between the stone members. This behaviour can be evaluated through centrifuge testing using scaled-down models [19]. The three-story stone pagoda (Seokgatap (Shakamuni Stupa)) of Bulguksa Temple and the five-story stone pagoda of Jungnimsa Temple, which are representative of the architectural style of the time, were evaluated along with their seismic behaviours using multiple dynamic centrifuge tests with a shaking table [20]. The dynamic centrifuge tests were conducted for scaled models of Cheomseongdae, an astronomical observatory in Gyeongju that has unique structural features [21]. The results of this study were used in 2012 as basic data to establish an architectural heritage seismic performance and risk assessment method and to propose standards for seismic risk assessments.

The M 5.8 earthquake that hit Gyeongju on 12 September 2016 is analysed in this study from an engineering point of view and the related damages to the stone architectural heritage are reported with a focus on Cheomseongdae. Dynamic centrifuge model tests that apply the Gyeongju Earthquake motions are performed to prove that it is not a coincidence that Cheomseongdae, a masonry structure composed of nearly 400 stone members, survived numerous seismic events for over 1300 years. Based on the assessment of the damage to the stone architectural heritage and the results of earthquake monitoring and dynamic centrifuge model tests for Cheomseongdae, the efforts before and after the Gyeongju earthquake and its reliability are verified from an engineering viewpoint.

2. Gyeongju Historic Areas and Cheomseongdae

The Gyeongju Historic Areas, the millennium-old capital of the Silla Kingdom (57 B.C.–935 A.D.) and UNESCO World Heritage Sites, are the representative heritage areas, as they are the cradle and pivot of the development of Buddhism in the Korean peninsula. The vibrant Buddhist culture in the area resulted in the creation of numerous stone heritages such as pagodas, lanterns, bridges and revetments. Many of the existing stone heritage have structures that are seemingly susceptible to collapse under lateral loads such as those produced during earthquakes. However, despite the numerous earthquakes documented, most of the aforementioned architectural heritage remained intact for over a millennium.

The Gyeongju Historic Areas are adjacent to the Yangsan Fault, and many seismic events in this area have been documented. In Samguksagi (a historical record of the Three Kingdoms), there is a record of an earthquake in 779 that killed more than 100 people and is the largest documented estimation of casualties due to an earthquake. The scroll, whose deciphered content began to be released in 2007, was discovered in 1966 while dismantling the three-story stone pagoda in the Bulguksa Temple. It contains records of great damage to the Bulguksa Temple facilities and Seokgatap pagoda due to earthquakes in 1013 and 1036. The scroll also describes the entire process from dismantling to rebuilding of the Seokgatap pagoda.

Park et al. [18] summarised and systemised the pre-existing site investigation databases of 50 sites, the results of in situ experiments conducted at 32 architectural heritage sites in the Gyeongju area and the results of the site-specific ground response analyses for scenario earthquakes. Using these data and site-specific response analysis results, they generated a GIS-based visualisation of the bedrock depths, natural ground periods, peak ground accelerations and ground amplification ratios for an earthquake scenario. Gyeongju is located in a basin of plains and low hilly areas with settlements and paddies/fields surrounded by mountains, through which the Hyeongsan River flows. The river is fed by various tributaries, such as Bukcheon (north stream) and Namcheon (south stream). The ground here is predominantly composed of a gravelly alluvial soil layer below a ~1–4 m-deep gravelly fill layer in which weathered soils, residual soils, weathered rocks and soft rock layers are locally distributed. The depth of the soft rocks varies greatly according to the drilling location and are distributed 8 to 50 m below the ground surface.

Located between the Bukcheon and Namcheon streams in the centre of the Gyeongju Historic Areas, Cheomseongdae (National Treasure 31) stands on a typical sedimentary basin terrain covered with a thick flood-induced alluvial deposit layer. Core samples taken from the subsoils near Cheomseongdae revealed a vertical soil profile (1–16 m) of sandy gravel and boulders. The depth-dependent stiffness distribution was derived from in situ down-hole and surface wave tests performed; these results were used to determine the seismic motions considering the site amplification factors. In other words, the surface motion at Cheomseongdae ground in each earthquake scenario was analysed. As a result, the ground around Cheomseongdae was found to incur a peak ground acceleration amplification of approximately 1.85 times.

The Cheomseongdae observatory was constructed during the reign of Queen Seondeok (632–646). Its original form survived over 1300 years of exposure to the natural and manmade disasters such as earthquakes. This stone masonry structure is comprised of nearly 400 stone members. Its materials and structure are maintained in their original state, and it is a monument of high artistic value with a mixture of harmoniously contrasting curves, straight lines, circles and squares.

Cheomseongdae's smoothly curved cylindrical main body consists of 27 layers of stone members piled up on a square stylobate and pedestal as shown in Figure 1 [21]. Two layers of interlocking headstones are placed on top of the main body. There is a small three-layer-high (13th to 15th layer) entrance or window facing south. The inside of the wall from the first to the 12th layer is densely packed with irregularly shaped and sized rubble and soil, whereas the inside is empty from the 13th layer onwards (Figure 1a). Unlike the smooth exterior of the main body, the interior is composed of roughly touching stone members up to the 18th layer (Figure 1b). The 19th layer comprises two stone beams with rectangular cross-sections that are placed in the east-west direction, piercing the cylindrical plane with their ends protruding. The 20th layer comprises two identically shaped stone beams placed in the north-south direction across the stone beams below (19th layer) without piercing the wall (Figure 1c). This structure is considered to enhance the frictional and adhesive forces between the members. The 21st—24th layers comprise stone members that are piled in the same manner as in the 12th—18th layers. The 25th and 26th layers show the same pattern as in the 19th and 20th layers, but in the opposite direction, i.e. with the two stone beams piercing the cylindrical plane in the north-south direction in the 25th layer and another two beams placed across them in the 26th layer. The 27th layer has a stone slab slanted to the east. The top part of Cheomseongdae is square. The 28th layer has two stone beams with quasi-square cross-sections that are placed in the east-west direction and another two identically shaped beams that are placed in the north-south direction inserted into the grooves made on the members underneath them. The same interlocking pattern is applied to the 29th layer with the reverse directions, i.e. the east-west members are inserted into the north-south members using a half-lap joint (Figure 1d).

Figure 1. The photos of exterior and interior of Cheomseongdae: (**a**) Filler beneath the 13th layer; (**b**) Inner irregular-shaped stones; (**c**) Long-shaped horizontal tie stones of 19th, 20th, 25th and 26th layers; and (**d**) Grid of interlocking stones and half-lap joint [21].

In a previous study [21], dynamic centrifuge tests were performed on two Cheomseongdae models to compare their seismic responses. One model was an exact duplicate of the actual structure and the other was a duplicate without the interlocking headstones and intersecting stone beams. As a result, the roles of the interlocking headstones and stone beams in enhancing Cheomseongdae's seismic performance were verified. Additionally, the rough interior members in contrast to the smooth exterior, the densely packed filler materials in the space below the window, the shapes of the stone members from the 19th layer onward and the shape of the joints lead to the belief that the Cheomseongdae structure is resistant to dynamic lateral loading such as earthquakes. In the dynamic centrifuge model tests on the Cheomseongdae models, the headstones underwent lateral displacements of up to 1.46 mm from a 1 s earthquake excitation in a 15 g centrifugal gravitational field under a seismic wave similar to that in the Ofunato earthquake with a return period of 1000 years (0.333 g of surface peak ground acceleration). By applying this result to the ground response analysis, a lateral displacement of 21.9 mm was estimated to occur under a 15 s earthquake excitation based on scaling factors from the dynamic centrifuge test.

3. Gyeongju Earthquake on 12 September 2016 and Damage to Architectural Heritage

There are 75 records of seismic events from 64 to 932 when Gyeongju was the capital of the Silla Dynasty. The Korea Meteorological Administration set up an analogue seismic monitoring network in 1978 after the Hong Sung earthquake (M 5.0), and a nationwide digital seismic monitoring network in 1997 after the Gyeongju Earthquake (M 4.2). A total of 21 earthquakes were recorded in the Gyeongju area between 1978 and 2016 prior to the 2016 Gyeongju earthquake. On 12 September 2016, a M 5.1 foreshock hit a region 8.2 km south-south-west of Gyeongju at 19:44, followed by a M 5.8 earthquake at 20:32. It was the largest earthquake after the creation of the seismic observation network in Korea, and vibrations were felt in most parts of the country [22].

Despite being the largest-scale earthquake ever recorded in the Korean peninsula, the damage of the 9–12 Gyeongju Earthquake was not as extensive as expected because it lasted only 5–7 s, with an approximately 1–2 s strong motion duration, and mostly high-frequency components were observed. The majority of the damage was incurred by medium-high and low structures, stone fences and stone structures, including architectural heritage, which were more susceptible to the high-frequency components.

At the MKL station in Gyeongju Myeonggye-ri (35.7322 °N, 129.2420 °E), the earthquake monitoring station with rock-outcrop condition closest to the epicentre (epicentral distance: 5.68 km), the peak ground acceleration values measured within the time domain sampled at 100 Hz were 0.346 g for the E-W component and 0.275 g for the N-S component. At USN station in Ulsan Bokan-ri (35.7024 °N, 129.1232 °E), the earthquake monitoring station with soil condition closest to the epicentre, the measured peak acceleration was 0.400 g for the E-W component and 0.430 g for the N-S component.

The 9–12 Gyeongju Earthquake damaged 52 designated cultural heritage sites. Bulguksa Temple suffered considerable damage. For example, roof tiles fells from the Main Buddha Hall (Daewoongjeon), walls in the west corridor were cracked, the fence of Guanyeumheon was destroyed and the banister joints in Dabotap Pagoda were displaced. In Cheomseongdae, the interlocking headstones moved and were partially displaced. Figure 2 illustrates the typical damages to the stone pagoda caused by the 9–12 Gyeongju Earthquake. Table 1 provides an overview of the seismic damage to the stone pagodas due to the Gyeongju Earthquake (modified from Kim et al. [23]). Most of the damaged stone pagodas have three stories and a main body composed of massive stones. Various parts of the pagodas suffered damage including the base, main body and steeple. Serious structural damage was not observed, and most damage included cracking, separation and displacement of non-structural members.

(a)   (b)   (c)   (d)

Figure 2. Typical damages to stone pagoda caused by 9–12 Gyeongju Earthquake (Photo by NRICH): (**a**) Separated third-story body stone, three-story Stone Pagoda at Cheollyongsa Temple Site in Namsan Mountain; (**b**) Rotated stone member, displaced third-story body stone, three-story Stone Pagoda of Yongjanggye Jigok Valley in Namsan Mountain; (**c**) Rotated body and roof stones, three-story Stone Pagoda of Bipagok Valley in Namsan Mountain; and (**d**) Displaced third-story body and roof stones, rhree-story Stone Pagoda of Jiamgok Valley in Namsan Mountain.

Table 1. Seismic damage incurred by stone pagodas due to the 9–12 Gyeongju Earthquake.

Classification	Name of Cultural Heritage	Seismic Damage	Height (m)
National Treasure 20	Dabotap Pagoda of Bulguksa Temple	Displaced banister joints	10.3
National Treasure 30	Stone Brick Pagoda of Bunhwangsa Temple	Cracked lower part	
Treasure 168	East and West Three-story Stone Pagodas in Cheongun-dong	Destroyed steeples of the east and west pagodas	East 6.73, West 7.72
Treasure 678	East and West Three-story Stone Pagodas of Unmunsa Temple	Loosen east pagoda roof stone, tilted west pagoda, cracked main body	
Treasure 908	Three-story Stone Pagoda in Yongmyeong-ri	Chipped upper end of main body	5.6
Treasure 1188 (Figure 2a)	Three-story Stone Pagoda at Cheollyongsa Temple Site in Namsan Mountain	Separated third-story body stone	7.16
Treasure 1429	East and West Three-story Stone Pagodas at Wonwonsa Temple Site	Broken east pagoda roof stone	6.04
Treasure 1867	Three-story Stone Pagoda at Changnimsa Temple Site of Namsan Mountain	Cracked base, separated main body, partially broken roof stone	7.0
Treasure 1935 (Figure 2b)	Three-story Stone Pagoda of Yongjanggye Jigok Valley in Namsan Mountain	Rotated stone member, displaced third-story body stone	
Historic Site 311	Archaeological Area of Namsan Mountain (Three-story Pagoda of Yeonbulsa Temple Site)	Loosened third-story body stone	7.0
Province-Designated Tangible Cultural Heritage 448 (Figure 2c)	Three-story Stone Pagoda of Bipagok Valley in Namsan Mountain	Rotated body and roof stones (counter clockwise)	
Province-Designated Tangible Cultural Heritage 449 (Figure 2d)	Three-story Stone Pagoda of Jiamgok Valley in Namsan Mountain	Displaced third-story body and roof stones	

In the case of Cheomseongdae, continuous monitoring has been carried out since 2009 with regard to member separation and displacement, separation of interlocking headstones due to destroyed joints and tilting of the structure [24]. Before the earthquake, the centre axis of Cheomseongdae was tilted towards the north by ~200 mm, and the southeast and northwest corner of headstones had partially lost their original joint.

As predicted in the previous model tests [21], the seismic damage of Cheomseongdae during the Gyeongju Earthquake on 12 September 2016 was more intense in the upper part than in the lower part filled with filler materials, and the separation and detachment of the stone members in the upper part and the displacement of the southeast headstones was clearly visible. A precise analysis of the results of a 3D survey conducted immediately after the Gyeongju Earthquake showed that the upper headstones were twisted and rotated clockwise, as shown in Figure 3 [25].

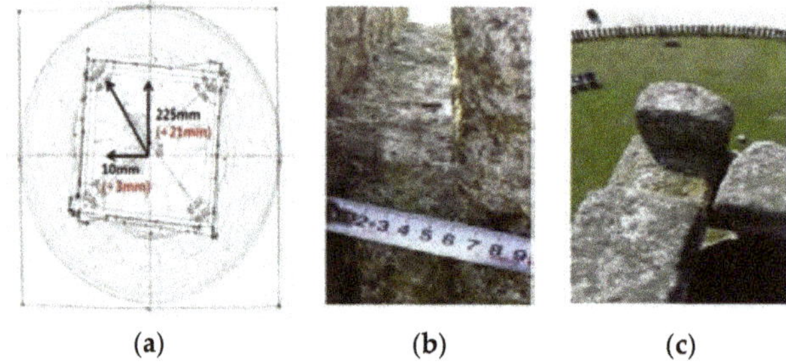

(a)   (b)   (c)

Figure 3. Displacement of Cheomseongdae headstones caused by the 9–12 Gyeongju Earthquake: (a) Displacement and rotation of headstones; (b) Displacement measurement of headstones; and (c) Separation of headstones on southeast corner.

The southern upper headstone moved 38 mm to the north after joint failure caused by the M4.6 aftershock on September 19, 2016. The overall behaviour of the structure was evaluated by analysing the extent of member separation and tilting. No damages were reported other than the cracking and separation of the upper headstone and separation of some stone members.

4. Dynamic Centrifuge Tests for Cheomseongdae Model

As mentioned, a seismic performance assessment for architectural heritage sites begins with obtaining an understanding of the engineering characteristics of the subsoil and architectural structure. First, a site investigation and seismic response analysis are performed at an architectural heritage site to predict the level of ground shaking for each earthquake scenario. Seismic loading exerted on an architectural heritage triggers different seismic behaviours and performances depending on the structure type and style and the load transfer characteristics. Therefore, an empirical or analytical approach based on experimental or numerical modelling of each architectural heritage site is necessary to understand its seismic behaviour.

In the previous study [19], a dynamic centrifuge test was proposed to evaluate the seismic behaviour of stone block structures. If appropriate scaling factors are applied and the model material has the same density as the real structure, the seismic behaviour can be observed under the same stress conditions as reality. Based on the study, the dynamic centrifuge model test was proposed for seismic performance assessment to evaluate how the stone architectural heritage resists seismic loads and whether it can resist damage from future earthquakes [20]. Table 2 shows the scaling factors for the dynamic centrifuge model test.

Table 2. Scaling factors of dynamic centrifuge test for stone block and architectural heritage model.

Quantities	Scaling Factors (Prototype/Model)
Displacement, Length	N
Acceleration, Gravity	N^{-1}
Mass	N^3
Density	1
Stress	1
Strain	1
Time (Dynamic)	N

The Cheomseongdae ground response was analysed based on an actual seismic waveform measured on 12 September 2016, at MKL station. Dynamic centrifuge model tests were performed on the Cheomseongdae model using the surface response obtained from the site-specific ground response analysis, which enabled Cheomseongdae's seismic performance to be re-assessed.

4.1. Ground Response Analysis and Generation of Input Motion for Shaking Table

The ground response for the Gyeongju Earthquake accelerograms measured at the MKL station was analysed using the shear-wave velocity profile and the modulus reduction and damping curve of each ground material at Cheomseongdae determined in a previous study [18]. The peak acceleration value of the E-W component measured at the MKL station was higher than that of the N-S component. The E-W component contained a higher proportion of high-frequency components whose frequency exceeded 25 Hz than the N-S component. Given that components with frequencies lower than 25 Hz are more important in ground engineering and structural engineering and considering the excitation frequency band of the shaking table, the N-S component of MKL was used in this study [26].

Figure 4 shows the results of the Cheomseongdae ground response analysis for the Ofunato Earthquake motion and the Gyeongju MKL (N-S component) motion. The waveforms of the latter had a shorter duration of strong motion than those of the former with a similar peak acceleration value. Based on a threshold acceleration of 0.05 g, the bracketed durations T_d were 11.35 s and 3.23 s for the Ofunato and Gyeongju earthquakes, respectively. Moreover, the Arias intensity I_a values [27] used for a quantitative calculation of the duration of strong motion were 0.0335 and 0.0039 m/s for the Ofunato and Gyeongju Earthquake motions, respectively.

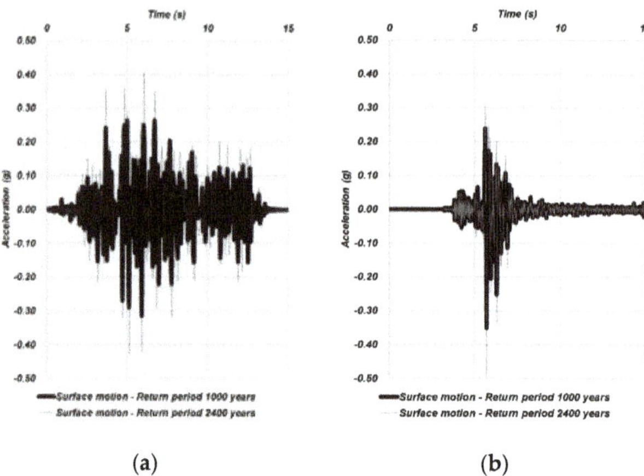

Figure 4. Ground response analysis for Cheomgeongdae site: (**a**) Ofunato earthquake motion; and (**b**) Gyeongju MKL motion.

The frequency range of the shaking table mounted on the centrifuge was 30–300 Hz for the scale model. Hence, the tests were performed at a centrifugal gravitational field of 15 g using a 2–20 Hz prototype-scale component. The time history from the ground response analysis results was converted using the bandpass filter of the 2–20 Hz component. As a result, the T_d values of the input seismic wave used for actual testing were 11.15 s and 2.46 s for the Ofunato and Gyeongju Earthquake motions, respectively.

4.2. Preparation of Cheomseongdae Models and Centrifuge Model Testing

To analyse Cheomseongdae's seismic behaviour characteristics, 1/15-scale models were created, and dynamic centrifuge model testing was performed using a centrifuge device and shaking table from the KAIST Geo-Centrifuge Testing Center [28,29]. The scaling factors were considered by applying a 15 g centrifugal force vertically, which is 15 times the gravitational acceleration, to the 1/15-scale Cheomseongdae model. Hwangdeung granite was used for the model members because its density and stiffness are closest to those of the material at Cheomseongdae. This enhances the plausibility of reproducing the actual seismic behaviour of Cheomseongdae.

All the stone members of the model were fabricated manually to produce models as accurately as possible. Cheomseongdae is composed of over 380 stone members, of which the long members of the 19th, 20th, 25th and 26th layers, and the two headstone layers were proven to contribute to bearing larger lateral loads. Shaking table tests were performed within centrifuge to compare the seismic behaviour of the Cheomseongdae models with and without the four headstones that are supposed to enhance the seismic performance, as shown in Figure 5.

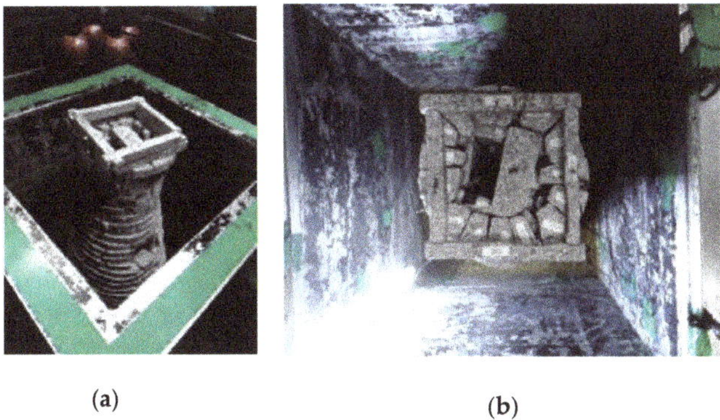

(a)   (b)

Figure 5. Cheomseongdae model: (**a**) Cheomseongdae model with headstones; and (**b**) Cheomseongdae model without headstones.

Accelerometers were installed at eight heights on the east side of the Cheomseongdae model, and the model was subjected to Gyeongju Earthquake motions in the east-west direction. The accelerometers were placed in the vibration shaker of the centrifuge model tester, Cheomseongdae base, 12th layer (on the filler surface), 13th layer (southern window above the filler), 19th layer (stone beam), 25th layer (stone beam), 26th layer (stone beam), 28th layer (lower headstone) and 29th layer (upper headstone). The 28th and 29th layers were used for the prototype model only. The Ofunato and Gyeongju Earthquake motions were applied during the tests, starting from a low acceleration that was gradually increased. For both Cheomseongdae models, seismic waves were applied six times per waveform. The acceleration results were processed after converting them into the actual Cheomseongdae prototype unit and taking the scale factor into account. These were used to explain the actual phenomena that may occur at Cheomseongdae during an earthquake, considering the scale factor for the 1/15 scale and 15 g conditions with respect to size, duration and frequency.

4.3. Acceleration Amplification With Height at Cheomseongdae

Figure 6 shows the peak acceleration values measured at various heights for the two Cheomseongdae models (with and without headstones) exposed to Ofunato and Gyeongju Earthquake motions. Almost no acceleration amplification was measured in the lower parts under the window, i.e. the part packed with filler materials, and different acceleration values were measured in the upper part depending on the absence or presence of headstones. Neither model collapsed, with only slight headstone displacement occurring at the 0.5 g level of strong motion. The movement of headstone members was similar to that reported after the actual Gyeongju Earthquake. A similar tendency was demonstrated when the model was exposed to Gyeongju Earthquake motion with a higher proportion of high-frequency components, and considerable differences were observed in the amplification characteristics of the upper part depending on the absence or presence of headstones.

During the dismantling of the models upon the completion of testing, a stone beam in the 19th layer was found to be broken in the centre, as shown in Figure 7. This proves that the stone beams play an important role in enhancing seismic performance by functioning as tie bars.

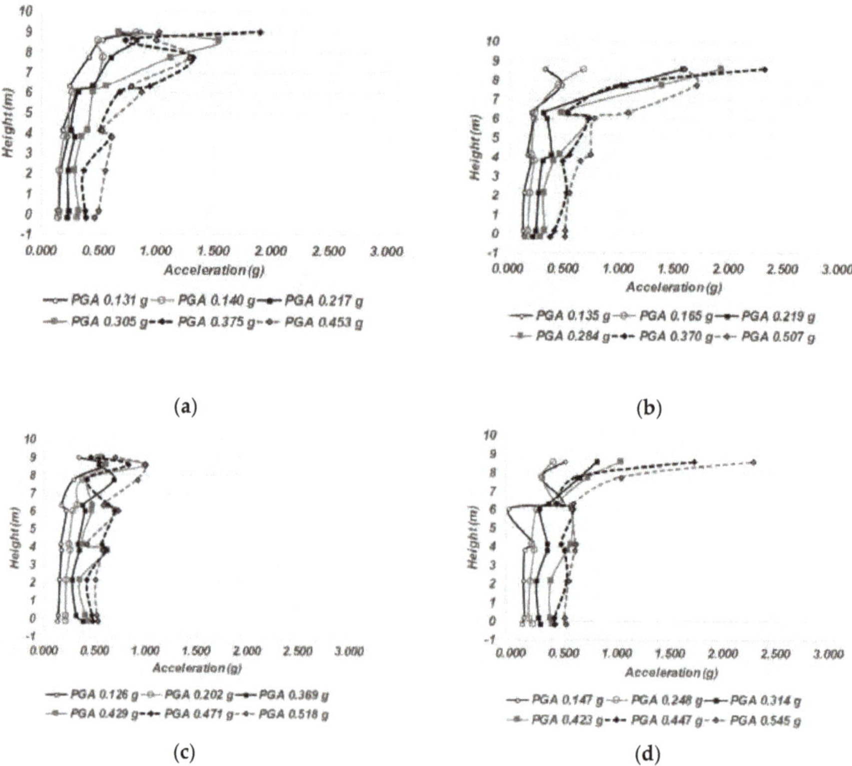

Figure 6. Peak acceleration values measured at various heights in the two Cheomseongdae models: (**a**) Ofunato earthquake motion in original model; (**b**) Ofunato earthquake motion in Cheomseongdae model without headstones; (**c**) Gyeongju Earthquake motion in original model; and (**d**) Gyeongju Earthquake motion in Cheomseongdae model without headstones.

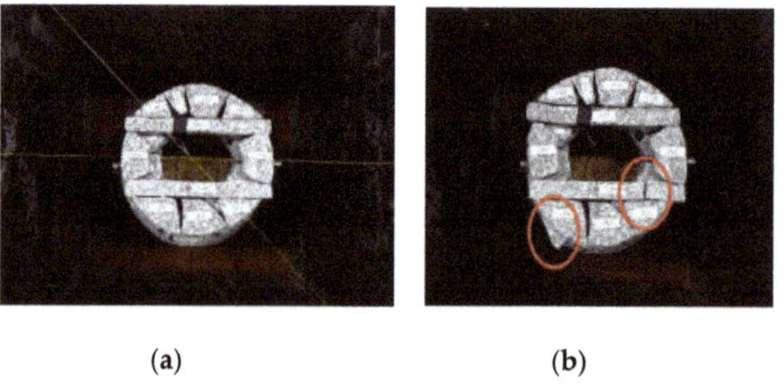

Figure 7. Broken stone beam of 19th layer after seismic excitations: (**a**) Before testing; and (**b**) After testing.

4.4. Frequency Response of Structure During Earthquake Excitation

Figure 8 shows the frequency response of the structure exposed to the Ofunato earthquake motion with a 0.137 g-level surface peak acceleration. A peak acceleration of 0.660 g was measured at the upper headstone (29th layer), and partial amplification occurred in the 2.4–3.9 Hz and 5.1–6.3 Hz bands. The amplification profile was not clearly discernible. This may be due to the complex mode shape of the structure, which was composed of nearly 400 members. Figure 9 shows the frequency response of the structure exposed to the Gyeongju Earthquake motion with a 0.126 g-level surface peak acceleration. A peak acceleration of 0.329 g occurred at the topmost layer, and partial amplification occurred at 2.7 and 4.9 Hz.

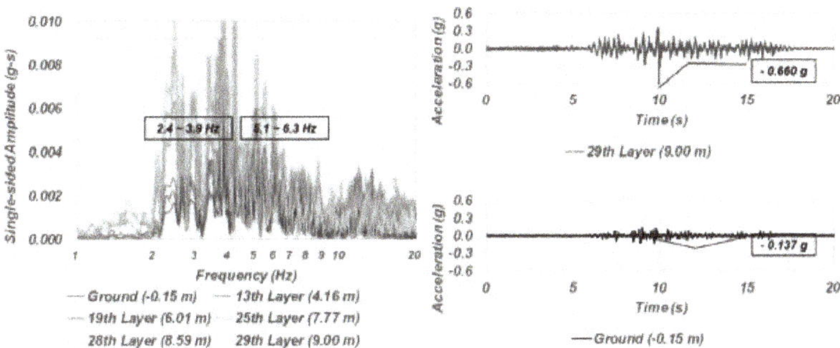

Figure 8. Frequency response of structure exposed to Ofunato earthquake motion.

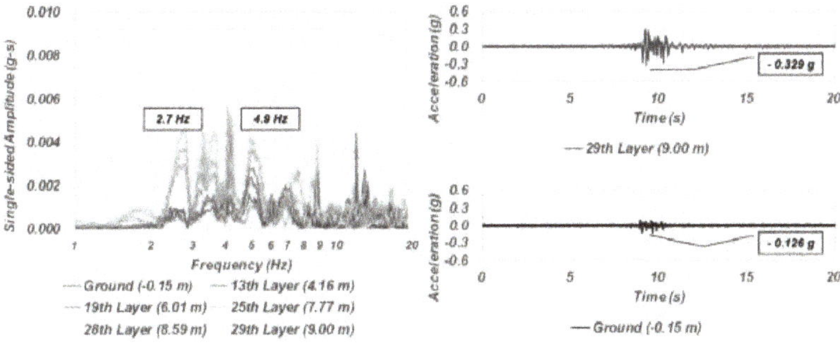

Figure 9. Frequency response of structure exposed to Gyeongju Earthquake motion.

It is difficult to determine the natural frequency of Cheomseongdae, which has almost 400 members [21]. In the Cheomseongdae prototype, acceleration amplification was measured in the 3.23 Hz and 7.1–11.07 Hz bands under excitation in the east-west direction. From the test results, it was concluded that Cheomseongdae has a high seismic performance with a harmony between the local behaviour of each member and the overall behaviour of the structure. Additionally, the behaviour of the lower part under the window level packed with filler materials exhibited rigid motion, and the intersecting long beams and interlocking headstones were found to enhance the seismic performance of the upper part with a large empty space.

5. Conclusions

This paper summarised and categorised studies conducted between 2009 and 2012 on the seismic risk assessment of architectural heritage in Korea's historic areas in relation to the evaluation of the engineering characteristics of architectural heritage sites, the site-specific ground response analysis for earthquake scenarios, the GIS-based seismic microzonation according to geotechnical earthquake engineering parameters, the reliability assessment of the dynamic centrifuge model testing of stone masonry structure and the seismic performance of architectural heritage. The M 5.8 earthquake that hit Gyeongju on 12 September 2016 was analysed from an engineering point of view and the related damages to stone architectural heritage were categorised according to the type of damage. The reliability and importance of the research achievements presented in this study were highlighted in the Cheomseongdae damage report written after the 2016 Gyeongju Earthquake. The reliability of the seismic performance assessment method for architectural heritage based on dynamic centrifuge model testing was also proven using a Gyeongju Earthquake records that have rarely been used in Korea thus far.

The findings of the dynamic centrifuge tests for the Cheomseongdae models using actual Gyeongju Earthquake motion were (1) the different acceleration values with height in the upper part depending on the absence or presence of headstones; (2) the observation of movement of headstone members similar to that reported after the actual 2016 Gyeongju Earthquake; and (3) a broken stone beam in the 19th layer during dynamic centrifuge tests. These findings show that it is not a coincidence that Cheomseongdae, a masonry structure composed of nearly 400 stone members, survived numerous seismic events for over 1300 years. The structural characteristics of Cheomseongdae, such as the well-compacted filler materials in its lower part, the rough inside wall in contrast to the smooth exterior, intersecting stone beams and interlocking headstones, were shown to contribute to its overall seismic performance based on the results of dynamic centrifuge model testing. It is interesting that Cheomseongdae maintained its original form given the ground conditions since the seventh century. It is a great cultural heritage in which we can still feel our ancestors' breaths and rediscover their outstanding seismic design technology.

Author Contributions: Conceptualization, H.-J.P. and J.-G.H.; Formal analysis, J.-G.H.; Funding acquisition, H.-J.P.; Investigation, S.-H.K.; Methodology, H.-J.P., S.-H.K. and S.-S.J.; Resources, S.-H.K. and S.-S.J.; Software, J.-G.H.; Supervision, H.-J.P.; Validation, S.-S.J.; Writing—original draft, H.-J.P.; Writing—review & editing, H.-J.P.

Funding: This research was also supported by the Basic Science Research Program through the National Research Foundation of Korea (NRF) funded by the Ministry of Education (grant number NRF-2014R1A6A3A04056405).

Acknowledgments: This study, which forms a part of the project, has been achieved with the support of national R&D project, which has been hosted by National Research Institute of Cultural Heritage of Cultural Heritage Administration. The authors would like to thank Kyung-Sik Seo, Sung-Bok Lee, Tae-Sic Oh, Moon-Gyo Lee, Yeong-Hoon Jeong, Kil-Wan Ko, Hye-Lim Lee, and Sunji Park for their advice and support for centrifuge model tests.

Conflicts of Interest: The authors declare no conflict of interest.

References

1. Calabresi, G. 1st Kerisel lecture: The role of Geotechnical Engineers in saving monuments and historic sites. In Proceedings of the 18th International Conference on Soil Mechanics and Geotechnical Engineering, Paris, France, 2–6 September 2013; pp. 71–83.

2. Viggiani, C. 2nd Kerisel lecture: Geotechnics and Heritage. In Proceedings of the 19th International Conference on Soil Mechanics and Geotechnical Engineering, Seoul, Korea, 17–22 September 2017; pp. 119–140.

3. Cid, J.; Susagna, T.; Goula, X.; Chavarria, L.; Figueras, S.; Fleta, J.; Casas, A.; Roca, A. Seismic zonation of Barcelona based on numerical simulation of site effects. *Pure Appl. Geophys.* **2001**, *158*, 2559–2577. [CrossRef]

4. Gizzi, F.T. Identifying geological and geotechnical influences that threaten historical sites: A method to evaluate the usefulness of data already available. *J. Cult. Herit.* **2008**, *9*, 302–310. [CrossRef]

5. Glatron, S.; Beck, E. Evaluation of socio-spatial vulnerability of citydwellers and analysis of risk perception: Industrial and seismic risks in Mulhouse. *Nat. Hazards Earth Syst. Sci.* **2008**, *8*, 1029–1040. [CrossRef]

6. Salamon, A.; Katz, O.; Crouvi, O. Zones of required investigation for earthquake-related hazards in Jerusalem. *Nat. Hazards* **2010**, *53*, 375–406. [CrossRef]

7. Psycharis, I.N.; Lemos, J.V.; Papastamatiou, D.Y.; Zambas, C.; Papantonopoulos, C. Numerical study of the seismic behaviour of a part of the Parthenon Pronaos. *Earthq. Eng. Struct. Dyn.* **2003**, *32*, 2063–2084. [CrossRef]

8. Konstantinidis, D.; Makris, N. Seismic response analysis of multidrum classical columns. *Earthq. Eng. Struct. Dyn.* **2005**, *34*, 1243–1270. [CrossRef]

9. Kounadis, A.N.; Papadopoulos, G.J.; Cotsovos, D.M. Overturning instability of a two-rigid block system under ground excitation. *ZAMM-J. Appl. Math. Mech.* **2012**, *92*, 536–557. [CrossRef]

10. Aguilar, R.; Marques, R.; Sovero, K.; Martel, C.; Trujillano, F.; Boroschek, R. Investigations on the structural behaviour of archaeological heritage in Peru: From survey to seismic assessment. *Eng. Struct.* **2015**, *95*, 94–111. [CrossRef]

11. Maria, B.D. Decision Making Based on Benefit-Costs Analysis: Costs of Preventive Retrofit versus Costs of Repair after Earthquake Hazards. *Sustainability* **2018**, *10*, 1537. [CrossRef]

12. Kim, J.K.; Ryu, H. Seismic test of a full-scale model of a five-storey stone pagoda. *Earthq. Eng. Struct. Dyn.* **2003**, *32*, 731–750. [CrossRef]

13. Peña, F.; Lourenço, P.B.; Alfredo, C.C. Experimental Dynamic Behavior of Free-Standing Multi-Block Structures Under Seismic Loadings. *J. Earthq. Eng.* **2008**, *12*, 953–979. [CrossRef]

14. Santi, M.C.; Vincenzo, S.; Irene, L.; Simona, M.C.P. Fiber-Reinforced Polymer Nets for Strengthening Lava Stone Masonries in Historical Buildings. *Sustainability* **2016**, *8*, 394. [CrossRef]

15. Zheng, L.; Zhang, H.R.; Jia, C.G.; Peng, Z.M. Seismic Performance of a New Type of Fabricated Tie-Column. *Sustainability* **2018**, *10*, 1716. [CrossRef]

16. National Research Institute of Cultural Heritage (NRICH). *Research Planning on the Disaster Mitigation of Architectural Heritage*; NRICH: Daejeon, Korea, 2008; p. 282. (In Korean)

17. Park, H.J.; Kim, D.M.; Kim, K.S.; Ahn, H.Y.; Kim, D.S. Noninvasive geotechnical site investigation for stability of Cheomseongdae. *J. Cult. Herit.* **2012**, *13*, 98–102. [CrossRef]

18. Park, H.J.; Kim, D.S.; Kim, D.M. Seismic risk assessment of architectural heritages in Gyeongju considering local site effects. *Nat. Hazards Earth Syst. Sci.* **2013**, *13*, 251–262. [CrossRef]

19. Park, H.J.; Kim, D.S. Centrifuge modelling for evaluation of seismic behaviour of stone masonry structure. *Soil Dyn. Earthq. Eng.* **2013**, *53*, 187–195. [CrossRef]

20. Park, H.J.; Kim, D.S.; Choo, Y.W. Evaluation of the seismic response of stone pagodas using centrifuge model tests. *Bull. Earthq. Eng.* **2014**, *12*, 2583–2606. [CrossRef]

21. Park, H.J.; Kim, D.S. Evaluation of seismic behaviour of Cheomseongdae using dynamic centrifuge model test. *Earthq. Eng. Struct. Dyn.* **2015**, *44*, 695–711. [CrossRef]

22. Ministry of Public Safety and Security. *9.12 Earthquake Whitepaper*; Ministry of Public Safety and Security: Sejong, Korea, 2017; 406p. (In Korean)

23. Kim, N.; Hong, S.G.; Koo, I.Y. Study on seismic evaluation for stone heritage pagodas. In Proceedings of the Earthquake Engineering Society of Korea (EESK) Conference, Incheon, Korea, 23 March 2018; pp. 143–144.

24. Yu, H.R.; Park, C. A study on the damaged architectural heritage—Cheomseongdae Observatory-in Gyeongju's from an earthquake. *J. Soc. Cult. Herit. Disaster Prev.* **2016**, *1*, 83–89. (In Korean)

25. Jo, S.S.; Kim, D.M.; Kim, D.S. Study on Stone Cultural Heritage's Structural Stability Management-with a Focus on Cheomseongdae. In *Structural Analysis of Historical Constructions*; Aguilar, R., Torrealva, D., Moreira, S., Pando, M.A., Ramos, L.F., Eds.; RILEM Bookseries; Springer: Basel, Switzerland, 2019; Volume 18, pp. 652–660.

26. Lee, C.H.L.; Park, J.H.; Kim, T.; Kim, S.Y.; Kim, D.K. Damage potential analysis and earthquake engineering-related implications of Sep.12, 2016 M5.8 Gyeongju Earthquake. *J. Earthq. Eng. Soc. Korea* **2016**, *20*, 527–536. [CrossRef]

27. Arias, A. A measure of earthquake intensity. In *Seismic Design for Nuclear Power Plants*; Hansen, R.J., Ed.; MIT Press: Cambridge, MA, USA, 1970; pp. 438–483.

28. Kim, D.S.; Kim, N.R.; Choo, Y.W.; Cho, G.C. A newly developed state-of-the-art geotechnical centrifuge in Korea. *KSCE J. Civ. Eng.* **2013**, *17*, 77–84. [CrossRef]

29. Park, H.J.; Ha, J.G.; Kwon, S.Y.; Lee, M.G.; Kim, D.S. Investigation of the dynamic behaviour of a storage tank with different foundation types focusing on the soil-foundation-structure interactions using centrifuge model tests. *Earthq. Eng. Struct. Dyn.* **2017**, *46*, 2301–2316. [CrossRef]

 *sustainability*

Article

Serpentinite from Moeche (Galicia, North Western Spain). A Stone Used for Centuries in the Construction of the Architectural Heritage of the Region

José Nespereira [1], Rafael Navarro [2], Serafín Monterrubio [1,2], Mariano Yenes [3] and Dolores Pereira [2,3,*] 

[1] Department of Geology, University of Salamanca, Avd. Requejo 33, 49022 Zamora, Spain; jnj@usal.es (J.N.); seramp@usal.es (S.M.)
[2] CHARROCK Research Group. University of Salamanca, Plaza de los Caídos s/n, 37008 Salamanca, Spain; rafanavarro74@gmail.com
[3] Department of Geology, University of Salamanca, Plaza de los Caídos s/n, 37008 Salamanca, Spain; myo@usal.es
*   Correspondence: mdp@usal.es

Received: 1 April 2019; Accepted: 10 May 2019; Published: 12 May 2019

Abstract: Serpentinites are characterized by highly variable mineralogical, physical, and mechanical properties. Serpentinites from Moeche (North Western Spain) have been studied to establish their mineralogical, petrographic, and textural characteristics, as well as their physical and mechanical parameters and the factors influencing rock failure, to evaluate the possible use of these rocks either for new construction or for conservation-restoration of the architectonic heritage of the region. In this paper, we highlight the importance of a detailed mineralogical and petrographic characterization in the fracture zones, which will determine the viability of quarrying the stone. A strong correlation between the petrographic features and the uniaxial compression strength values has been observed. The most important aspects were found to be the rock texture, the mineralogical composition of the fracture area and foliation, although mineralogy was also found to be involved (% of carbonates) in the strength of the stone. An important preliminary result of the study was the low asbestos content of these serpentinites, which will help in the potential re-opening of the quarries.

Keywords: serpentinites; rocks characterization; dimension stone; architectural heritage

1. Introduction

Serpentinites are metamorphic rocks formed from the alteration of ultramafic rocks. These alteration processes lead to the hydration of original mineral phases that were formed prior to serpentine and other minerals [1]. Macroscopically, foliation, which bestows a clear anisotropy to these rocks, brecciated areas, and carbonates, filling fractures or replacing serpentine minerals, are quite common features of serpentinites around the world. As a consequence of their mineralogical and structural variability, serpentinites are very heterogeneous rocks. They have been used as construction materials since ancient times; many monuments and historical buildings are made of these important heritage stones [2–4]. Serpentinites appear recurrently in commercial catalogs as green marbles due to the frequent replacement of serpentine minerals for carbonates [5,6]. However, they are not marbles, and a detailed characterization from several points of view (mineralogical, physical and mechanical) is needed if the rocks can be applied either in construction or conservation-restoration of the architectonic heritage.

A few published papers refer to the physical and mechanical properties of serpentinites. Coumantakis [7] observed that in ultramaphic rocks, geomechanical behavior depends on the degree of serpentinization. Other authors [8] report data from two construction projects in which serpentinites are used as construction material: The Hepp damn (Indonesia) and the Berke powerhouse access tunnel (Turkey). Oszoy et al. [9] characterize serpentinites, among other stones, for the Yakakayi dam site (Turkey), highlighting their different behavior depending on the degree of alteration. Marinos et al. [10], who also study serpentinites from the perspective of rock mass, report that the value of the Geological Index Strength (GSI) considerably differs depending on the petrographic and structural variations of serpentinites. The results of these studies have led to a common conclusion: The behavior of serpentinized rock masses depends on the degree of alteration and on the structure or foliation affecting the rock.

Other studies have approached the characterization of serpentinites without taking into account rock mass, but rather focus on the rock as dimension stone, both for conservation-restoration and new building. In this sense, Christensen [11] describes their findings concerning the elastic properties of serpentinites and peridotites obtained from P and S wave velocities. Other authors [12] have studied serpentinites from Egypt used as dimension stones in heritage, and Diamantis et al. [13] have proposed several correlations among the physical and mechanical properties of Greek serpentinites used in heritage. Kurtulus et al. [14] have determined dynamic engineering values, as well as the geotechnical and mechanical properties of serpentinized ultrabasic rocks in NW of Turkey, showing some correlations between different parameters. Navarro et al. [15] studied the serpentinites from Macael (S of Spain) used as dimension stone, obtaining varied results with respect to their mechanical properties, owing to the mineralogical differences between samples. Other authors [16] have compared two serpentinites from the same quarry in Macael (SE of Spain) with different degrees of carbonation. In addition, they remarked on the differences found in the main parameters used in the dimension stone industry among the samples tested. More recently, Navarro et al. [6] have carried out a complete study with six varieties of carbonated and non-carbonated serpentinites from the SE of Spain. They showed how carbonation of serpentinites greatly influences the behavior of these rocks when used as dimension stone. Pereira et al. [5] have remarked on the need for petrographic and mineralogical data in order to be able to decide on the utility of serpentinites as dimension stones.

A significant amount of architectonic heritage found in this part of Galicia was built using this kind of rock, which is known locally as "Pedra de Toelo" [3]. The reason of their extensive use in the past in this area is that this stone presents the quality of being much softer and ductile to the chisel of the stonemason than the granite, another frequently used stone in the area. This characteristic allows a more refined sculpts as well as an easier finishing and polishing. Its crass shine and its color, varying from greenish to dark gray, pinkish and bluish, as well as the presence of abundant veins and a particular mottled, gives a pleasant polychrome appearance acquiring a variable tone with the changes of luminosity, originating several kinds of works. However, all these features trigger a greater degradation of the material over the passage of time [17]. This is why most of the architectural heritage sites built on this stone are in very bad states of conservation at present, and in need of an intense conservation-restoration, including replacement of some blocks. Also for this reason, it is very important to analyze the content of asbestos in these rocks, because a high content of the fibrous minerals should prevent any intention of re-opening the quarries, to obey the very strict European laws (e.g., Directive 2009/148/EC on the protection of workers from the risks related to exposure to asbestos at work).

The quarries in this area were mined for a long period in the past centuries. In fact, they are found in the commercial catalogs under the commercial name of "Verde Pirineos" by Marmolera Gallega, S.L.; currently, a mining company is carrying out new studies and is considering the possibility of resuming mining activity. The name of the company is omitted for confidential reasons.

Preliminary studies on stones from Moeche focused on the mineralogy and the physical and mechanical characteristics of different varieties of serpentinites of the region and their use as a dimension stone in the region, which also forms part of the architectural heritage of the area [3,18–20]. The heterogeneity of serpentinites was made clear, even within a single outcrop. The only geotechnical standard that contemplates the requirements of physical and mechanical behavior for serpentinites as natural stone is the American C1526-02 standard [21]. In this document, the range of accepted values for water absorption (0.2% in exteriors and 0.6% in interiors) and the minimum values for uniaxial compression strength (69 MPa) and bulk density (2560 kg/m^3) are specified as the requirements.

The present paper studies the mineralogical, physical, and mechanical properties of one of the main bodies of serpentinites in Spain, located in the area surrounding Moeche (Galicia, NW of Spain). The uniaxial compression strength (UCS), elastic moduli (E_s, v_s), bulk density (ρ), and water absorption (a) have been determined, as well as the characterization of the mineral composition carried out using X-ray diffraction and the study of thin sections. This work also includes a detailed description of the failure surfaces developed during the UCS tests. Accordingly, ultrasonic wave velocities were measured and the results were compared to bulk density, the UCS, and the elastic moduli in search of connections that could help to explain their behavior. Furthermore, this characterization could also facilitate the commercial use of these stones, both for construction and conservation-restoration purposes.

The objectives of this work are focused on describing the main properties of serpentinites from Moeche and to determine the mechanisms of rock failure to facilitate the evaluation of these serpentinites as potential construction and conservation-restoration material.

2. Geological Context

The study area is located in the NW of Spain and belongs to the Cabo Ortegal Complex, one of the five allochthonous complexes outcropping in the NW of the Iberian Massif and framed within a context of collision between Gondwana and Laurasia [22–25] (Figure 1).

The ultramafic rock outcroppings in this complex are in large part allochthonous units, the stacking of nappes being the consequence of subduction and collision in the early phases of the Variscan orogeny in Northwestern Iberia. The different units are fragments of the lithospheric mantle at the base of the ophiolitic complexes and they have undergone several metamorphic events before being piled up and incorporated into the autochthonous terrane [19]. Serpentinization and tectonic structure is well known [23,24,26,27], and three different units can be distinguished at the complex: The Basal Unit, The Ophiolitic Unit, and the Upper Unit. Within the Ophiolitic Unit, the Somozas Mélange has been described, where two sub-units can be differentiated. The upper sub-unit, 800 m thick, is made up of a highly sheared matrix of serpentinites surrounding tectonic blocks and slices of variable size and continuity, with gabbros, diabases, granitoids and volcanic rocks, and intercalations of phyllites and phyllonites. In this sub-unit, inclusions of sandstones, conglomerates and marbles can also be found, but are less common. In the case of the lower sub-unit, which can reach a thickness of 1000 m, is a mélange with a matrix of ocher-colored or blue phyllites [24]. Our samples were collected in quarries located in Moeche, in one of the massifs from the uppermost unit of this complex.

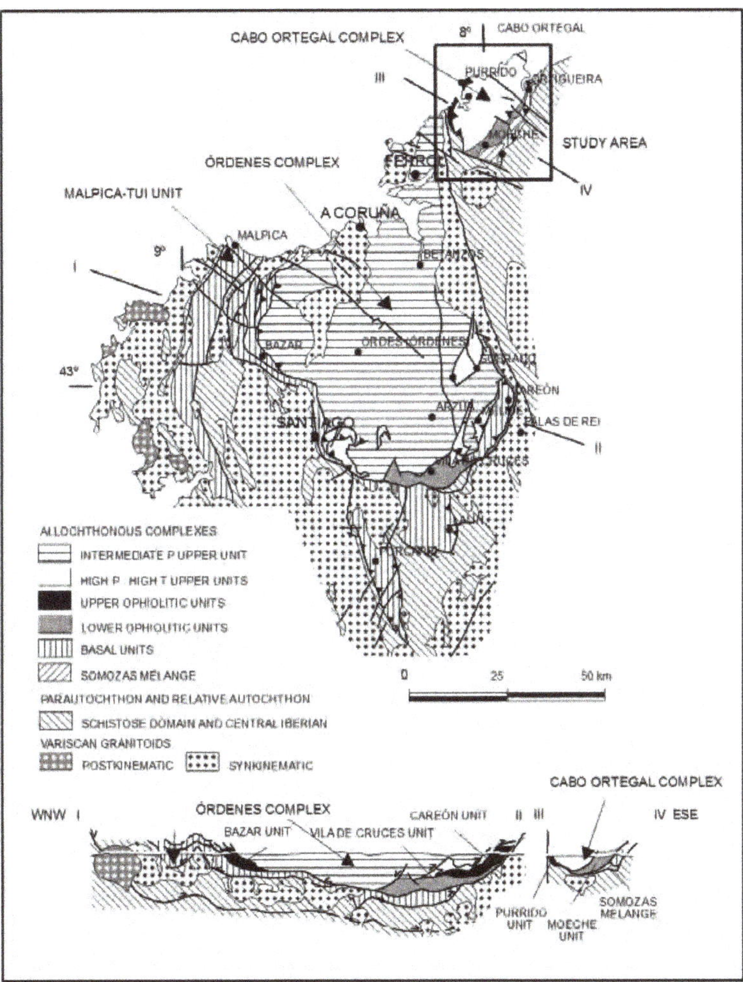

Figure 1. Geological map of the area according to Albert et al. [25]. The solid black arrow shows the Moeche study area.

3. Materials and Methods

3.1. Sampling

Seven block samples were collected from abandoned quarries close to the village of Moeche (Figure 2a), measuring approximately 30 × 30 × 20 cm (Figure 2b). The quarries are located in Monte Ferrerías (samples from blocks 2, 3, 4, 5, 7, 8), and Monte Gradoy (samples from block 10). The samples were collected directly from the outcrops and also from abandoned blocks from previous quarry stonework. Although the structure of these stones is very chaotic, with heterogeneous presence of foliation, fractures, etc . . . , the general direction of the quarry beds follows the principal direction of the regional structure.

A laboratory core drill and cutting machines were used to prepare cylindrical specimens, with diameters between 50.7 and 51.1 mm and lengths between 101 and 104 mm. Thus, the length-to-diameter ratio was 2.0, in accordance with the accepted suggestions for uniaxial compression tests in rocks [28].

The samples analyzed and the tests carried out are shown in Table 1.

Figure 2. (**a**) Abandoned quarry in Monte Ferrerías (Moeche) where sampling was carried out. The high of quarry beds is around 1 m; (**b**) serpentinite block sample from the same quarry.

Table 1. Samples used in this study and the tests and analyses carried out. In the Petrography column, the number of thin sections prepared is indicated in parentheses. The last row shows the total number of samples, tests, and analyses performed. (Id): Sample identification; (UCS): Uniaxial compression strength test; (V_P/Vs), the ratio between P-wave and S-wave velocities.

Id	Bulk Density	Petrography	UCS	V_P/Vs	Strain Gauge	XRD	Water Absorption
M-2d	X	X (2)	X	X	X	X	-
M-3a	X	X (1)	X	X	X	X	-
M-3b	X	X (1)	X	X	X fail	X	-
M-3c	X	-	-	X	-	X	X
M-3e	X	X (1)	X	X	X	-	X
M-4a	X	X (1)	X	X	X fail	X	-
M-4c	X	X (1)	X	X	X	X	-
M-5a	X	X (1)	X	X	X	X	-
M-5b	X	X (1)	X	X	X	X	-
M-5c	X	X (1)	X	X	X	-	X
M-7a1	X	X (1)	X	X	X	X	-
M-7a2	X	-	-	X	-	-	-
M-7b1	X	-	-	X	-	-	-
M-7b2	X	X (1)	X	X	X	X	X
M-7c1	X	-	-	X	-	X	X
M-7c2	X	X (1)	X	X	X	-	X
M-8a	X	-	-	X	-	X	-
M-8c	X	X (1)	X	X	X	X	-
M-8d	X	X (1)	X	X	X	X	X
M-10a	X	X (2)	X	X	X	X	-
M-10b	X	-	-	X	-	X	X
Total	21	17	15	21	15	16	8

The working hypothesis was that the serpentinites would fail across bands or discontinuous zones where highly variable mineralogy might appear, and mechanical tests have proved this to be true in most cases.

Some extra samples were analyzed to describe the possible content of asbestos in these rocks.

3.2. Physical Properties

Bulk densities were obtained for each cylindrical sample, and geometrical volume was determined. Thus, in each specimen, the average diameters (D) from six measurements were recorded (two at the upper end, two in the middle, and another two at the lower end, and the average height (H) from three measurements taken along the probe axis and equally spaced at approximately 120°). Specimen weights were determined using a scale after oven drying to constant mass.

Water absorption (a, measured as %), the ratio between the water mass in a saturated sample and its dried mass, were determined after the samples had been subjected to 48 h of immersion in water, after which they were weighed on a precision balance. Dry mass measurements were taken after oven drying at 60 °C until constant mass, using some of them to test the water absorption.

ASTM C-1526-02 was followed to test properties such as water absorption (a) or bulk density (ρ) [20].

3.3. Ultrasonic Test

The study of sound waves was performed at the Laboratory of Applied Petrology of the CSIC-University of Alicante Associated Unit, following the ASTM C1721-15 standard. This involved the coupling of two piezoelectric sensors on opposite faces of the sample. A viscoelastic gel was used to achieve a good connection between the sample and the transducers. The travel time was registered using Panametric non-polarized transducers (1 MHz) connected to an oscilloscope to detect and identify the wavefronts. The length and travel time in each sample allowed the wave propagation velocities (V_P and V_S) to be calculated.

Specimens of 50 mm diameter and 100 mm length were tested after being oven dried and subsequently cooled in a desiccator. The same specimens were later used to perform the uniaxial compression tests. Nine samples were also tested after they had been soaked in water for 48 h.

Elastic dynamic moduli were obtained indirectly from the ultrasonic tests [29] using the following expressions:

$$v_d = \frac{\left(\frac{V_P}{V_S}\right)^2 - 2}{2 \cdot \left[\left(\frac{V_P}{V_S}\right)^2 - 1\right]} \tag{1}$$

$$E_d = 2 \cdot \rho \cdot V_S^2 \cdot (1 + v_d) \tag{2}$$

v_d: Dynamic Poisson coefficient.
V_P: Compressional ultrasonic wave velocity
V_S: Shear ultrasonic wave velocity
E_d: Dynamic Young modulus
ρ: Bulk density

3.4. Uniaxial Compression Strength and Elastic Modulus (Static)

Uniaxial compression tests were carried out with the same specimens used for the ultrasonic tests. The load rate applied was 0.1 MPa/s, and failure was reached after 11.3 min (average value).

Simultaneously, the elastic parameters E_s and V_s—the static Young modulus and the Poisson coefficient—were measured with the help of two pairs of electrical strain gauges (for axial and diametric strain) appended to the middle portion of the curved surface of the specimens with epoxy resin adhesive after cleaning the surface with acetone and polishing to get a homogeneous surface. The Young modulus was obtained as the average slope of the straight portion of the stress-strain curve, one of the three methods included in the standard EN 22-950-90/3 [30].

3.5. Mineralogy and Petrographic Analyses

The mineralogical composition was determined by X-ray powder diffraction analysis (Bruker D8 Advance) of bulk rock with Cu Kα radiation and a velocity of 2θ/min. The semiquantitative determination of the different proportion of the detected phases was done following Rietveld's method [31]. The equipment used is a Siemens D-500 (40 KV and 30mA) diffractometer using CuKa radiation (λ = 1.5437Å) Ni-filter, equipped with the Diffract ATV3 software package, at the University of Salamanca.

The petrographic study was carried out on thin sections from the fractured area after testing samples for compression strength. Careful reconstruction using adhesive tape was required for each broken sample. Then, thin sections perpendicular to the specimen axis were obtained after embedding the specimen in resin to regain a single unit. Samples M-2d and M-10a were also cut parallel to the axis. A complete petrographic examination was done following Reference [32] to describe the mineralogy and textures as recommended by Reference [33]. A Leica DM2500P microscope under transmitted light was used for this purpose.

Selected samples were also studied through Scanning Electron Microscopy (SEM) in order to detect a suspicious content of fibrous minerals, which would immediately suggest avoiding any further quarrying. For SEM, samples were prepared in aqueous suspension in an ultrasonic bath, and SEM was recorded with a Zeiss EM 900 instrument.

4. Results

The results obtained for the physical and mechanical properties are summarized in Table 2.

Table 2. Statistical results of the mechanical properties evaluated. (CV), coefficient of variation (CV is the ratio of the standard deviation to the mean; (ρ), bulk density, (a), water absorption; (V_P), P-wave velocity; (V_S), S-wave velocity; (V_P SAT), P-wave velocity in saturated samples; (V_S SAT), S-wave velocity in saturated samples; (E_d), dynamic elastic modulus; (UCS), uniaxial compression strength; (E_s), static elastic modulus.

Statistic Parameter	Max	Min	Average	Median	Standard Deviation	CV	Total Number of Measurements
ρ (kg/m^3)	2838	2597	2747	2759	75	3%	21
a (%)	0.7	0.1	0.4	0.5	0.2	50%	8
V_P (m/s)	5914	1859	4673	4731	1098	23%	21
V_S (m/s)	3342	1264	2657	2722	567	21%	21
V_P/V_S	2.05	1.35	1.75	1.23	0.18	10%	21
V_P (m/s)(Sat)	5983	3258	5079	5082	837	16%	9
V_S (m/s)(Sat)	3257	1608	2738	2728	499	18%	9
E_d (GPa)	78	9	51	49	20	39%	21
Poisson (v_d)	0.34	0.07	0.25	0.25	0.07	28%	20
UCS (MPa)	119.6	24.2	57.1	59	24	42%	15
E_s (GPa)	66	12	39	44	16	41%	15
Poisson (v_s)	0.48	0.08	0.22	0.18	0.14	64%	13

Concerning the physical properties, the average bulk density was 2747 kg/m^3. The highest values found corresponded to the highest values determined for V_P, which were associated with sample M-7b2. The water absorption value was 0.4%, ranging between 0.1%–0.7%. One of the lowest values obtained also had a low bulk density and the lowest V_P (sample M-10b), although there were no significant differences with respect to the other samples.

With regard to the mechanical properties, the uniaxial compression strengths (UCS) values were highly discordant, where the lowest value was observed for sample M-5c (24.2 MPa) and the highest for sample M-2d (119.6 MPa). The average value for the fifteen specimens tested was 57.1 MPa, the coefficient of variation, defined as the ratio of the standard deviation to the mean, being especially significant and reaching 41%. This reflects the enormous variation in the values obtained for this parameter, and the lack of meaning of averages in this context (Table 2). Sample breakdown was reached after an axial strain between 0.07%–38% (mean 0.20%), and the Es from 12 (specimen M-10a) to 66 GPa (specimen M-7b2), with a mean of 39 GPa. Additionally, values ranged from 0.08–0.48 for the Poisson ratio, with a mean of 0.22. Most of the breakdown occurred following previous discontinuity surfaces, frequently with light colors in hand sample. The angle between the perpendicular line to the failure surface and the axis of the specimen varied between 60° and 70°.

This scattered behavior occurs quite frequently and is well known in foliated rocks [34–36] as its strength is a function of the angle between the loading axis and the orientation of anisotropy planes (β). The strength is at its maximum when β is equal to 90° and at its minimum when the β angle is around 30° (45° -φ/2, being φ the friction angle of discontinuity in the Mohr-Coulomb breakdown criterion). Taking this into account, it is possible to evaluate rock anisotropy using the anisotropy ratio [37,38], defined as the ratio of compressive strength at β = 90° and the minimum strength observed ($\sigma_{c90}/\sigma_{cmin}$). The maximum UCS obtained in the Moeche serpentinites was 119.6 MP (sample M-2d) in a nodulous serpentine with scarce planes of foliation normal to the direction of the load (β = 90°). A foliated serpentinite, on the other hand, with β = 30° and a fracture surface in a talc rich area, showed the minimum UCS (24.2 MPa, sample M-3b). This sample would be classified as high anisotropy since the ratio $\sigma_{c90}/\sigma_{cmin}$ is between 4 and 6. Other samples with similar foliation orientations ($\beta \approx 30°$) show a lower anisotropy ratio, which highlights that the mechanism involved in the failures also depends on the intensity of foliation, the amount of nodulous areas, and the mineralogical composition of the failure planes. Even still, the typical behavior of a foliated rock is clearly present.

According to the Reference [39] classification for the uniaxial compression strength in rocks (Figure 3), the samples were classified in groups that included weak to very strong rocks, with most of the samples being classified as strong (50–100 MPa). In most cases, the relative Es/UCS ratio [40] was medium, although occasionally the ratios were high.

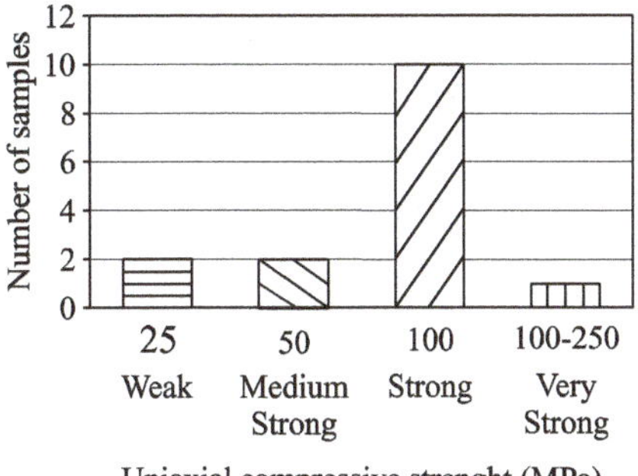

Figure 3. Samples from Moeche grouped according to their UCS based on the ISRM classification system [39]: Weak (UCS > 25 MPa), medium-strong (25–50 MPa), strong (50–100 MPa), and very strong (>100 MPa).

As for the ultrasonic velocities, V_P ranged from a low of 1859 to a maximum of 5914 m/s in dried specimens. The average was 4673 m/s and the relation between the standard deviation and the average was 23%. In the case of saturated samples, the mean V_P was slightly higher, 5079 m/s. The V_S in dry specimens was between 1264–3342 m/s, with an average of 2627 m/s and a coefficient of variation of 21%. Under wet conditions, the average was also slightly higher: 2738 m/s. The V_P/V_S ratio ranged between 1.35–2.05, with a mean of 1.75.

Samples M-10a and M-10b had the lowest V_P under dried conditions, obtaining values below the average, with less than 2000 m/s. Samples M-7, M-5, and M-2d displayed the highest V_P values, all of them above 5000 m/s. We continue to study the correlation between ultrasonic velocities and the other parameters to understand the results, but the preliminary explanation seems to be related to the different mineralogical composition of samples, from only serpentine (antigorite) to a mixture with talc and carbonates.

The dynamic elastic modulus ranged between a minimum of 9 GPa and a maximum of 78 GPa, with a mean of 49 GPa. Evidently, the highest and the lowest values were associated with the samples with the maximum and minimum V_P values.

With respect to mineralogy, the results of the x-ray diffraction (XRD) (Table 3) showed that serpentine polymorphs (antigorite, lizardite, and traces of chrysotile) were the main minerals in all of the samples, followed by magnesite, whereas talc and dolomite appeared in smaller proportions. The latter was detected in only four samples, but these data are very important in terms of conservation-restoration, as this carbonated stone, locally called Pedra de Toelo or Pedra de Moeche, was used for the sculptures in façades of many buildings and sculptures of the region [3,17] (Figure 4).

Table 3. Percentage (%) of the phases detected by XRD. Semi-quantitative analysis.

Sample	Serpentine	Magnesite	Talc	Dolomite
M-10a	100	0	0	0
M-2d	89	0	0	11
M-8c	84	12	4	0
M-4c	79	17	4	0
M-7c2	79	14	7	0
M-4a	78	16	6	0
M-7b2	71	23	6	0
M-8d	68	25	6	0
M-7a1	62	29	9	0
M-5a	53	38	9	0
M-3e	50	34	12	3
M-5b	50	24	25	0
M-5c	50	24	25	0
M-3b	49	41	8	2
M-3a	48	40	10	3
Max	100	41	25	11
Min	48	0	0	0
Average	67	22	9	1
Standard Deviation	16.3	12.4	7.3	2.9

Figure 4. Details of the use of the carbonated serpentinite (Pedra de Toelo or Pedra de Moeche) in the architectural heritage of Galicia. This stone can be found in monuments and historical buildings from the Middle Ages. (**a**) Door lintels, and other architectonic ornaments in the façade of Labacengos church, in Moeche (16th century); (**b**) coat of arms of the Monastery of Oseira (oso means bear in Spanish).

Different textures were found in the Moeche serpentinites depending on the original mafic mineral and on the degree of recrystallization. In the thin sections analyzed, it was possible to distinguish the following:

- Preferentially mesh texture and sometimes hour-glass texture. This is a typical texture of olivine serpentinization and is frequently observed in nodulous areas (not affected by foliation or fractures) (Figure 5a).
- Bastite texture. Serpentine mineral aggregates elongated and with similar orientation. This is a serpentine that has replaced old pyroxenes, and in these cases, the presence of oriented magnetite following pyroxene exfoliation surfaces is quite frequent (Figure 5b).
- Thorn texture. This has a higher degree of crystallization than the previous textures and is developed from them. Original minerals may not be clear, although there is a tendency for this texture to appear normal to olivine, and parallel to the pyroxenes (Figure 5c).
- Fibrous serpentine (chrysotile) growing normal or parallel to fractures (Figure 5d).

Figure 5. Textures found in the Moeche serpentinites (Crossed Nichols): (**a**) Mesh texture (M-8c); (**b**) bastite (Ba) (M-3b); (**c**) thorn texture beginning to recrystallize on bastite (M-2d); (**d**) fibrous serpentine (possibly chrysotile) (M-10a).

In addition, most of the samples included chromite spinels and chromites as accessory minerals, whose borders are frequently replaced by magnetite.

The Moeche serpentinites usually showed an irregular foliation with varying directions, and wavy surfaces marked by veins filled either with carbonates or by slices parallel to them, where talc and/or serpentine was predominant. Frequent fractures running in all directions and filled with carbonates were also observed.

Foliation and fractures often affected the whole rock, but sometimes they were less persistent, leaving nodular areas with no discontinuities. These areas were of a dark green color in the hand specimens and under the petrographic microscope, serpentine, talc and carbonate could be distinguished. Some textures were the remains of the original mineralogy (olivine and pyroxene) (Figures 6 and 7).

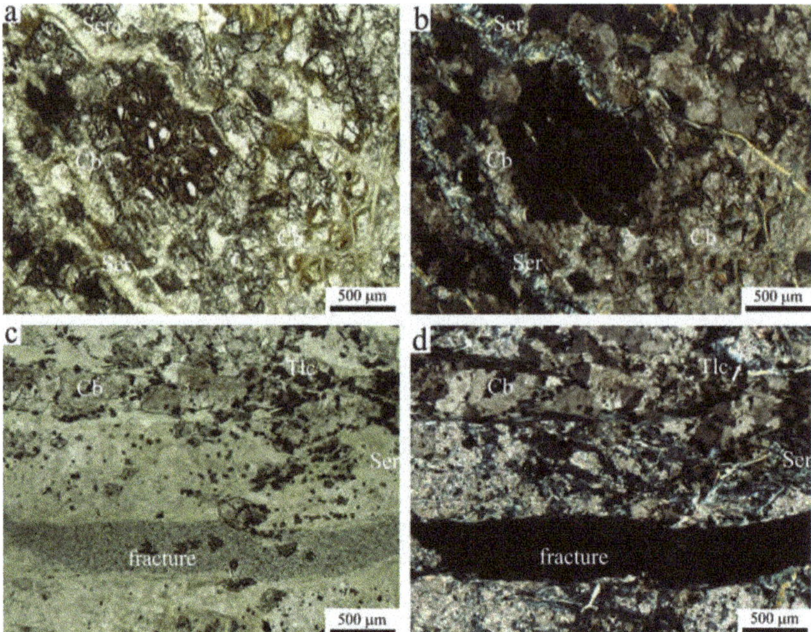

Figure 6. (**a,b**) Sample M-4a. Carbonation of olivine in an early stage of serpentinitization with serpentine with mesh texture (a, Parallel Nichols and b, Crossed Nichols); (**c,d**) sample M-3a showing clean carbonates filling fractures, fracture with predominant talc and carbonate crystals aligned following the foliation and parallel to the fracture (c, Parallel Nichols and d, Crossed Nichols). Ser: Serpentine; Cb: Carbonates; Tlc: Talc.

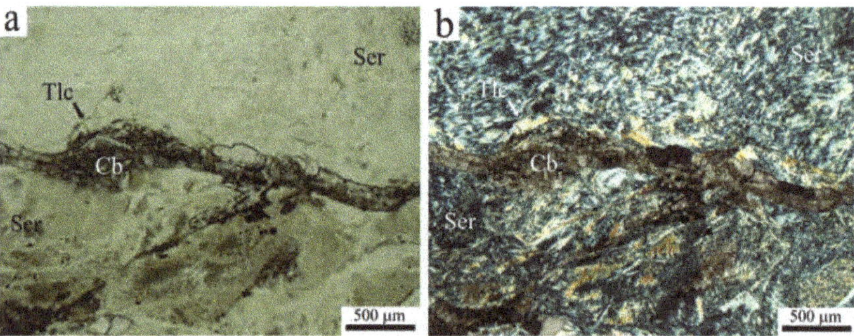

Figure 7. The microscopic aspect of a nodulous sample, not foliated, with serpentine minerals, talc and carbonates, the latter filling a fracture. (**a**) Parallel Nichols and (**b**) Crossed Nichols. Sample M-2d. Ser: Serpentine; Cb: Carbonates; Tlc: Talc.

In cases where a relation between rock failure and the mineralogy was established in the uniaxial compression strength test, it was observed that most of the fractures occurred following foliation or in bands parallel to this foliation. In the latter case, the predominant mineralogy was talc, a talc-serpentine association and, occasionally, carbonates filling these discontinuities. In general, the failures were not brittle, except in the case of the nodular samples, in which the fracture surface developed more irregularly and encircled porphyroblasts.

Weak specimens failed following the foliation planes defined by talc (Figure 8a–c) or chrysotile (Figure 8d–f), with an irregular morphology that sometimes seemed to avoid carbonated areas. This is an indication of the higher resistance present in these areas.

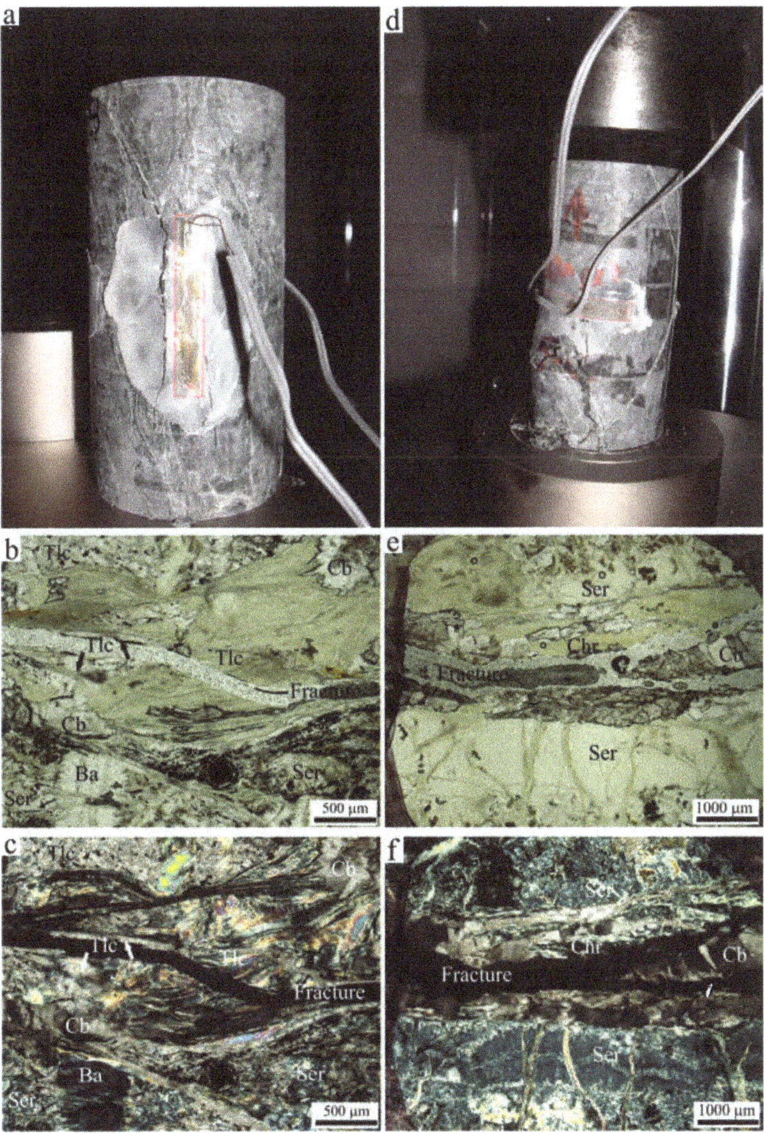

Figure 8. Fractured weak rock samples. Left (**a–c**): Sample M-3b (UCS = 24 MPa), with the failure surface following foliation with abundant talc in the surrounding areas. Apparently, the fracture does not pass through the most carbonated zones. (**a**) Sample after UCS testing; (**b**) Parallel Nichols, and (**c**) Crossed Nichols. Right (**d–f**): Sample M-5c (UCS = 25 MPa), with the failure surface following an undulated foliation defined by chrysotile and replacement carbonate in relict olivine where serpentinization has taken place. (**d**) Sample after UCS testing, (**e**) Parallel Nichols, and (**f**) Crossed Nichols. Ser: Serpentine; Cb: Carbonates; Tlc: Talc; Chr: Chrysotile.

In moderately strong samples, the foliation was not as penetrating as in the weak samples, but was more regular (Figure 9a–c). Unlike the weak specimens, the fractures sometimes passed through areas where carbonate recrystallization had occurred (Figure 9d–f).

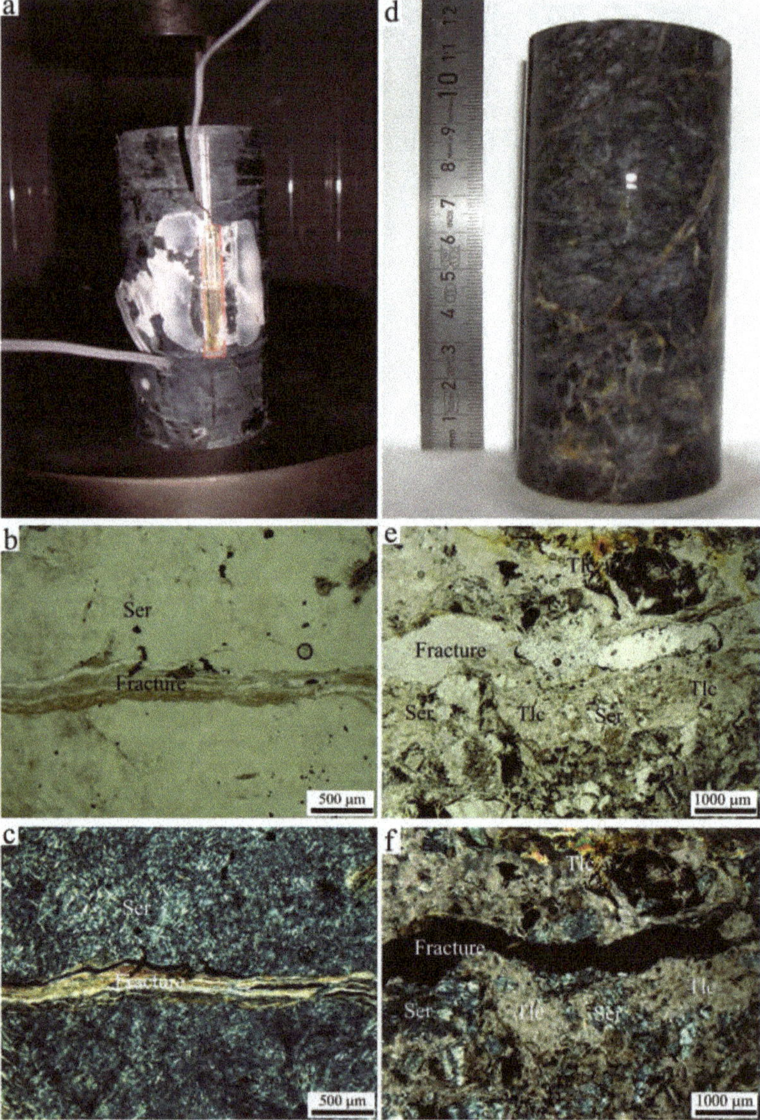

Figure 9. Fractured medium-strong rock samples. Left (**a–c**): Sample M-10a (UCS = 35.1 MPa) with the failure following the serpentinite foliation in a specimen with few previous fractures. (**a**) Sample after UCS testing; (**b**) Parallel Nichols, and (**c**) Crossed Nichols. Right (**d–f**): Sample M-8d (UCS = 35.8 MPa) with the fracture following the foliation, defined by intense carbonation and relevant talc contents; (**d**) Sample before testing; (**e**) Parallel Nichols, and (**f**) Crossed Nichols. Ser: Serpentine; Tlc: Talc.

Strong and very strong samples shared a brecciated aspect, but the size of the rock fragments in the strong samples was smaller (Figure 10a,d). Also, in the strong samples, the fractures developed on foliation planes, filled with carbonate and talc (Figure 10b,c). Moreover, when these foliation planes were not continuous, the fracture tended to surround areas containing relict porphyroblasts, usually completely serpentinized (Figure 10e,f).

Figure 10. Fractured strong rock. Left (**a**–**c**): Sample M-7a1 (UCS = 52.2 MPa) showing the fracture following the foliation defined by talc and carbonates. (**a**) Sample before testing; (**b**) Parallel Nichols and (**c**) Crossed Nichols. Right (**d**–**f**): Sample M-7c2 (UCS = 70 MPa), with the failure surface surrounding a serpentinized nodule with talc and carbonates. (**d**) Sample before testing; (**e**) Parallel Nichols, and (**f**) Crossed Nichols. Ser: Serpentine; Cb: Carbonates; Tlc: Talc.

Regarding the very strong samples, their fractures were brittle. It was found that neither the foliation nor the previous fracture orientations were followed, although the latter were clearly visible at the macroscopic scale, but with scarce continuity and irregular orientation (Figure 11).

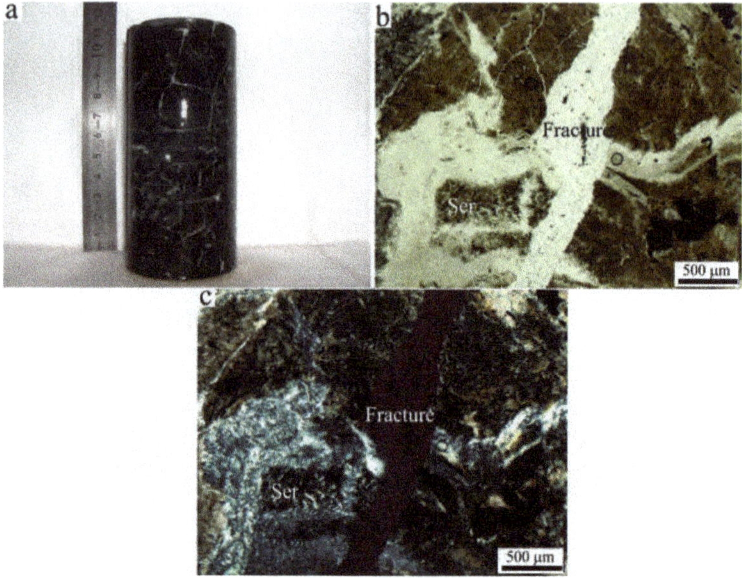

Figure 11. The fractured very strong rock sample. Sample M-2d (UCS = 119.6 MPa). The failure surface does not follow previous discontinuities. (**a**) Sample before testing, (**b**) Parallel Nichols, and (**c**) Crossed Nichols are shown. Ser: Serpentine.

5. Discussion

Moeche serpentinites show a complete transformation of primary minerals in low-grade metamorphism phases, such as minerals from serpentine, carbonate, talc and magnetite, preserving the remains of the chromium rich original spinel [41]. All of these changes cause the mechanical properties of the rock to change, as reflected in the highly variable results obtained during the uniaxial compression strength testing carried out in this study. This parameter is significant for the determination of the behavior of dimension stones, and therefore it is important to establish the factors governing it and other properties, which can be used as an index to predict the uniaxial compression strength of these rocks.

The serpentinites of Moeche generally have low UCS, with an average value of 57 MPa, which is even below the minimum threshold required by ASTM C-1526-02 for susceptible serpentinites to be used as ornamental stones [21]. The maximum water absorption value obtained, 0.7%, is also outside the limit values of the above standard for both outdoor (0.2%) and indoor (0.6%) use. The only ASTM requirement that these rocks accomplish is the minimum bulk density, which was 2600 kg/m^3, above the minimum value required of 2560 kg/m^3.

Table 4 compares the properties of serpentinites from different countries reported in the literature and the ones studied here. From this data it can be observed that only the carbonated serpentinites from Macael (SE of Spain) studied by Navarro et al. [6] fulfill all of the requirements set out by the ASTM standard (UCS > 69 MPA; ρ > 2560 kg/m^3 a < 0.2% (exterior) and a < 0.6% (interior)).

The UCS values obtained are similar to the serpentinites studied in Diamantis et al. [13] and in Kurtulus et al. [14] from the ophiolite complex in Mount Kallidromo, in central Greece and from NW Turkey, respectively. However, differences can be found with respect to the serpentinites from Granada and Macael (SE of Spain) that have an average UCS above 227 MPa. Serpentinites can also have very high values, such as 346 MPa, as is the case of serpentinites from Granada [6] and Egypt [12]. But it is important to highlight that the UCS values obtained in a laboratory setting are highly dependent on multiple factors. Some of these factors are intrinsic (mineralogy, fabric or discontinuities), but others are extrinsic such as shape, size, the relation between width and length and the speed and the direction of the application of the load [42,43]. UCS values obtained with different protocols (e.g., EN, ASTM) vary even with the same stone, therefore, the comparison between different stones must be approached carefully [44]. The low resistance and significant dispersion found for this property could be explained by the macrostructure of foliated rocks such as serpentinites. The UCS in these rocks depends largely on the relative orientation between the microstructure and the axial load, and therefore the range of dispersion obtained is usually broad [40]. This circumstance often leads to a misinterpretation of the shear strength values of discontinuity planes, where they are thought to correspond to failures in the rock matrix [10]. There are important implications when the global behavior of serpentinized rock masses is analyzed. Upon studying several outcrops, the same authors proposed a serpentinite matrix uniaxial compression strength of around 40 MPa, which is slightly lower than the average obtained in this study (57 MPa). However, on measuring and comparing the inclinations of the principal planes of the macroscopic Moeche foliation samples tested, it was found that apparently not only the relative orientation between the load applied and the foliation control the uniaxial compression strength of these rocks (Table 5), since there are very weak samples that, macroscopically, are not foliated, such as sample M-3b. This apparently contradicts our hypothesis, but this is in fact not the case, because under the petrographic microscope a microscopic foliation can be distinguished. Therefore, macroscopic observations alone can lead to incorrect conclusions being drawn in cases where apparently foliated specimens are analyzed, as previously noted in Hawkins et al. [28] and references therein. By contrast, the Moeche sample M-2d is weakly foliated at both microscopic and macroscopic scale and its UCS scored the highest of all the specimens tested: 120 MPa. Therefore, as a first conclusion, it can be inferred that foliated samples are weaker than unfoliated ones. In unfoliated samples, although they are sometimes highly fractured, the presence of minerals in the fracture filling—typically carbonates—could hamper the progression of failed surfaces, resulting in a higher UCS.

Table 4. Parameters of different serpentinites compared with the Moeche serpentinites. (m) minimum; (Av) average; (M) maximum; (SD) standard deviation.

Origin			PARAMETER				
			UCS (MPa)	ρ (kg/m^3)	a (%)	V_P (m/s)	V_S (m/s)
Moeche (Spain) (this paper)		m	24	2600	0.10	1859	1264
		Av	57	2750	0.40	4673	2657
		M	129	2840	0.70	5914	3342
		SD	24	70	0.20	1098	567
Egypt [12]		m	89	2480	0.01	-	-
		Av	152	2520	0.10	-	-
		M	189	2590	0.20	-	-
		SD	32	30	0.06	-	-
Greece [13]		m	19	2490	0.14	4842	2425
		Av	60	2610	0.57	5344	2783
		M	126	2730	1.84	5789	3109
		SD	27	60	0	225	170
Turkey [14]		m	22	2430	0.16	4265	-
		Av	54	2564	0.46	5018	-
		M	77	2660	1.33	5461	-
		SD	16	58	0.39	341	-
South of Spain [6]	Granada	m	331	2658	0.19	5532	-
		Av	346	2662	0.26	5646	-
		M	361	2666	0.33	5589	-
		SD	51	18	0.10	457	-
	Macael (Carbonated)	m	139	2704	0.10	6024	-
		Av	227	2796	0.13	6078	-
		M	315	2888	0.16	6132	-
		SD	43	32	0.03	309	-
	Macael (Non-carbonated)	m	246	2650	0.42	5308	-
		Av	279	2949	0.48	5203	-
		M	263	2800	0.53	5378	-
		SD	61	134	0.10	673	-

Table 5. Comparative data on uniaxial compression strength obtained for each sample, ISRM classification, macroscopic features, and main general aspects, singularities of fracture and failure angle. Foliation and angle of failure measured from the base of the specimen and typology of the failure surface. ISRM Classification: (W) Weak, (M) Medium-Strong, (S) Strong, (VS) Very Strong.

SAMPLE	UCS (MPa)	ISRM CLASSIF	FOLIATION	MACROSCOPIC FEATURES	FRACTURE SURFACE	FAILURE ANGLE
M-2d	120	VS	No foliation	Nodulous, scarce foliation. Anisotropy due to fractures. Brecciated aspect.	Does not follow previous orientations. Breaks minerals.	30°
M-3a	51	S	60–90°	Foliated, nodulous aspect locally. Anisotropic aspect.	Following foliation, within an area where talc is predominant and aligned carbonates appear.	75°
M-3b	24	W	60–90°	Foliated, nodulous aspect locally. Anisotropic aspect.	Foliation in talc-rich area.	75°
M-3e	52	S	50–60°	Foliated, nodulous aspect locally.	Following foliation.	90°
M-4a	60	S	70°	Nodulous and irregular fractures. Scarce foliation. Anisotropic aspect.	Following fractures.	70°
M-4c	60	S	20–45°	Nodulous and irregular fractures. Scarce foliation. Anisotropic aspect.	Subvertical, locally following foliation.	80°
M-5a	68	S	80°	Irregular. Anisotropic aspect.	Affects both foliated and unfoliated areas.	60°
M-5b	60	S	60–70°	Two main foliations easily recognized in the specimen. Important failure at 60°. Anisotropic aspect.	Previous carbonated plane.	60°
M-5c	25	W	60–90°	Nodulous areas with mineral alignments (80°–85°) and variable fissures (60°–90°). Anisotropic aspect.	Previous carbonated plane.	50°
M-7a1	52	S	70°	Foliated, locally nodulous aspect. Brecciated.	Carbonated foliation with talc-rich bands.	70°
M-7b2	87	S	40–80°	Nodulous, scarce foliation. Anisotropy due to fractures. Anisotropic aspect.	Partially follows foliation.	75°
M-7c2	70	S	45–70°	Foliated, locally nodulous aspect. Brecciated aspect.	Talc in the foliation and parallel to carbonated bands.	65°

Table 5. *Cont.*

SAMPLE	UCS (MPa)	ISRM CLASSIF	FOLIATION	MACROSCOPIC FEATURES	FRACTURE SURFACE	FAILURE ANGLE
M-8c	59	S	40–65°	Foliation marked by fractures and nodulous zones in between. Brecciated and foliated area.	Foliation, but not always.	45°
M-8d	36	M	No foliation	Irregular due to several orientations. Anisotropic aspect.	Foliation	55°
M-10a	35	M	0–50°	Irregular due to several orientations. Massive aspect.	Foliation	40°

In the light of the dominant role of foliation in these rocks, the question arises as to whether it might be possible that mineralogical differences between the samples could account for the wide dispersion of the UCS obtained. Figure 12 shows the mineralogy of each sample, ordered from the highest to the lowest UCS, and no clear relationship can be obtained between the UCS and the percentage of serpentine, talc, dolomite or magnesite.

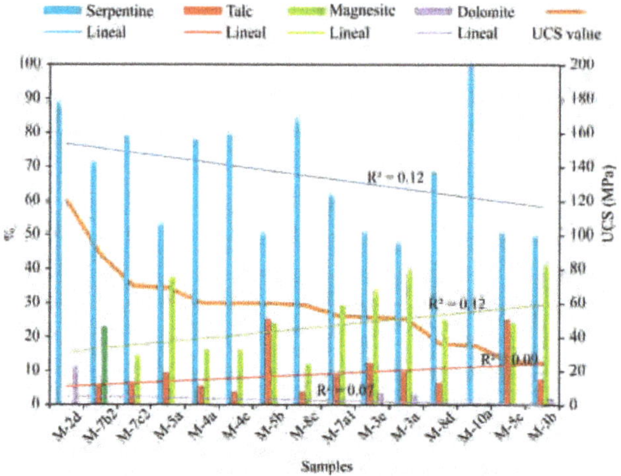

Figure 12. Mineralogy of the Moeche serpentinites according to the semiquantitative results obtained in the XRD analyses. Samples are organized from left to right according to their UCS (MPa). Straight lines represent the linear correlation between each mineral and its UCS.

Nevertheless, the relationship between the percentage of serpentine minerals and the other minerals seems to increase in the strongest samples (Figure 13a), except for sample M-10a, which is comprised of 100% serpentine minerals but has a UCS of 36 MPa. In this sample, the orientation of the load applied and the influence of foliation prevail over mineralogy. Some authors [6,12,15,18] have explained that in similar rocks the carbonate content seems to be responsible for the greater UCS. The trajectories of failure planes in weak rocks were analyzed under a petrographic microscope, and it was observed that the fractures did not progress following the path determined by carbonates. In contrast, they progressed through the carbonates in medium-strong and strong rocks (Figure 13b). This can be explained by the alteration processes that accompany carbonation, which sometimes lead to the formation of microfractures due to the loss of material in these areas [2]. This was also observed by Navarro et al. [6]. In the case of carbonated serpentinites from Macael, the UCS value in the less carbonated samples was much higher than the more carbonated serpentinites. They concluded that in the initial phases of carbonation, carbonates act as cement, increasing the strength. As carbonation progresses, the pressure of crystallization produces fractures, causing the stone to become weak.

The serpentinites from Moeche vary from weak, medium strong, strong and very strong rocks, whose behavior cannot be explained only by their mineralogy. In weak samples there exists slight anisotropy defined not only by the existence of discontinuity planes, but also by a very strong preferred orientation associated with mineralogy (mainly talc); medium strong and strong samples are foliated but this feature is less intense than in the weak samples. Finally, the very strong samples display macroscopic discontinuity surfaces, but these are less continuous than in the previous cases. This fact favors a tendency towards isotropic behavior of the rock, observed microscopically in the irregular shape of the fractures formed without following any foliation or preferred mineral alignment. Glawe and Upreti [8] classified serpentinites with discontinuities of little persistence as very strong rocks, and our data from Moeche corroborates their conclusion.

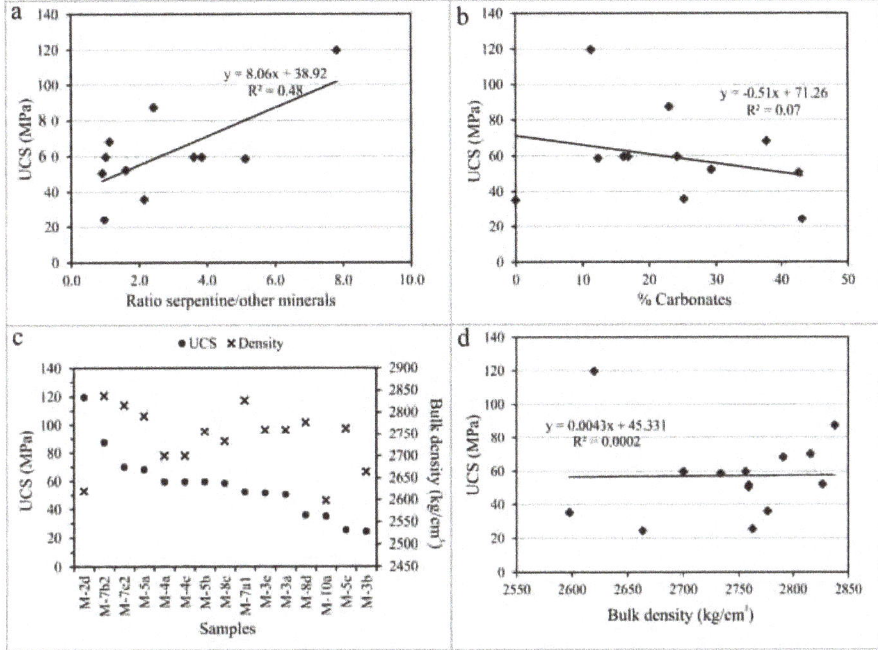

Figure 13. Correlation between UCS and some of the parameters studied: (**a**) UCS and serpentine/other minerals ratio; (**b**) UCS and % carbonates; (**c**) UCS and density bulk for each sample, from the highest value (**left**) to the lowest (**right**); (**d**) UCS test and bulk density.

The enormous difficulties involved in obtaining samples for conducting uniaxial compression tests underscore the need for other types of much simpler tests which may provide relevant information on different mechanical aspects. The correlations between UCS and other tests or analyses are relatively abundant in the literature [45–48], but they do not address serpentinites. One of the most studied correlations is bulk density and the UCS; this correlation is shown in Figure 13c,d for the serpentinites collected from Moeche. The results are not satisfactory at all, showing a very low correlation coefficient (R^2), and disagree with the observations reported by Diamantis et al. [13] for Greek serpentinites, Nespereira et al. [49] for tertiary sandstones, and those of ISRM [39]. Our results are in agreement with the data reported by Reference [50] in sedimentary rock and Reference [51] in basic rocks, where the relationship between the two parameters is very low. Bulk density is directly related to mineralogy, and the fact that they are not well correlated reinforces the idea that the mechanical behavior of serpentinites usually depends on several factors and is much more complex than in other types of rocks.

Other correlations considered in other works are related to the propagation velocities of ultrasonic waves. V_P and V_S correlate well with each other (Figure 14a), but do not correlate well with the mechanical properties obtained from the uniaxial compression tests. It was apparent that the higher the V_P or the V_S, the higher the UCS, but the value of R^2 ranges between 0.17–0.29 (Figure 14b). The same can be said for the correlation between the velocities and bulk density ($R^2 = 0.60$ to 0.59) (Figure 14c) and between the static elastic modulus (E_s) and the velocities (Figure 13d) ($R^2 = 0.74$ to 0.72), in this case using an exponential correlation. An $R^2 = 0.60$ is obtained for E_d and E_s (Figure 14e). Comparing these results with the results for other rocks of different origin, it is common to obtain good correlations between UCS and E_s in granites, basalts and quartzites [52], in sedimentary rocks [46], and even in some serpentinites and peridotites [11]. V_P and UCS are also correlated positively with each other in serpentinites [13,14], but much more pronounced (with R^2 above 0.80 in both cases) than those observed in the Moeche rocks or in the case of serpentinites from Southern Spain, where no correlation was observed [6]. Finally, Figure 14e also shows the values of E_s and E_d. Their ratio lies between 1 and 2.3, with only two data points below 1. The correlation between both is similar to that obtained by Martínez-Martínez et al. [53] for a wide range of carbonate rocks, and follows the general trend observed by Reference [54]: As the value of E_d increases, so does the E_d/E_s ratio. A good correlation between V_P and Es was observed by Kurtulus et al. [14], but in this case, the authors observed a linear correlation.

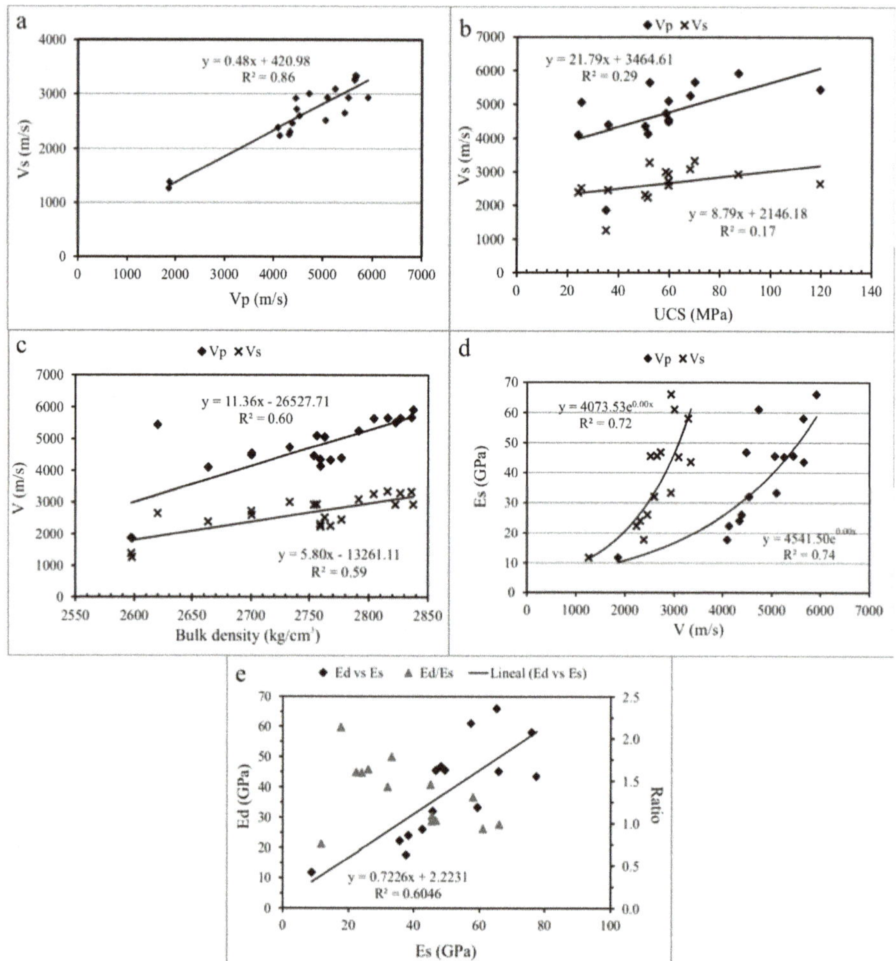

Figure 14. Correlation between propagation velocities of ultrasonic wave and other parameters: (**a**) V_P and Vs; (**b**) V_P and Vs with UCS; (**c**) V_P and Vs with bulk density; (**d**) V_P and Vs with static modulus Es; (**e**) Elastic modulus ratio.

Because careful petrographic studies showed the existence of some mineral fibers in these rocks, corresponding to chrysotile, and because the increasing awareness regarding the content of asbestos in quarrying materials, we analyzed extra samples from the same quarries by SEM to clarify the major component of the serpentine phases in the rocks from Moeche. The preliminary results are compatible with all the analytical methods used to characterize the studied serpentinites: SEM shows that the major serpentine phase is antigorite (Figure 15). A small percentage is shown as lizardite, and very few fibers of chrysotile were detected, but in a small amount that could be compatible with the potential extraction of the stone. Further studies on this subject are being carried out to study the safe handling of a potential re-opening.

Figure 15. Scanning electron micrograph of a serpentinite sample from Moeche. There are predominant globular morphologies identified as antigorite, which is in agreement with the x-ray diffraction analysis. The few prismatic shapes correspond to lizardite crystals, and also a few particles with acicular shapes observed in the micrographs are assigned to the chrysotile phase of the serpentine [55].

6. Conclusions

Serpentinites from Moeche are very heterogeneous rocks. It is difficult to establish relatively well-defined intervals for their mechanical properties. They are rocks whose strengths range from weak to very strong, and this property seems to be reflected by the textural appearance of the rock. It is confirmed that in most cases, and in contrast to the initial general idea, the most resistant samples are those with a more brecciated aspect. This type of sample has significant carbonation that progresses, in conjunction with talc, through a network of irregular fractures, some of them coinciding with the foliation of the rock, but having poor continuity. Strongly-carbonated rocks are of interest for the potential conservation-restoration of architectural heritage that used that stone, locally named as "Pedra de Toelo" or "Pedra de Moeche".

Continuous and persistent foliation weakens these rocks, even when detected only at a microscopic scale.

The bulk density of the Moeche serpentinites does not correlate with either the UCS or the velocity of ultrasonic waves, and hence they cannot be recommended for indirect estimations. However, ultrasonic data can be useful for obtaining a first approach to the static elastic parameters, which are usually lower than the dynamic parameters, the E_d/E_s ratio increases with E_d.

The general trend observed in previous studies addressing serpentinites from other locations cannot be confirmed in the Moeche serpentinites, but, similar to the carbonated serpentinites from

Macael (SE Spain), the higher the percentage of carbonates, the lower the UCS. In any case, it seems that carbonates govern the fracture morphology. What might possibly further determine the UCS in the serpentinites from Moeche is their origin; the UCS increases if there is a carbonate replacement of serpentine phases, and decreases if carbonate has filled the fractures.

A direct relationship between the mineralogical composition and the compressive strength of these rocks has not been determined. Nevertheless, a general tendency of the latter to be higher when the proportion of serpentines is also higher can be observed.

The complex relationship between mineralogy, structure, and microstructure is responsible for the large spread of values of resistance of these rocks, which in turn makes it difficult to establish a threshold value for gauging its suitability as a dimension stone based on this property. For such thresholds, these rocks do not meet American standards (ASTM), owing to their high water absorption and their low minimum UCS, which is in agreement with serpentinites from other currently operating quarries. Consequently, a draft outlining the European standards for these rocks is necessary and should include, among others, the petrographic characteristics of the rock and its mineral composition, both of which are factors that, according to the results obtained in this study, determine the geomechanical behavior of serpentinites.

According to the study presented in this paper, the serpentinites from Moeche need a complete and specific mineralogical, physical, and mechanical characterization before they can be used properly, as an ornamental stone or as conservation-restoration material, to avoid undesirable results for construction and conservation-restoration activities. Preliminary positive results associated to our research are related to the very low content of asbestos, in any of the possible phases of amphibole or chrysotile, detected in these rocks, which is an advantage is terms of quarrying even for further testing.

The present paper will be a step forward in the preservation of cultural and architectonic heritage of Galicia.

Author Contributions: J.N., R.N., S.M. and D.P. were responsible for the conceptualization and validation of data, as well as for the writing and editing of the manuscript. All authors were involved in sampling. D.P. was responsible for the funding acquisition.

Funding: This research was funded by the projects: CGL2004–03048, CGL2006–05128/BTE and BIA2017-85897-R from the Spanish government, and the University of Salamanca through project TCUE 2015 and the University's own financial program.

Acknowledgments: The authors wish to thank the Laboratory of Applied Petrology of the CSIC-University of Alicante Associated Unit for their help with the ultrasonic tests, and the Language Services of the University of Salamanca for the language revision. D. Pereira acknowledges the research teams of Inorganic Geochemistry of the universities of Salamanca and Complutense of Madrid for SEM analysis and discussion of results. Two anonymous reviewers are acknowledged for their help in improving this paper.

References

1. O'Hanley, D.S. *Serpentinites: Records of Tectonic and Petrological History*; Oxford University Press: New York, NY, USA, 1996; p. 277.
2. Meierding, T.C. Weathering of serpentine stone buildings in the Philadelphia region: a geographic approach related to acidic deposition. In *Stone Decay in the Architectural Environment*; Turkington, A.V., Ed.; Geological Society of America: Denver, CO, USA, 2005; Volume 399, pp. 17–25. [CrossRef]
3. Pereira, M.D.; Peinado, M.; Yenes, M.; Monterrubio, S.; Nespereira, J.; Blanco, J.A. Serpentinites from Cabo Ortegal (Galicia, Spain): A search for correct use as ornamental stones. In *Natural Stone Resources for Historical Monuments*; Prikryl, R., Török, A., Eds.; Geological Society: London, UK, 2010; Volume 333, pp. 81–85.
4. Navarro, R.; Pereira, D.; Rodríguez-Navarro, C.; Sebastian-Pardo, E. The Sierra Nevada serpentinites: The serpentinites most used in Spanish heritage buildings. In *Global Heritage Stone: Towards International Recognition of Building and Ornamental Stones*; Pereira, D., Marker, B., Kramar, S., Cooper, B., Schouenborg, B., Eds.; Geological Society: London, UK, 2015; Volume 407, pp. 101–108.

5.   Pereira, M.D.; Blanco, J.A.; Peinado, M. Study on Serpentinites and the consequence of the misuse of natural stone in buildings for construction. *J. Mater. Civ. Eng.* **2013**, *25*, 1563–1567. [CrossRef]

6.   Navarro, R.; Pereira, D.; Gimeno, A.; Del Barrio, S. Influence of natural carbonation process in serpentinites used as construction and building materials. *Constr. Build. Mater.* **2018**, *170*, 537–546. [CrossRef]

7.   Coumantakis, J. Comportement des peridotites et des serpentinites de la grece en travaux publics et leur proprietes physiques et mecaniques. *Bull. Eng. Geol. Environ.* **1982**, *25*, 53–60. [CrossRef]

8.   Glawe, U.; Upreti, B. Better understanding of the strengths of serpentinite bimrock and homogeneous serpentinite. *Felsbau Rock Soil Eng.* **2004**, *22*, 53–60.

9.   Ozsoy, E.A.; Yilmaz, G.; Arman, H. Physical, mechanical and mineralogical properties of ophiolitic rocks at the Yakakayi dam site, Eskisehir, Turkey. *Sci. Res. Essays* **2010**, *5*, 2579–2587.

10.   Marinos, P.; Hoek, E.; Marinos, V. Variability of the engineering properties of rock masses quantified by the geological strength index: The case of ophiolites with special emphasis on tunnelling. *Bull. Eng. Geol. Environ.* **2005**, *65*, 129–142. [CrossRef]

11.   Christensen, N.I. Serpentinites, peridotites, and seismology. *Int. Geol. Rev.* **2004**, *46*, 795–816. [CrossRef]

12.   Ismael, I.; Hassan, M. Characterization of some Egyptian serpentinites used as ornamental stones. *Chin. J. Geochem.* **2008**, *27*, 140–149. [CrossRef]

13.   Diamantis, K.; Gartzos, E.; Migiros, G. Study on uniaxial compressive strength, point load strength index, dynamic and physical properties of serpentinites from central Greece: Test results and empirical relations. *Eng. Geol.* **2009**, *108*, 199–207. [CrossRef]

14.   Kurtulus, C.; Bozkurt, A.; Endes, H. Physical and mechanical properties of serpentinized ultrabasic rocks in NW Turkey. *Pure Appl. Geophys.* **2012**, *169*, 1205–1215. [CrossRef]

15.   Navarro, R.; Pereira, M.D.; Gimeno, A.; Del Barrio, S. Verde Macael: A serpentinite wrongly referred to as a marble. *Geoscience* **2013**, *3*, 102–113. [CrossRef]

16.   Navarro, R.; Pereira, D.; Gimeno, A.; del Barrio, S. Characterization of the natural variability of Macael serpentinite (Verde Macael) (Almería, south of Spain) for their appropriate use in the building industry. In *Engineering Geology for Society and Territory. Volume 5: Urban Geology, Sustainable Planning and Landscape Exploitation*; Lollino, G., Manconi, A., Guzzetti, F., Culshaw, M., Bobrowsky, P., Luino, F., Eds.; Springer International Publishing AG: Cham, Switzerland, 2015; Volume 5, pp. 209–211.

17.   Burgoa-Fernández, J.J. El patrimonio etnográfico y el arte popular: cruceros y petos de ánimas de los municipios de Moeche y San Sadurniño. *Anuario Brigantino* **2000**, *23*, 477–494.

18.   Pereira, M.D.; Blanco, J.A.; Yenes, M.; Peinado, M. Las serpentinitas y su correcta utilización en construcción. *Roc Maquina* **2005**, *9*, 24–27.

19.   Pereira, M.D.; Peinado, M.; Blanco, J.A.; Yenes, M. Geochemical characterization of serpentinites at Cabo Ortegal, northwestern Spain. *Can. Mineral.* **2008**, *46*, 317–327. [CrossRef]

20.   Blanco, J.; Peinado, M.; Pereira, D.; Yenes, M.; Nespereira, J.; Monterrubio, S. Mineralogy of serpentinites: A clue for their use as ornamental stones. *Geophys. Res. Abstr.* **2007**, *9*, 05494.

21.   ASTM. *C-1526-02 Standard Specification for Serpentine Dimension Stone*; American Society for Testing and Materials International: Pennsylvania, PA, USA, 2002; p. 2.

22.   Matte, P. The Variscan collage and orogeny (480–290 Ma) and the tectonic definition of the Armorica microplate: A review. *Terra Nova* **2003**, *13*, 122–128. [CrossRef]

23.   Arenas, R.; Martínez-Catalán, J.R.; Sánchez-Martínez, S.; Díaz-García, F.; Abati, J.; Fernández-Suárez, J.; Andonaegui, P.; Gómez-Barreiro, J. Paleozoic ophiolites in the Variscan suture of Galicia (northwest Spain): Distribution, characteristics, and meaning. In *4-D Framework of Continental Crust*; Hatcher, J.R.D., Carlson, M.P., McBride, J.H., Martínez-Catalán, J.R., Eds.; Geological Society of America: Denver, CO, USA, 2007; Volume 200, pp. 425–444.

24.   Arenas, R.; Sánchez-Martínez, S.; Castiñeiras, P.; Jeffries, T.E.; Díez-Fernández, R.; Andonaegui, P. The basal tectonic mélange of the Cabo Ortegal Complex (NW Iberian Massif): A key unit in the suture of Pangea. *J. Iberian Geol.* **2009**, *35*, 85–125.

25.   Albert, R.; Arenas, R.; Sánchez-Martínez, S.; Gerdes, A. The eclogite facies gneisses of the Cabo Ortegal Complex (NW Iberian Massif): Tectonothermal evolution and exhumation model. *J. Iberian Geol.* **2012**, *38*, 389–406. [CrossRef]

26.   Ordóñez-Casado, B.; Gebauer, D.; Schäfer, H.J.; Gil Ibarguchi, J.I.; Peucat, J.J. A single Devonian subduction event for the HP/HT metamorphism of the Cabo Ortegal complex within the Iberian Massif. *Tectonophysics* **2001**, *332*, 359–385. [CrossRef]

27. Mendia, M.; Ibarguchi, J.I. Petrografía y condiciones de formación de las rocas ultramáficas con granate asociadas a las eclogitas de Cabo Ortegal. *Geogaceta* **2005**, *37*, 31–34.

28. Hawkins, A.B. Aspects of rock strength. *Bull. Eng. Geol. Environ.* **1998**, *57*, 17–30. [CrossRef]

29. Benavente, D.; Martínez-Martínez, J.; Jáuregui, P.; Rodríguez, M.A.; García del Cura, M.A. Assesment of the strength of building rocks using signal processing procedures. *Constr. Build. Mater.* **2006**, *20*, 562–568. [CrossRef]

30. AENOR. *Geotecnia: ensayos de campo y de laboratorio*; Asociacion Española De Normalizacion Y Certificacion: Madrid, Spain, 1999; p. 385.

31. Rietveld, H.M. A profile refinement method for nuclear and magnetic structures. *J. Appl. Crystallogr.* **1969**, *2*, 65–71. [CrossRef]

32. ASTM. *C1721-15 Standard Guide for Petrographic Examination of Dimension Stone*; American Society for Testing and Materials International: Pennsylvania, PA, USA, 2015; p. 2.

33. Jaeger, J.C. Shear failure of anisotropic rock. *Geol. Mag.* **1960**, *97*, 65–72. [CrossRef]

34. Wicks, F.J.; Whittaker, E.J.W. Serpentinite textures and serpentinization. *Can. Mineral.* **1977**, *15*, 459–488.

35. Donath, F.A. Strength variaton and deformational behaviour of anisotropic rocks. In *State of Stress in the Earth's Crust*; Judd, W.R., Ed.; Elsevier: New York, NY, USA, 1964; pp. 281–298.

36. Hoek, E. Fracture of anisotropic rock. *J. South. Afr. Inst. Min. Metall.* **1964**, *64*, 510–518.

37. Singh, J.; Ramamurthy, T.; Venkatappa Rao, G. Strength anisotropies in rocks. *Ind. Geotech. J.* **1989**, *19*, 147–166.

38. Ramamurthy, T. Strength, modulus responses of anisotropic rocks. In *Comprehensive Rock Engineering*; Hutson, J.A., Ed.; Pergamon: Oxford, UK, 1993; Volume 1, pp. 313–329.

39. ISRM. Rock characterization. Testing and monitoring. In *Suggested Methods Prepared by the Commission on Testing Methods*; Brown, E.T., Ed.; International Society for Rock Mechanics; ISRM Turkish National Group; Pergamon Press: Ankara, Turkey, 1981; p. 221.

40. Deere, D.U.; Miller, R.P. *Engineering Classification and Index Properties for Intact Rock*; U.S. Air Force Systems Command Air Force Weapons Lab.: Kirtland Air Force Base, NM, USA, 1966; p. 324.

41. Pereira, M.D.; Neves, L.; Pereira, A.; González-Neila, C. Natural radioactivity in ornamental stones: An approach to its study using stones from Iberia. *Bull. Eng. Geol. Environ.* **2011**, *70*, 543–547. [CrossRef]

42. Benavente, D. Propiedades físicas y utilización de rocas ornamentales. In *Utilización de Rocas y Minerales Industriales. Seminarios de la Sociedad Española de Mineralogía*; García del Cura, M.A., Cañaveras, J.C., Eds.; Sociedad Española de Mineralogía: Madrid, Spain, 2006; Volume 2, pp. 123–153.

43. Romana, M.; Vásárhelyi, B. A Discussion on the decrease of unconfined compressive strength between saturated and dry rock samples. In Proceedings of the 11th ISRM Congress, Lisbon, Portugal, 9–13 July 2007.

44. Siegesmund, S.; Dürrast, H. Physical and mechanical properties of rocks. In *Stone in Architecture: Properties, Durability*, 4th ed.; Siegesmund, S., Snethlage, R., Eds.; Springer: Berlin, Germany, 2011; pp. 97–225.

45. Kahraman, S. Evaluation of simple methods for assessing the uniaxial compressive strength of rock. *Int. J. Rock Mech. Min. Sci.* **2001**, *38*, 981–994. [CrossRef]

46. Tsiambaos, G.; Sabatakakis, N. Considerations on strength of intact sedimentary rocks. *Eng. Geol.* **2004**, *72*, 261–273. [CrossRef]

47. Sonmez, H.; Gokceoglu, C.; Medley, E.W.; Tuncay, E.; Nefeslioglu, H.A. Estimating the uniaxial compressive strength of a volcanic bimrock. *Int. J. Rock Mech. Min. Sci.* **2006**, *43*, 554–561. [CrossRef]

48. Yesiloglu-Gultekin, N.; Gokceoglu, C.; Sezer, E.A. Prediction of uniaxial compressive strength of granitic rocks by various nonlinear tools and comparison of their performance. *Int. J. Rock Mech. Min. Sci.* **2013**, *62*. [CrossRef]

49. Nespereira, J.; Blanco, J.A.; Yenes, M.; Pereira, D. Irregular silica cementation in sandstones and its implication on the usability as building stone. *Eng. Geol.* **2010**, *115*, 167–174. [CrossRef]

50. Ulusay, R.; Türeli, K.; Ider, M.H. Prediction of engineering properties of a selected litharenite sandstone from its petrographic characteristics using correlation and multivariate statistical techniques. *Eng. Geol.* **1994**, *38*, 135–157. [CrossRef]

51. Moos, D.; Pezard, P.A. Relationships between strengths and physical properties of samples recovered during Legs 137, 140, and 148 from Hole 504B. In *Proceedings ODP, Scientific Results*; Alt, J.C., Kinoshita, H., Stokking, L.B., Michael, P.J., Eds.; Ocean Drilling Program: College Station, TX, USA, 1996; Volume 148, pp. 401–407.

52. Gupta, A.S.; Seshagiri Rao, K. Weathering effects on the strength and deformational behaviour of crystalline rocks under uniaxial compression state. *Eng. Geol.* **2000**, *56*, 257–274. [CrossRef]
53. Martínez-Martínez, J.; Benavente, D.; García del Cura, M.A. Comparison of the static and dynamic elastic modulus in carbonate rocks. *Bull. Eng. Geol. Environ.* **2011**, *71*, 263–268. [CrossRef]
54. Al-Shayea, N.A. Effects of testing methods and conditions on the elastic properties of limestone rock. *Eng. Geol.* **2004**, *74*, 139–156. [CrossRef]
55. Cressey, B.A.; Whittaker, E.J.W. Five-fold symmetry in chrysotile asbestos revealed by transmission electron microscopy. *Mineral. Mag.* **1993**, *5*, 729–732. [CrossRef]

 *sustainability*

Article

"Trachytes" from Sardinia: Geoheritage and Current Use

Nicola Careddu [1],* and Silvana Maria Grillo [2]

1   Department of Civil, Environmental Engineering and Architecture (DICAAr), University of Cagliari, via Marengo 2, Cagliari 09123, Italy
2   Department of Chemistry and Geology (DSCG), University of Cagliari, via Trentino 51, Cagliari 09127, Italy
*   Correspondence: ncareddu@unica.it; Tel.: +39-070-675-5561

Received: 11 June 2019; Accepted: 4 July 2019; Published: 6 July 2019

Abstract: Sardinia was affected by an intense igneous activity which generated calc-alkaline products during the Oligo-Miocene period. The volcanic substance shows large variations, ranging from pyroclastic flow deposits, lava flows and domes. By composition, the deposits are all primarily dacites and rhyolites, with subordinate andesites and very scarce basalts. The rhyolite lavas show porphyritic and ash-flow tuffs. Ignimbrite structures are found in the dacitic domes and rhyolitic lavas. These rocks—commercially known as "Trachytes of Sardinia"—used to be quarried in all historical provinces, mainly in the central part of the island to be used as ornamental and building stone. They continue to be commonly used nowadays, but their use dates back to the prehistoric age. They are easily found in many nuraghi, "domus de janas", holy wells, Roman works (mosaics, paving, roads, bridges), many churches built in Sardinia and practically in all kinds of structural elements in public and private buildings, such as walls, houses, and bridges. Contrary to the granitoid rocks, whose appearance is largely influenced by the mineralogical composition, the aesthetic feature of volcanic rocks is rather affected by the widest range of colors, structure and texture, i.e., shape, size and distribution of mineral components, porphyric index, etc. "Trachyte" is quarried opencast with the "single low step" method, with descending development, with prevalent use of double-disc sawing machines. Whenever the stone deposit allows higher steps, the chain cutting machine, in combination with diamond wire, becomes the preferred extraction solution. This study aims to at look Sardinian "trachytes" from a geoheritage perspective. After a geological-petrographic framework, the paper discusses the historical uses of "trachyte" in Sardinia. The current state of the art of "trachyte" quarrying, processing and usage in the Island is also described. An analysis of the "trachyte" production has been carried out. Finally, a consideration about how to enhance geotourism in the area is suggested.

Keywords: pyroclastic rock; dimension stone; Sardinia; geoheritage; market

---

1. Introduction

"Trachyte" is the term which was used to indicate clear, rough rocks outcropping in Greece in the area of Thrace in ancient times [1]. Another etymological interpretation of the word might derive from the Greek word "τυραχζ" which means "rough". From the petrographic point of view, trachyte is the volcanic equivalent of syenites, which are rocks formed with K-feldspar, subordinate plagioclase oligo-andesinic and mafic, such as green hornblende and biotite. Quartz is generally missing.

Differently, "Trachite auctorum" is the latin term which, in the past, indicated a multiplicity of vulcanites, from acidic to neutral (rhyolites, riodaciti, dacites and andesites), found in Central and North-Central Sardinia (Ottana, Sedilo, Allai, Fordongianus, Bosa, Ozieri, Oschiri and Osilo), southern (Serrenti) and southwestern (Sulcis) as shown in Figure 1. This term is still in use in the commercial field to identify this wide range of stone materials. Volcanic rocks are very different in texture (lavas, ignimbrites, pyroclastic rock welded in different ways, stagnation dome), easy to carve, they have variable porosity, and present very different color features, often characterized by a wide variable intensity: Pink, red, yellow, gray, dark gray, green, green-blue.

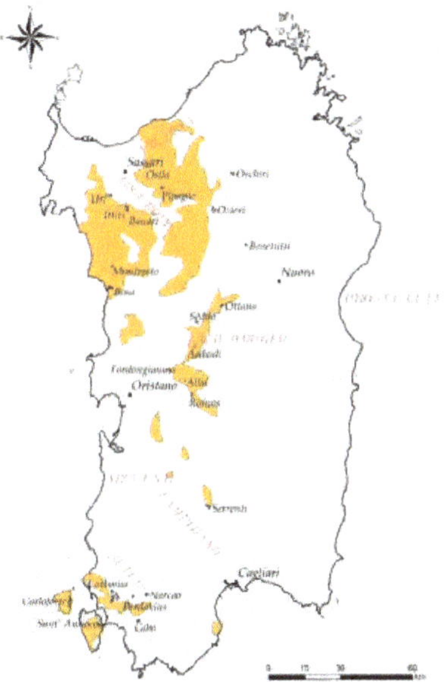

Figure 1. Lithological map of Oligo-miocenic volcanite outcrops and localities mentioned in the text (elaborated by Authors from [2]).

The aim of this study is to highlight how the "trachytes" of Sardinia can be regarded as a heritage stone resource (HSR). HSR is a natural stone, which was used in the construction of historical buildings and monuments over an extended period of time (sometimes centuries) and that should be involved in a great deal of consideration for its use in the restoration of those same historical buildings—when necessary—even when local quarries may no longer be active [3,4]. It is exactly for this reason that an international category has been created to include those global natural heritage stone resources (GHSR) that have achieved widespread utilization in human culture [5].

The present study is considered a geoheritage subtopic; geoheritage can indeed be defined as a the branch of geosciences that studies "Globally, nationally, state-wide, the regional features of geology such as its igneous, metamorphic, sedimentary, stratigraphic, structural, geochemical, mineralogic, paleontologic, geomorphic, pedologic, and hydrologic attributes at all scales, that are intrinsically important or culturally important, which offer information or insights into the formation and evolution of the Earth, or into the history of science, or which can be used for research, teaching or reference" [6]. For these reasons, heritage stone is a branch of geoheritage, which deals with the conservation of building natural stone resources that had an important role in human culture and also supports the preservation of historical quarries that once were the source of such stones.

Historical and Current Uses

These rocks have been widely used since ancient times, and can be found in numerous constructions located throughout the island. A great use of it was done during the Roman age: For example, the ancient thermal baths of Fordongianus, shown in Figure 2, which were built with squared grey boulders of "trachyte" [7], as well as the seven-arches bridge over the river Tirso, which was built with "trachyte" on top of an older bridge; Sardinian "trachyte" was also used by the Romans for the composition of mosaics and flooring in Nora. There is extensive footage of ancient "trachyte", used in prehistory to build defensive walls building, "Domus de Janas" (i.e., fairies' house), stone huts, nuraghi (i.e., Santu Antine, shown in Figure 3, which is one of the biggest and best preserved nuragic settlements of the island), carved vases (currently exhibited at the local Sardinian museums), mostly dating back to what is generally referred to as the "cultura di Ozieri" (middle Neolithic), etc.

Figure 2. Forum Traiani, Ancient Roman thermal baths in Fordongianus (I–III centuries a.c).

Figure 3. Santu Antine nuraghe, near Torralba (North-West Sardinia). It was built from 1800 to 1450 b.c.

Important examples of the use of these volcanic rocks, which are used in construction as well as by architects, can be typically seen in the civic center of several Sardinian towns: "Trachytes" are used to make different architectural parts in both public and private buildings, as ashlars, pillars, capitals, architraves, jambs, statues, friezes and etc. A special mention should be made of the many churches built in the 12th and 13th centuries, that can be considered a showcase where the visitor can observe the "trachytes" in all shapes and sizes; some examples are shown in Figure 4a–c.

(a)   (b)   (c)

Figure 4. Ancient Roman bridge in Sant'Antioco (**a**). It was restored in 1858, 1893 and 1920; The Church of San Pietro extra muros in Bosa (north-west Sardinia, XI-XIII centuries) (**b**). It was built in Romanesque style by using various shades of pink and red rhyolites; (**c**) Fort Sabaudo, Sant'Antioco (built in 1813–1815).

The main areas where "trachytes" are currently quarried and processed are practically near to historic sites, which follow the geological distribution of outcrops. The main production facilities for both ornamental use and construction are located in Fordongianus (near Oristano) and Serrenti (Campidano) where the quarrying activity has started again about twenty years ago.

With regard to the latter, which is currently being reassessed, it is quarried to obtain the stone known as "Pietra di Serrenti", a light-grey pyroclastic andesite, which is easy to be carved and split; it has been quarried and processed by skilled stonemasons since the 19th century from the quarry of Monte Atzorcu and used to build the traditional portals of both Serrenti houses including the houses (Figure 5), the old market hall and the local Court of Justice in Cagliari (shown in Figure 6), a part of the portico of Via Roma, the churches of San Francesco (in Cagliari, too), squares, and an array of monuments all over Campidano, etc.

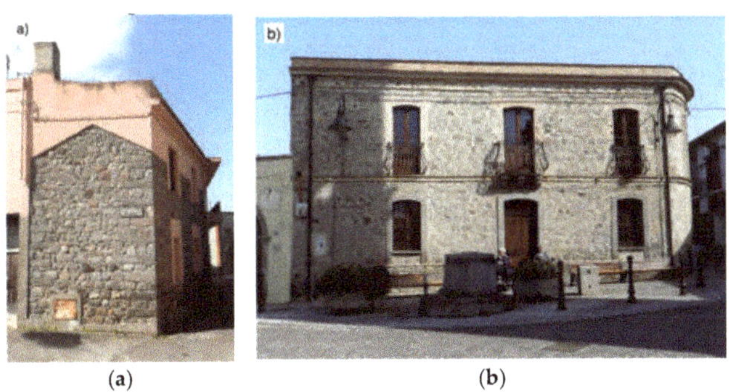

(a)   (b)

Figure 5. (**a**) and (**b**): two examples of houses, (**a**) and (**b**), built by using "Pietra di Serrenti".

Figure 6. Court of Justice in Cagliari, built in 1938 by using also the "Pietra di Serrenti".

It should be noted that all the different Sardinian "trachytes" started to be utilized at different moments throughout history. For example, Fordongianus "trachytes" have been used since Roman times and the Pietra di Serrenti since 19th century, while the use of Red Montresta in building is relatively recent.

2. Geological Framework

Figure 1 shows the lithological map of volcanics in Sardinia, which are related to two different volcanic cycles: A Plio-Pleistocenic one and a Oligo-Miocenic the second.

The first cycle occurred abundantly in the mid-western part of Sardinia and started when the island began to separate from the European continental mass during the Lower Oligocene. The subsequent opening of the Balearic basin activated a subduction zone to which the main phase of this volcanism is linked. This volcanic phase is composed of two main complexes both starting with mainly andesitic breccias and lavas, followed by more felsic product as tuff, lavas and ignimbrites. Most of these products are subaerial. However, during the rest of Miocene, a marine transgression with deposition of sediments as conglomerate, sandstone, clay and limestone covered most of these outcrop [8–11].

These volcanic activity can be shared in different phases:

- A first sequence from 32 My: Outcrops in Central-West and South Sardinia, is characterized by lava with an intermediate to basic chemistry. These rocks generally present a dome structure, which is often associated with pyroclastic sediments.
- The second acid explosive from 23 My: Characterized by volcanic products essentially dacitic to rhyolitic, ignimbrites and subordinate lavas. This sequence outcrops in several parts of Central Sardinia.
- The third sequence goes from 19 to 16 Ma: Prevalent outcrops in Western, Central and Southern Sardinia. Here, volcanic rocks are composed by andesites and basalts that have been deposited by means of lava flows and pyroclastic sediments, both subaqueous and subaerial, or as lava domes and associated explosive phreatic-magmatic products.
- The last activity between 17 and 13 Ma is characterized by ignimbritic, pyroclastic and lava-like rocks. These products are riolitic and dacitic (with rare episodes of pyroclastic sedimentation with sanidine), sometimes commenditic (Sulcis).

All these volcanic products have subalkaline affinity as shown in the Middlemost diagram of Figure 7. This diagram has been successfully, and often compared to the original definition of Middlemost [12], which reported only alkaline series [13].

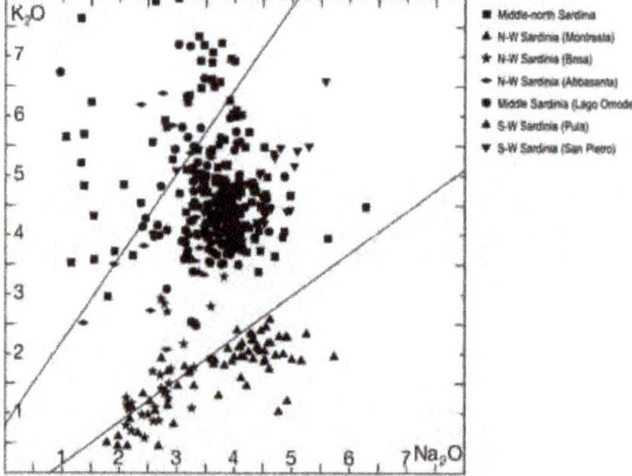

Figure 7. Middlemost (1975) summary diagram of all oligo-Miocene volcanics of Sardinia (elaborated by Columbu et al. [13]).

In summary, dacitic to rhyolitic ignimbrites are the prevailing products of volcanic activity, followed by andesitic, basaltic andesitic and, finally, basaltic lavas [11].

Figure 8a–f shows the micro-photographies under crossed polars of six different Sardinian "trachytes"; their differences, explained in the figures caption, are clear.

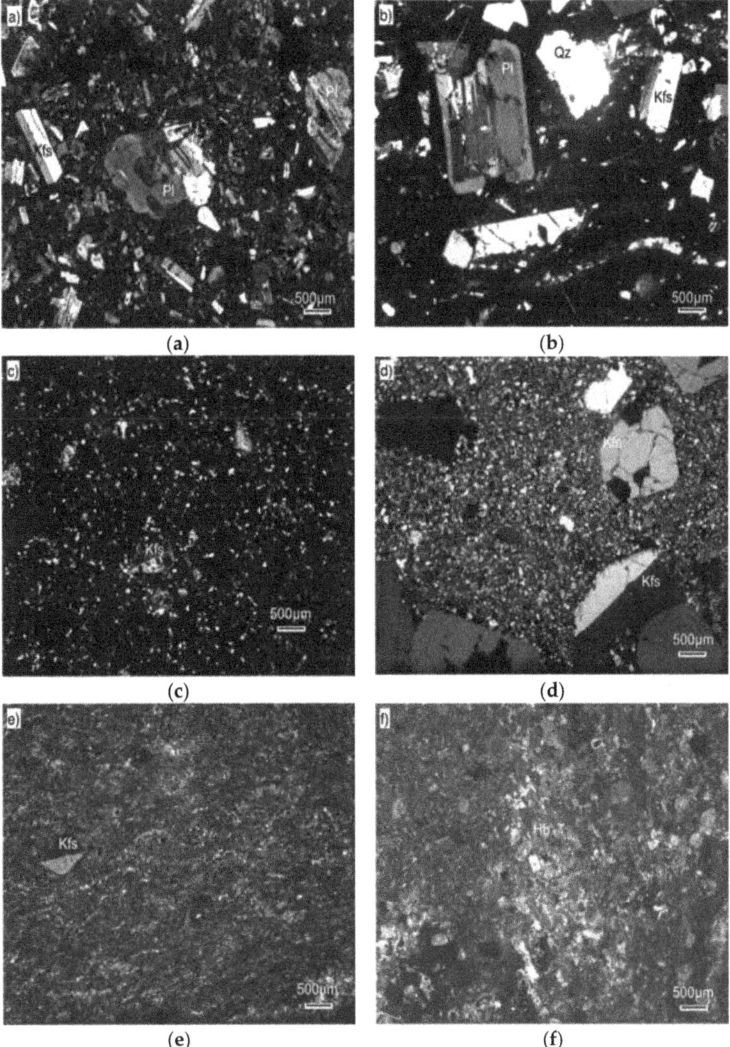

Figure 8. Thin section micro-photographies for some representative samples of Early Miocene volcanic rocks from Sardinia. The observed mineral phases are euhedral to subhedral K feldspar, plagioclase, quartz, horneblende in a groundmass from hypohyaline (Ploaghe (**a**), Sant'Antioco (**b**)) to hypocrystalline (Fordongianus red (**c**), Carloforte (**d**), Allai (**e**)) showing a porphyritic texture and different porphyric index. In the Serrenti area, the andesitic rocks, which include domes and volcaniclastic products, is interested by epithermal phenomena, so the outcrops are interested by propilitic alteration and silicification (**f**). Legend: pl = plagioclase; Kfs = alkali feldspar:sanidine; qz = quartz, hb = horneblende.

3. State of Art

Pyroclastic rock are widely quarried in all the historical provinces of the Island, mainly in the areas of Fordongianus-Allai, Ottana, Sedilo, Ozieri and Serrenti as ornamental stone and building stone both as aggregates and ashlars.

The beauty of Sardinian "Trachyte" lies in its warms colors, whose intensity may vary even within a single quarry run, a single slab and therefore a single product, enhancing the unique character of this natural stone.

Color variability of Sardinian "trachytes" is neither a defect, nor a limit, but a very peculiar feature of this type of material that is skillfully used for building and ornamental purposes. It should be noted that the variability of color is extremely common; there are so many intermediate and softened shades amongst a vast range of colors [14].

According to the various genetic processes, appearance and texture may vary, not only from type to type, but even within the same commercial type. This is made possible by the orientation in which the quarryman makes the cuts, which can be enhanced, emphasized or minimized the role of constituents of each material (crystals, lithic fragments, veins etc.). A good example is provided by the "trachyte" Red Montresta, which looks very different depending on the direction of the cut (Figure 9a,b).

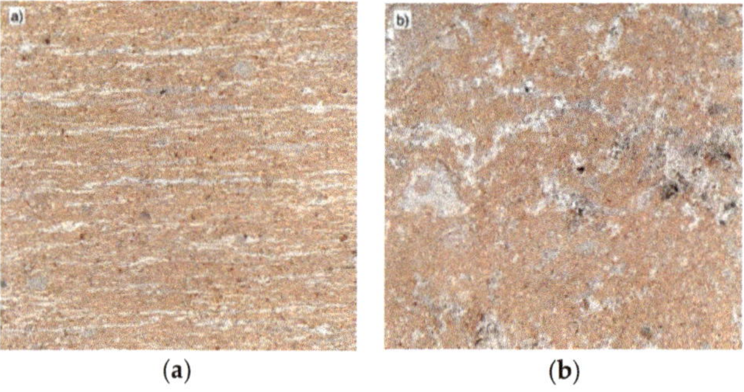

(a)   (b)

Figure 9. Appearances of "trachyte" Red Montresta: (**a**) Sample cut parallel to the flow planes, (**b**) sample cut orthogonal to the flow planes. Both the samples have a saw-plane surface.

The use of "trachyte" has its roots in ancient times, when it was used to make external paving, which has coupled to the use of cladding in contemporary times (as shown in Figure 10).

Figure 10. A detail of Teatro Massimo in Cagliari; the cladding, installed following a "run" pattern with runs in variable heights, was built using Fordongianus red "trachyte" (renewed in 2009).

Using this pyroclastic rocks to build stairs best enhances the properties of various types; in fact, it is representative of all the various surface finishes that can be made, ranging from the mild thicknesses of the covered steps to the one of the solid steps in any shape (rectangular, triangular, fan-shaped, etc.) and any edge profiling.

Another important use is beautifully summed up both by celebratory and commemorative works of art, such as monuments and sculptures (Figure 11a,b), and by other works that are able to combine functionality, design and sensitivity for natural stone, as is the case with some fountains.

(**a**)   (**b**)

Figure 11. (**a**) Parco delle Rimembranze (Cagliari), the monument was built with Bosa red "trachyte"; (**b**) two sculptures in Pietra di Serrenti at San Sperate (Cagliari).

Sardinian "trachytes" are also used in a wide range of architectural components as arches, jambs, columns, pillars (Figure 12).

Figure 12. Some examples from a private building showing a combination of Red Montresta "trachyte" with Orosei marble (courtesy of Marmi & Pietre I.C.S. s.r.l. [15]).

In recent years, production has proved to be particularly prolific in the segment of furnishing accessories, for which has supported the development of a whole range of interesting lines of production. It includes both objects with a strong aesthetic vocation, such as rose windows, cornices, inlay works, fretworks, etc., which combine aesthetics with functionality, such as clocks (wall, table), mirrors, tables of all sizes, fireplaces, bathroom sets, tops and elements for kitchens.

Among the furnishing accessories, a new concept of mosaic was developed in bathroom fittings, kitchen countertops, tables, fountains and with stone elements.

Sardinian "trachytes" are often associated with other stones, including basalt and travertine, a material particularly appreciated by the antiqued finishing.

3.1. Quarrying

As stated above, the commercial name "Sardinian Trachyte" covers a wide variety of pyroclastic rocks that henceforth will all be referred to by their common trade name.

Currently, the Sardinian market offers a number of "trachyte" stone types, which come in different grains, hardness, compactness and colors (pink, red, yellow, gray, brown; less frequently, green-blue); Figure 13 shows some Sardinian "trachytes" which are important from an historical and commercial point of view. Moreover, as we look at Figure 13, it is interesting to see how geodiversity, regarded as the natural range (diversity) of geological rocks [16] and relative properties, is embedded in geoheritage.

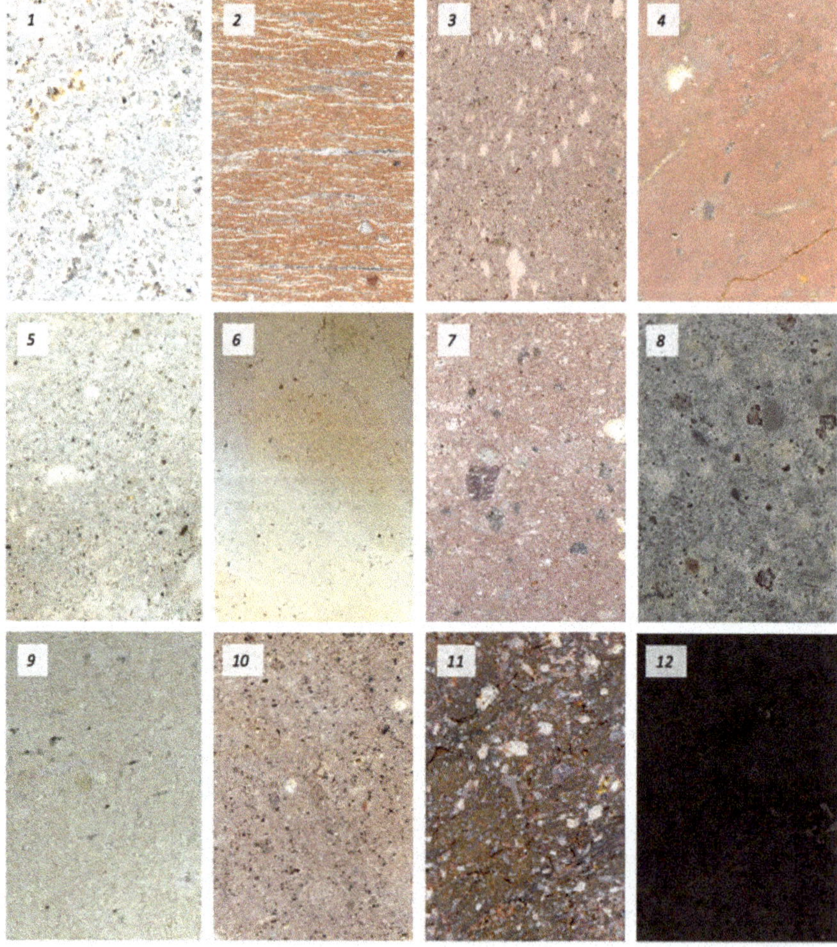

Figure 13. Some hystorical and commercial Sardinian "trachytes": **1)** Carloforte, **2)** Red Montresta, **3)** Fordongianus red, **4)** Allai, **5)** Fordongianus green, **6)** Ozieri, **7)** Bosa, **8)** Pietra di Serrenti, **9)** Fordongianus gray, **10)** Carbonia-Perdaxius, **11)** Sant'Antioco, **12)** Ploaghe.

"Trachyte" is quarried and worked by companies located in the provinces of Oristano (Ardauli, Bosa, Fordongianus, Ruinas), Sassari (Benetutti, Ittiri, Oschiri, Ozieri, Uri), Cagliari (Serrenti, whose local quarry is shown in Figure 14), and Nuoro (Sedilo). Sardinian deposits are usually extensive and only slightly fractured, with exploitable depths ranging from 10 up to 65 m [17].

Figure 14. "Pietra di Serrenti" quarry in Monte Atzorcu.

The most common and effective quarrying method is by "single slice and horizontal step". Quarrying motion leads downward, using quarry cutters mounted on tracks and equipped with two toothed sawing disks—one vertically, and the other one horizontally. Cutting speed is about 1–2 m/min, with a maximum daily output of about 200 m^2. After an adequate preparation of the quarry yard, parallel lines are traced on the surface of the rock slice with the vertical blade. The cut is then made simultaneously by the two blades: The vertical disk has a maximum cutting depth of 40 cm, which determines the size of the primary blocks cut out of the quarry loaf (maximum size: 40 × 60 cm, and length according to requirements) as shown in Figure 15.

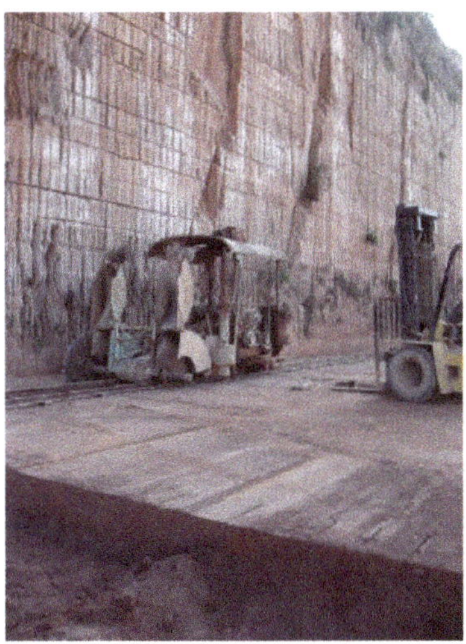

Figure 15. Partial view of a quarry of red "trachyte" excavated by using the "single slice and horizontal step" method (courtesy of Primavori, 2011).

Following this cutting method, other stone units as well as blocks can be obtained already during primary cutting in the quarry: Solid stone for masonry work, strips of different thicknesses, or slabs for paving applications. Strips and slabs can be obtained also in the fabrication plant, by cutting the block with a bridge sawing machine with diamond-tipped disk.

During the quarrying of larger blocks, a higher step is required. Its height should be at least equal to what will be the height of the quarried blocks; in this case, more advanced technologies such as chain-cutters and/or diamond-wire cutting machines have to be used (Figure 14).

Moreover, the stone deposit could have portions of rock of considerable volume which are already partially or completely detached from the rest of the rock mass; to isolate them, it is necessary to use earthmoving machines, such as shovels and excavators. The block is safely separated, perhaps with some drilling and metal wedges, and despite having irregular/non-geometric shapes, nevertheless provides significant volumes from which it is possible to produce large slabs dimensions, thicknesses and artifacts of complex shape.

3.2. Stone-Processing and Products

Similar to basalts [18], the processing of "trachytes" follows different techniques, depending on the quarry availability and the use of the stone material.

When dealing with a medium-large block, it is preferred to saw it into large slabs, obtained by means of diamond-wire plants, multi-blade gangsaw, giant disk cutters.

Some of the finishings that leave the surface smooth or semi-smooth, the saw-plane, the honed, the gloss and the antiqued are the more requested.

Bush-hammering, shot-blasting, brushing and splitting are the most popular surface finishings which have a more traditional rustic appeal.

Regarding the cut-to-size phase, this is carried out with modern CNC cutters, often equipped with devices that expand its performance and which can be used for profiling, shaping, lathes, and with traditional cutters (as jib mill or bridge milling).

The production of building elements, masonry blocks and brick-sized blocks represents a segment of particular importance for the "trachytes"; these are mainly parallelepiped-shaped artifacts, used during construction, and they are used as load-bearing elements in enclosure and containment walls, structural works and replacement restorations of architectural and artistic works, as shown in Figure 16.

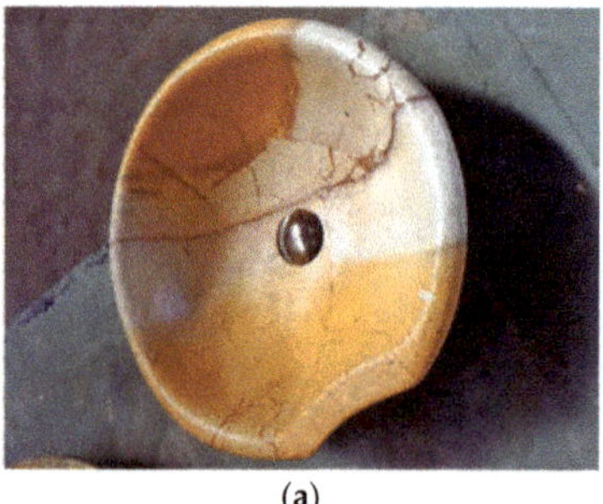 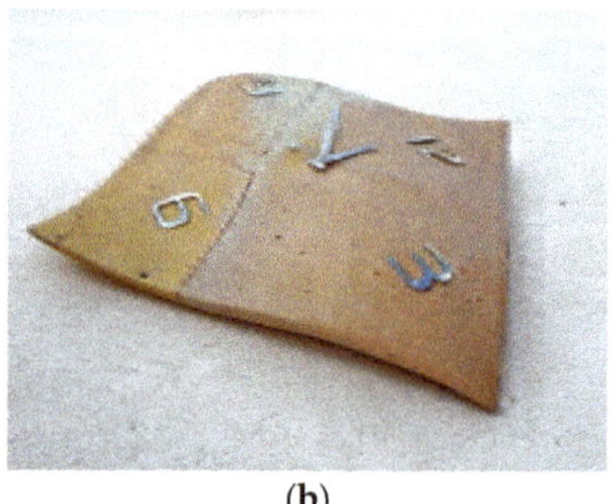

(a)   (b)

Figure 16. Sink (**a**) and clock (**b**) in Ozieri "trachyte" (courtesy of Pedra Noa di Gavino Mulas [19]).

3.3. Notes on the Wheatering of Sardinian Volcanic Calkaline Rocks

The different mineral composition (with some presence of glass, and some of different phenocrysts), a wide range of physical structure (due mainly to the porosity, density, hardness which are summarized in Table 1), can encourage the process of decay. In fact, the decay of stone depends on the combined action of weathering (e.g., solar radiation, rainwater, stream water, carbon dioxide, sulfate, seawater, dust etc.) and on its mineral composition, texture and structure of stone that is used in monuments and engineering structures. In particular, the weathering of volcanic rocks is mainly due to the combined action of the wind with sea salt spray in Sardinia. For example, in the facies with porphiric structure and high capillary porosity (i.e., "trachytes" of Fordongianus), salt from the seawater produces new crystal growths and salt accumulation in the pore, extensive efflorescence and subefflorescence, with consequent crumbling, flaking, detached crust, color change phenomena. On the contrary, in the well compacted ignimbrites and tuffs facies, that are characterized by very low porosity and permeability (as "trachyte" of Sant'Antioco) the wheatering is very weak and no changes are visible and the stones used are very resistant to the action of atmospheric agent.

Table 1. Physical and mechanical properties of some Sardinian trachytes currently on the market.

Stone Properties—UE STANDARDS	UNIT	Pietra di Serrenti	Red Montresta	Fordongianus "Trachyte"
Petrographic denomination—EN 12407:2007		Pyroclastic rock	Ignimbrite	Pyroclastic rock
Real density—EN 1936:2007	kg/m^3	2,204	2290	1612
Apparent density—EN 1936:2007	kg/m^3	2,688	n.a.	2662
Total porosity—EN 1936:2007	%	18.00	n.a.	39.40
Open porosity—EN 1936:2007	%	17.00	n.a.	39.20
Water absorption at atmospheric pressure—EN 13755:2008	%	6.30	n.a.	21.70
Water absorption coefficient by capillarity EN 1925:2000	$g/m^2s^{0.5}$	23.30	n.a.	n.a.
Flexural strength under concentrated load—EN 12372:2007	MPa	18.90	5.4–12.31 (dry sample) 1.63–6.871 (wet sample)	8.20
Flexural strength under concentrated load (after 48 freeze-thaw cycles)—EN 12372:2007 + EN 12371:2003	MPa	15.50	n.a.	not frost-resistant
Compressive strength—EN 1926:2007	MPa	93.00	93.2–104.51 (dry sample) 67.7–74.91 (wet sample)	49.00
Compressive strength (after 48 freeze-thaw cycles)—EN 1926:2007 + EN 12371:2003	MPa	90.10	n.a.	34.00
Resistance to ageing by thermal shock—EN 14066:2004	%	n.a.	n.a.	$\Delta E = -7.9$
Abrasion resistance—EN 14157:2005	mm	25	1.75–1.871 (dry sample) 1.91–2.041 (wet sample)	23.00
Resistance to ageing by SO2 action in the presence of humidity—EN 13919:2004	%	$\Delta m = -0.94$ significant variation of the color	n.a.	n.a.
Slip resistance by means of the pendulum tester—EN 14231:2004		77 (sawplane, wet sample)	n.a.	76 (sawplane, dry sample) 63 (sawplane, wet sample)
Linear thermal expansion coefficient—EN 14581:2005	$\mu m/m/^\circ C$	n.a.	n.a.	5.94
Rupture energy EN 14158:2005	J	8	n.a.	n.a.
Breaking load at dowel hole—EN 13364:2003	mm (d_1, b_A) kN (F)	$d_1 = 11$ $b_A = 49$ F = 1.50	n.a.	$d_1 = 11{,}1$ $b_A = 37$ F = 0.86

3.4. Market

Unlike marble and granite, whose market trend is widely documented, "trachytes" quarrying (raw production) can be difficult to track down and describe [20]. In fact, the quarried "trachytes" is often sold locally, in other words it doesn't cross the boundaries of the country where it was quarried and processed; this is due to the low market value when compared to more noble varieties of marbles and granites [21]. However, a little percentage of Sardinian "trachyte" is currently exported to Europe, especially to Germany and UK, and also to the USA. Some is exported from Sardinia to the Italian domestic market.

Statistics are shown in Table 2 and Figure 17; in Table 2, the production trend of Sardinian "trachytes and tuffs" for ornamental use is upward. Figure 16 shows the number of active quarries during the last 25 years.

Table 2. Sardinian "trachytes" and tuffs (used as ornamental stone) official quarrying production in 1994–2017 (in m^3). Elaboration from Regione Autonoma della Sardegna (R.A.S.) [22] and our own data [23,24].

	1994	1999	2002	2005	2007	2012	2014–2017
Quarrying Production [m^3]	21,300	5240	13,195	19,950	12,013	15,900	111,685

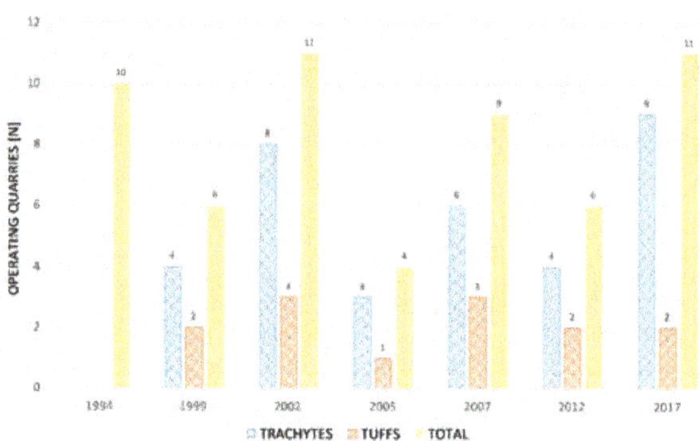

Figure 17. Number of "trachytes" and tuffs operating quarries in Sardinia. Elaboration from R.A.S. [22] and our own data [23,24].

It has been difficult to evaluate the productions before 1999 because "trachytes" and tuffs had been inserted in the group "stones", which included also sandstones and basalts in the previous statistics.

4. Geotourism Remarks

Geoethical values promote the right of citizens to access and receive transparent information pertaining to the intrinsic nature and meaning of geodiversity [25,26]. Geoheritage encourages the knowledge of the notions pertaining to the preservation of Earth Science principles [27].

We believe that the best way to represent geoethical values and geoheritage is the organization of a tour of historical towns and archaeological sites where Sardinian different types of "trachyte" have been used. This info-tour should be followed by the visit to historical quarries of "trachytes" as well as those quarries which are still fully operational. The tours will be led by qualified scientific guides with a degree in geoheritage topics.

A Possible Example: Bosa Heritage Stone Tour

Bosa is a municipality built from a medieval town on the western coast of Sardinia (see Figure 1). This municipality received a tourism award as Italy's second most attractive town in 2014 (tourism award 'I borghi più belli d'Italia' [28]). Our proposal consists of a walking tour to view a number of monuments which were built using the local Bosa "Trachyte" (Figure 13(7)) such as the castle of Serravalle—i.e., a XIII century fortification that dominates the town of Bosa from above—and the Church of San Pietro extra muros (already shown in Figure 4b). The tour shall end at the historical quarry "sas Pedraggias". It shall consist of about 6 km and last for about 4 hours. For the above mentioned sites, see Figure 18.

Figure 18. Aerial view of Bosa and location of principal monuments and historical "trachyte" quarry for the hypothetical Bosa heritage stone tour (from Google Earth).

5. Conclusions

Sardinian pyroclastic rocks, which are generally called Sardinian "trachytes" are stones which are widespread in the western part of Sardinia: These kinds of stone were used to build a lot of works that have featured the Island's history, from its prehistory (as witnessed by "domus de janas" and nuraghi), to the modern age, via the Roman and medieval ages.

During the last twenty years, the use of Sardinian "trachytes" has increased due not only to fashion, but also for the restoration of a high number of civic centers in several municipalities throughout the island.

Using commodity-related terms, Sardinian "trachyte" can show different colors from the pale yellow to the red through to pinkish, the green and the dark grey and has different mechanical characteristics. It is a fairly cohesive rock and shows remarkable elements of aesthetic value that have favored, its popularity as a building resource and ornamental stone in Sardinia.

At present, Sardinian "trachyte" is very much demanded for its compactness, homogeneity, color and because of its easy workability; it is usually utilized for flooring, covering, blocks for open-face constructions and buildings as well as for architectural and street furnishing elements. All this is highlighted by the growing production in the island.

Exactly for the above-mentioned reasons, it is necessary to raise awareness of people about the diversity, the history and use of Sardinian "trachytes" through guided tours.

Author Contributions: Conceptualization, N.C.; investigation, S.M.G.; resources, N.C. and S.M.G.; data curation, N.C.; writing—original draft preparation, N.C. and S.M.G.; supervision, N.C.; funding acquisition, N.C.

Funding: This research received no external funding.

Acknowledgments: The Authors wish thank to Antonio Dessena for his support in the cartographic elaboration.

Conflicts of Interest: The authors declare no conflict of interest.

References

1. Grillo, S.; Mocci, S.; Pia, G.; Spanu, N.; Tuveri, L. *Il Manuale Tematico Della Pietra*; Sanna, U., Atzeni, C., Eds.; Itaca (pub.): Fiumicino, Italy, 2009; pp. 20–34. (In Italian)
2. Carmignani, L.; Oggiano, G.; Barca, S.; Conti, P.; Salvadori, I.; Eltrudis, A.; Funedda, A.; Pasci, S. Note illustrative della Carta Geologica della Sardegna a scala 1:200.000. In *Memorie Descrittive della Carta Geologica d'Italia*; Servizio Geologico d'Italia, Ed.; ISPRA: Roma, Italy, 2001; Volume 60, 283p.
3. Cooper, B.J.; Marker, B.; Pereira, D.; Schouenborg, B. Establishment of the "Heritage Stone Task Group" (HSTG). *Episodes* **2013**, *36*, 8–10.
4. Pereira, D.; Baltuille, J.M. Documenting natural stone to preserve our cultural and architectonic heritage. *Materiales de Construcción* **2014**, *64*, 002.
5. Cooper, B.J. Towards establishing a "Global Heritage Stone Resource" designation. *Episodes* **2010**, *33*, 38–41.
6. Brocx, M.; Semeniuk, V. Geoheritage and geoconservation—History, definition, scope and scale. *J. R. Soc. West. Aust.* **2007**, *90*, 53–87.
7. Argiolas, S.; Carcangiu, G.; Floris, D.; Massidda, L.; Meloni, P.; Vernier, A. Le piroclastiti dell'antica Forum Traiani (Fordongianus)—Sardegna centrale: caratterizzazione, tecniche di estrazione e specificità di utilizzo nei secoli. In Proceedings of the Atti Conv. "Le Risorse Lapidee Dall'antichità a Oggi in Area Mediterranea", Canosa di Puglia, Italy, 25–27 September 2006; pp. 33–38. (In Italian).
8. Coulon, C. Le Volcanisme Calco-Alcalin Cénozoique de Sardaigne (Italie). Petrographie, Gèochimie et Genèse des Lavas Andèsitiques et des Ignimbrites. Ph.D. Thesis, Université St Jerome, Aix Marseille, France, 1977.
9. Conte, A.M. Petrology and geochemistry of Tertiary calcalkaline magmatic rocks from the Sarroch district (Sardinia, Italy). *Periodico di Mineralogia* **1997**, *66*, 63–100.
10. Lustrino, M.; Morra, V.; Melluso, L.; Brotzu, P.; d'Amelio, F.; Fedele, L.; Franciosi, L.; Lonis, R.; Petteruti Liebercknecht, A.M. The Cenozoic igneous activity of Sardinia. *Periodico di Mineralogia* **2004**, *73*, 105–134.
11. Guarino, V.; Fedele, L.; Franciosi, L.; Lonis, R.; Lustrino, M.; Marrazzo, M.; Melluso, M.; Morra, V.; Rocco, I.; Ronga, F. Mineral compositions and magmatic evolution of the calcalkaline rocks of northwestern Sardinia, Italy. *Periodico di Mineralogia* **2011**, *80*, 517–545.
12. Middlemost, E.A.K. The basalt clan. *Earth Sci. Rev.* **1975**, *11*, 337–364. [CrossRef]
13. Carboni, D.; Columbu, S.; Corazza, G.; Garau, A.M.; Ginesu, S.; Macciotta, G.; Marchi, M.; Marini, C. *Manuale Sui Materiali Lapidei Vulcanici Della Sardegna Centrale e Dei Loro Principali Impieghi Nel Costruito*; Tipografia Ghilarzese, Ed.; ISKRA: Ghilarza, Italy, 2011; p. 193. (In Italian)
14. Primavori, P. I materiali Lapidei della Sardegna. In *Sardegna Ricerche*; Regione Autonoma della Sardegna, Ed.; Sainas Industrie Grafiche: Villaspeciosa, Italy, 2011; pp. 261–307. (In Italian)
15. Marmi & Pietre I.C.S. s.r.l. Available online: www.marmiepietrevinci.it/index.htm (accessed on 10 June 2019).
16. Gray, M. *Geodiversity: Valuing and Conserving Abiotic Nature*, 2nd ed.; John Wiley & Sons: Chichester, UK, 2013.
17. Massacci, G.; Medici, G. Le piroclastiti Sarde come pietre ornamentali. In *Proceedings of the Atti Del Convegno "Le Materie Prime Minerali Sarde", Cagliari, Italy, 23–24 Giugno 1997*; Marini, C., Ed.; CUEC pub.: Cagliari, Italy, 1997. (In Italian)
18. Careddu, N.; Grillo, S.M. Sardinian Basalt—An Ancient Georesource Still En Vogue. *Geoheritage* **2019**, *11*, 33–45. [CrossRef]
19. Pedra Noa di Gavino Mulas. Available online: www.pedranoa.com/index.php (accessed on 10 June 2019).
20. Careddu, N.; Siotto, G.; Marras, G. The crisis of granite and the success of marble: Errors and market strategies. The Sardinian case. *Resour. Policy* **2017**, *52*, 273–276. [CrossRef]
21. Montani, C. *XXIX Report Marble and Stones in the World 2018*; Aldus: Carrara, Italy, 2018.
22. R.A.S, 2007. - Piano Regionale delle Attività Estrattive (PRAE). Regione Autonoma della Sardegna, ottobre 2007. Available online: http://www.regione.sardegna.it/speciali/pianoattivitaestrattive/ (accessed on 5 July 2019).
23. Careddu, N.; Scanu, M.; Desogus, P. (Eds.) *Map of Natural Stones from Sardinia*; Miniere di Sardegna: Cagliari, Italy, 2015.
24. Careddu, N.; Scanu, M.; Desogus, P. Notes on the poster "Map of natural stones from Sardinia (Italy)". *Key Eng. Mater.* **2019**, accepted.

25. Ferrero, E.; Giardino, M.; Lozar, F.; Giordano, E.; Belluso, E.; Perotti, L. Geodiversity action plans for the enhancement of geoheritage in the Piemonte region (north-western Italy). *Ann. Geophys.* **2012**, *55*, 487–495.

26. Careddu, N.; Di Capua, G.; Siotto, G. Dimension Stone Industry should meet the fundamental values of Geoethics. *Resour. Policy* **2019**. accepted.

27. Doyle, P.; Easterbrook, G.; Reid, E.; Skipsey, E.; Wilson, C. *Earth Heritage Conservation*; The Geological Society in Association with Open University: London, UK; City Print (Milton Keynes) Ltd.: Milton Keynes, UK, 1994.

28. Bosa è il secondo borgo più bello d'Italia. La Nuova Sardegna. Available online: http://www.lanuovasardegna. it/oristano/cronaca/2014/04/21/news/bosa-e-il-secondo-borgo-piu-bello-d-italia-1.9087617 (accessed on 2 July 2019).

 *sustainability*

Article

Houses Based on Natural Stone; A Case Study—The Bay of Kotor (Montenegro)

Dušan Tomanović *, Irena Rajković, Mirko Grbić, Julija Aleksić, Nebojša Gadžić, Jasmina Lukić and Tijana Tomanović

Faculty of Technical Sciences, Department of Architecture, University of Priština, ul. Knjaza Miloša 7, 38220 Kosovska Mitrovica, Serbia
* Correspondence: dusantomy@gmail.com; Tel.: +381-63-833-8170

Received: 26 May 2019; Accepted: 9 July 2019; Published: 16 July 2019

Abstract: The Bay of Kotor, in its exceptional natural conditions, thanks to its geographical location and influenced by historical events, saw the development of rural settlements that are historically, artistically and culturally worthy of recognition. These stone settlements were acquired completely spontaneously, keeping the same pace as the settling, and transformed to some degree due to contemporary social movement and migration. Up until the middle of the 20th century, structures on the coastline in general were built by applying the same verified methods, which remained unchanged for centuries. Unreinforced stone walls as load-bearing vertical elements, coupled with wooden floor joists attached in a traditional way are typically present in the stone architecture of the Adriatic region and karst areas in general. The construction characteristics of the stone houses built in such a way meet all needs in terms of strength, thermal insulation, and are suitable for the coastal climate of this region. The fast-paced development in the past 50 years, the inadequate legal protection of residential buildings in the Bay of Kotor, poverty, and the new rich have brought about the devastation of not only buildings built in traditional architecture styles themselves, but also the urban landscape of the bay. Throughout the Bay of Kotor, buildings built in traditional architecture styles are nowadays more and more rare to see in their original shape—houses outside of cities but which display all characteristics of civic coastal houses and buildings free of rigid style rules, even though closely in contact with them. Regardless of efforts to preserve the heritage inherited by our ancestors, cultural monuments and houses referenced here deteriorate on a daily basis due to troubles and neglect.

Keywords: houses based on natural stone; architectural heritage; the bay of Kotor

---

1. Introduction

This research tries to investigate traditional houses based on natural stone in the Bay of Kotor over a period of 200 years and describes the basic principles of these houses. Through a review of the old traditional architecture, the qualitative and functional advantages of the building were considered.

We selected the Bay of Kotor because it is the most interesting part of the Montenegrin coast and has a very interesting historical trajectory, geological formation and rich biodiversity (Figures 1 and 2).

This paper advocates the stance that rational action in further urban development can be planned only if we understand the meaning of the phenomena processes, condition and planning strategies of this site throughout history. It also highlights the significance and necessity to preserve these houses as well as the settlements.

A new wave of urbanization makes status of these houses more endangered, both on the Montenegrin coast and in the Bay of Kotor. In most cases, urban planners "bother" objects hundreds of years old of invaluable cultural significance.

Based on the results of a previous study on 95 buildings [1,2] that discussed the typology of houses in the Bay of Kotor in the 18th, 19th and 20th centuries, two basic types of houses were formed. In the first phase, the house was a ground-level type, while in the second phase it became a single-storey house. From these two main types, other subtypes of houses were developed. The representativeness of these buildings with respect to the total is 80%.

The age of certain rural settlements in the Bay of Kotor can be accurately determined based on the archival materials of the Decree of the City of Kotor and notary books from the 14th and 15th centuries [3]. The borders of villages were defined by a regulation in the Kotor Decree in the 15th century. Border setting was done using permanent natural terrain marks or artificial signage. This is telling us that the founding of village borders was planned. Even though the Decree only mentions maintaining security measures, a large part of the settlements at the time is referenced in the cited chapter. The name of the Vrmac (Vermez) peninsula is mentioned here for the first time. Progress into the settlements was spontaneous, keeping the same pace as the settling.

Figure 1. A map of Montenegro and the Bay of Kotor [4].

Figure 2. It consisted of the bays of Kotor, Risan and Tivat [5].

In the history of the Bay of Kotor, there are four major stages for which it is possible to reconstruct the settlement organisation and the attitude of humans towards the environment:

(a) The appearance of village settlements in the hills, far from the shore and main communication channels (the Middle Ages)

(b) Consolidation of settlements on inaccessible terrain with an intense use of the territory, deforestation and landscape transformation (the Venetian period)

(c)   Establishment of settlements in the lower terrain and development of new activities (the Austrian period)

(d)   Abandonment of upper villages, growth of urban centres and degradation of landscapes (the post-war period).

The settlements, the house formations and groupings reflect a traditional lifestyle and the principles of socio-economic organisation in their relationships to each other as well as in their attitude towards the landscape.

Borders of villages in the Bay of Kotor area are uneven as a result of their spontaneous formation and terrain configuration. When looking at village locations, climate and hydrographic factors are also important.

With the beginning of the 19th century and the cease of threats, coastal strips across the bay became attractive locations for settling, and the sea grew as the most important traffic artery. Population still depended on traditional economy for endurance. Land farming was still essential for survival. It was only with the strengthening of other activities (industry) at the end of the 19th century that the conditions were set for the changing of settlement structure within seemingly traditional relationships [6–9].

The construction of a coastal road by the Austro-Hungarian authorities at the beginning of the 20th century and the construction of the highway in the 1960s significantly determined the organisational way of the settlements in this period. Those events enabled a quick development of the coastal strip, while all other areas remained neglected and left to their own decline.

Between the two world wars, with the rise of industry in this period, the need for food production was reduced, thus transforming agricultural plots into construction sites. The landscape change in the Bay of Kotor continues with increased intensity. Intensive construction entirely disregards traditional architecture.

Existing and extinct settlements with reliable historical records behind them could be divided into two groups: lower and upper settlements.

Lower settlements are located in the coastal strip. Almost all of those lower settlements are located along the coast or roads and thus had no possibility of becoming a core settlement with a central role, since they were groupings of the same or similar levels clustered along the coast.

Upper settlements are located at an altitude of 250–450 m.

2. Materials and Methods

An analytical and comparative research method was used. Data was collected using direct survey. A part of the empirical material was derived from interviews with locals. General literature, previous studies and online material have also been reviewed.

The initial phase of the research included an analysis that was based on the extensive literature review of books, articles and reports covering topics related to the architectural heritage of the Bay of Kotor. In the second phase, we included the same research for nearby regions (Dalmatia in Croatia and Paštrovići region in southern Montenegrin coastline). Among the primary sources, the most important were archives, of which we will mention:

-    Serbian Academy of Sciences and Arts in Belgrade
-    State Archives of Montenegro in Cetinje
-    Historical Archives of Kotor

In our region, only a few scientists are engaged in researching stone houses in the Bay of Kotor and those are primarily: Kojić Branislav, Korać Vojislav, Đurović Vinko, Prof. Arch. M. Zloković, Prof. Arch. Zoran Petrović, Goran Božović and Prof. Arch. Lalošević Ilija.

In his two papers entitled "Izumiranje sela u Kotorskomom zalivu i na poluostrvu Vrmac" [7] and "Seoska arhitektura u Boki Kotorskoj" [10], Branislav Kojić writes about the architecture of the Bay of Kotor. The first paper concludes that extinct villages could hardly be reconstructed due to the position of the terrain in relation to the main roads, but the extinction of existing settlements only stop when conditions for intensive use of agricultural land are created. In the second paper, the author came to the conclusion that one of the characteristics of the rural house in the Bay of Kotor is erection of the kitchen on the highest level, which demonstrates an inclination to develop vertically.

In the paper entitled "Spomenici srednjovekovne arhitekture u Boki Kotorskoj" [8], Korac Vojislav came to the conclusion that early medieval architecture in the Bay of Kotor was almost completely unknown up to the middle of the 20th century.

Đurović Vinko pointed out in the paper "O konstrukcijama kuća od XVI do konca XIX veka u kotorskom zalivu i njihovim graditeljima" [11] that old walls were not strong because they were made of burnt lime.

Milan Zloković was particularly dedicated to the subject of proportional systems. He conducted research in the Bay of Kotor, together with certain collaborators, entitled "Građanska arhitektura u Boki Kotorskoj u doba Mletačke vlasti" [12]. Numerous palaces were technically recorded (Beskuća and Verona in Prčanj, Grgurina in Kotor, as well as palace Milošević in Dobrota).

Petrović Zoran put emphasis on the upper villages of the Vrmac peninsula in his paper, " Selo i seoska kuća u Boki Kotorskoj" [13]. He concluded that the depopulation of upper villages would not contribute to the sustainability of these settlements in the future.

Božović Goran provided an overview of preserved architectural heritage in the Bay of Tivat and gave a proposal to convert them into tourist facilities in the study, "Naselja i kuće Tivatskog zaliva" [6].

Ilija Lalošević in the paper, "Svojstva istorijskih zidanih konstrukcija u seizmičkim zonama (primjer Boke Kotorske)" [14], concludes that the tall stone structures in the Bay of Kotor have significant disadvantages in terms of stability during strong seismic shifts.

In the region, one of the most important publications is A. Freudenreich's "Narod gradi na ogoljenom krasu", where images and drawings of house typology in the Croatian coast are shown. The architecture of this part of the coast and that of the Bay of Kotor was also explored by Croatian authors Cvito Fisković and Katarina Horvat Levaj. Their experiences are similar to those of the above-mentioned authors.

Of studies on the southern part of the Montenegrin coast, an important publication is Stanko Gaković's "Paštrovska kuća", published in 1979. It provides a detailed overview of the typology of houses in the Paštrovići region. This region is neglected in the literature, although it is invaluable for both the southern part of the Montenegrin coast and the Montenegrin heritage in general.

2.1. Architectural Design and Formation of the Façade of a Stone House

The position of the house on the terrain, location of the house, slope of the terrain, sun exposure, line of sight, house entrance and function are factors that determine the shape and symmetry of the house. Houses in the Bay of Kotor characteristically face the sea, be it in upper or lower settlements or on the northern, western, southern or eastern side of the slope. The sea is therefore an important factor that determined the position of the entrance to the house.

The slope of the terrain is the second factor that determined what kind of house will be built and where. A milder slope offered easier access to the house, a symmetrical, calmer façade, but also better sun exposure, while a higher slope caused the construction of a larger number of asymmetrical buildings (Figures 3–6). Cascading terraced gardens should also be mentioned. They are made of roughly carved stone in a dry stone wall. Historically, the terraces represented a complex system for managing slope dynamics (from conservation of the soil to the triple function of runoff, drainage and collection of rain water) [15].

The styling of stone house façades was more modest in the hinterland than on the coast. It depended mainly on the wealth of its owners. The front façade was richer, made using better stone in comparison with the more-modest side façades. During heavy rains, joints tend to carry moisture through the walls into the interior of the house. This brought about the trend of plastering the side façades (in some places, all of them) with red hydraulic mortar (lime mortar with added milled brick); cement mortar was used later in the 20th century.

In this area, there is an abundance of limestone, thus dictating the shape of the house and garden. Over three-quarters of construction objects in Montenegro were built using limestone, which was processed by following a centuries-old tradition. Less regularly shaped blocks were obtained by splitting the rocks. This was done manually. The houses were made of fine and rough carved gray limestone, brought from local quarries.

Figure 3. Houses in Gornji Stoliv face both the slope and the sea (Photo: Dušan Tomanović).

Figure 4. The slope of the terrain was the reason behind the construction of a large number of asymmetrical buildings—Gornji Stoliv [16].

Figure 5. Lepetani—a row of stone houses located along the coastal road (Photo: Dušan Tomanović).

Figure 6. Muo—construction of the coastal road at the beginning of the 20th century caused the destruction of the urban core settlement, which was more oriented towards economic activity related to the sea (Photo: Dušan Tomanović).

The silhouette of the house depends on the shape of the base—which was spacious in the Bay of Kotor area. The external measures are 9 to 13 m in length, and 6 to 8 m in width. The number of floors is another important characteristic that influences the architectural shape of stone houses. In most cases, these houses consist of a ground floor, an upper floor, and a loft.

The biggest changes in the 20th century occurred in the overall horizontal and vertical dimensions, in materialisation, shapes of the roofs, the design of façades and details on them. The original structural system and applied elements as well as the traditional technique were not adequately assessed in this period.

2.2. Types of Stone Houses

There are two types of stone houses typical in this region:

A-   Single-storey stone house (a house with one floor only)
B-   Multiple-storey stone house (the developed kind)

2.2.1. Single-Storey Houses

There are no primal house types proven as old in the rural area of Kotor Bay. Archival materials do not provide any information on the shape of old rural houses, thus we cannot say with certainty what the oldest types of houses looked like. The single-storey house is at the core of rural architecture of this part of the Adriatic coast because of its shape, construction and organisation, since the arrival of the Slavs to this region until today (Figure 7). It is the primary cell from which today's rural architecture has developed [10].

The interior of this house was at the same time the sole eventful room in the whole household. It was not uncommon for the building to be divided into two with a low set stone or wooden bar. The partitioning of the single-storey house represents an interior development of the same type, the barriers were added afterwards and did not change the base of the unpartitioned type. The first part was intended for use by humans, while the other was used for livestock and, at the same time, as natural fertilizer storage space. Animals lived in the same house, divided from humans by a wall plastered with mud and bovine manure (Figures 8–10).

Figure 7. The single-storey stone house is at the core of rural architecture in this part of the Adriatic coast (Photo: Dušan Tomanović).

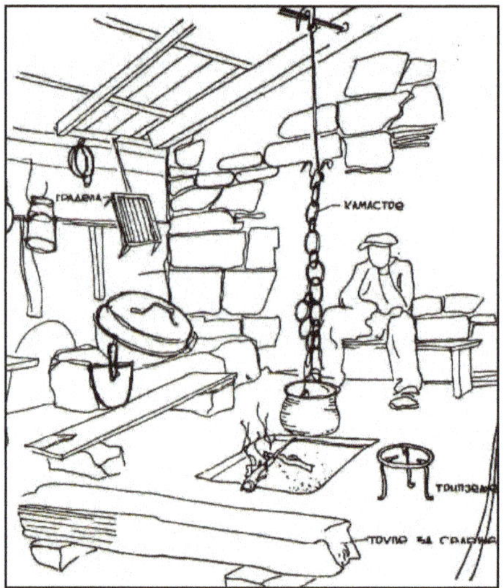

Figure 8. There was no furniture in the interior of the single-storey house [10].

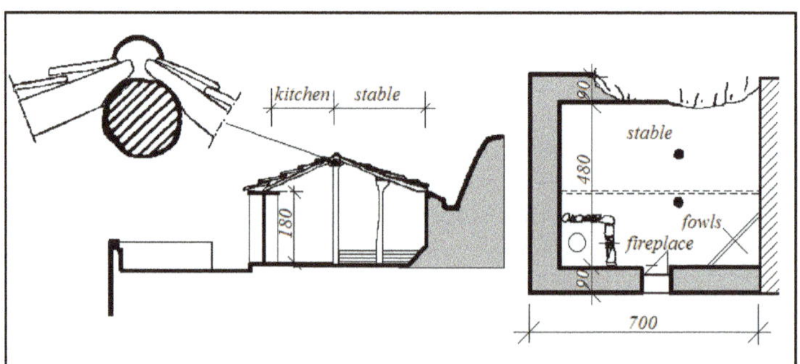

Figure 9. The plan and cross-section of an early (undivided) single-storey house [10].

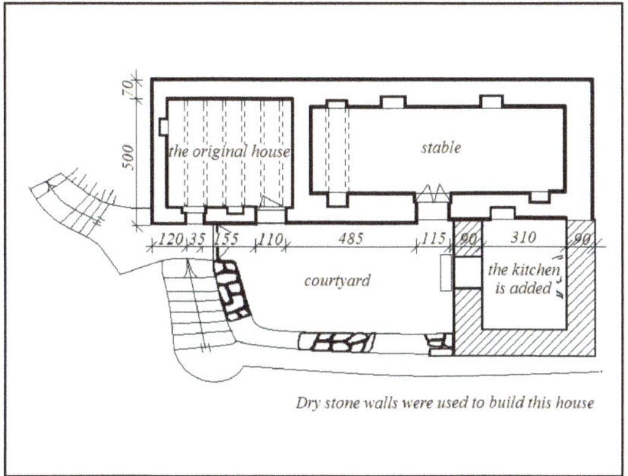

Figure 10. The area of the single-storey house extended horizontally at the beginning of the 20th century [10].

2.2.2. Multiple-Storey Houses

The multiple-storey house is a developed type of house in the Bay of Kotor area. A more rational construction stems from several reasons:

- ❖ A tendency to save space
- ❖ House compactness
- ❖ The construction is more rational (more space is obtained with the same foundation)
- ❖ Urban influences

Multiple-storey houses with ordinary lofts and kitchens located in separate buildings or outhouses in the yard can be found in the inner part of the bay. This kind of situation is typical for remote areas with low urban influence. The most common type of house consists of a ground floor, an upper floor, and an inhabited loft (Figures 11–13).

Figure 11. Original drawing (in colour) of the Tomanović family multi-storey house dating from 1826 in Lepetani (italian arch. F. Boriani) (Photo: Dušan Tomanović).

Figure 12. Typical seaside house with an upper floor and inhabited loft (Photo: Dušan Tomanović).

Figure 13. Typical multiple-storey house with an upper floor and inhabited loft in upper villages (Photo: Dušan Tomanović).

Natural factors were of essential importance when it comes to the position of the house and yard. In this case, slope was the factor, since all houses face the valley, i.e., the sea and the arable land. Cardinal directions did not play any role in the position of the houses, since they face different directions, influenced only by slope orientation of the land they were built on.

In the 18th, 19th and at the beginning of the 20th centuries, a large multiple-storey house consisted of three typical rooms: the lowest part is the konoba area (intended for use by men), the residential part is on the upper floor, while the loft area is occupied by a kužina (intended for use by women). Elevation of the kitchen to the uppermost level was the most striking feature of the vertical house development in this period.

2.3. Analysis of Construction Technique

2.3.1. Materials

Construction materials used for building houses and enclosures in the whole Bay of Kotor area, or even in a wider sense along the Adriatic coast, were locally provided, taken from nearby sites.

Natural stone has been used as a construction material since ancient times (e.g., since about 4700 BC in Egypt). Initially, these expensive materials were used for temples, tombs, palaces, civic buildings and major infrastructure as well as decoration and sculptures. Selection of stone was guided partly by suitability, but also influenced by personal prestige and social and mystic beliefs [17].

This region contains a lot of high-quality limestone. The Verige and Kamenare quarries contain plate stone, which was used then but also today for paving the floors of taverns, courtyards, city squares, and other public spaces (Figure 14). Straight blocks of limestone extracted in Strp and Lipci close to Morinj were used for facades on the houses (Figures 15 and 16).

Figure 14. Verige and Kamenare quarries contain gray and red plate stones (Photo: Dušan Tomanović).

Figure 15. Less regular shaped blocks on the lateral façade from Strp and Lipci quarries (Photo: Dušan Tomanović).

Figure 16. Front façade walls were built using the highly regarded fine stone from Strp and Lipci quarries (Photo: Dušan Tomanović).

Limestones are massive with thick-layered types of stone, grayish brown to light brown in color, mainly of Mudstone type with layers of pellet Wackestone and Packstone, and less of Grainstone. According to Folk's classification (Folk, 1959 and 1962), these are mainly micrites, pelletiferous micrites, completely or partially recrystallized micrites of the grumous structure, and less often pelmicrites to pelsparites. Limestone contains little fossil remains, from non-skeletal components, mainly pellets, to skeletal components of myliolide and ostracods. Based on the findings of *Favreina njegosensis* BRÖNNIMANN, these deposits are from the Neocomian age (Matičec & Fuček, 1999). Microfossil structures are mainly distinguished by a micrite or a fibrous calcite diameter of 0.02–0.8 mm [18].

Manual stone extraction is painstaking work. Pin holes were done in pairs, with one person holding the rod and the other one striking it. After a row of holes was lined up, planks were inserted between them and wedges were hammered in, one by one until the block breaks off. The oldest tools used for manual stone extraction and stone processing (levers, chisels and hammers) can be found across the whole karst region under different local names. Areas with softer stone types were easier to process with devices akin to a carpenter's tools (saw, drill, axe).

Sand for construction was brought from Meljine and Žanjic, later from Ulcinj, Cavtat, and the valley of the Neretva river [11].

There was always a lack of wooden materials in the construction of buildings on the seaside. The shore hardly contains any forests, while the modest timber did not yield adequate building material. In the 19th century, timber was imported and bought in Lepetane, and transported via direct sea pathways from Venice. After the World War II, the situation became a whole lot better. The industrialisation of a large country brought about the opening of many storage spaces and there were more diverse building materials being offered.

All sculptural elements were made of stone. They were prefabricated in local quarries or in those on the Brač and Korčula islands (Figure 17). By this, we are primarily referring to heavy house balconies borne by stone brackets (*regula*), frames around the windows and doors, small stone brackets underneath the cornices, and terrace fences made from stone balusters.

Figure 17. The Korčula stone is soft and easy to process; however, due to its lightness and brittleness, it is rarely used for construction, but only for decorating the facade (Photo: Dušan Tomanović).

When houses were being established, their roofs were covered by flagstones or by an old type of dry-layered imbrex. The oldest type of tile can still be found on some houses.

The materials mentioned here not only meet all requirements in terms of strength and thermal insulation, but also favour the coastal climate they adapt to [19].

2.3.2. Techniques of Constructing Single-Storey Stone Houses

These buildings were constructed in a primitive way. The base is square-shaped in most cases. The width and length of the building does not exceed overall dimensions larger than 3–5 m (Figure 18). Local barely regular stones were used for dry construction (with no binding agent). Dry construction (without a binding agent) is the oldest construction technique. It is still used today in the building of stone fences and retaining walls, and was being applied in traditional construction until almost the end of the 20th century.

The walls of these houses are 70 to 100 cm thick, 2–2.2 m in height. The height of the ridge is 3–4 m. In the interior of the wall, sporadic indentations (*panjega*) were left in several places by omitting a single stone in the wythes of the wall. The largest blocks were positioned at wall corners.

Blocks on the outer sides of the wall were always set first, followed by the inner ones. The outer side of the wall was built using larger and more regular-shaped stone blocks, while smaller ones were used for the inner sides, so row heights of the face and back of the wall did not match. All cavities between blocks on the face and back of the wall were filled by small pebbles (*sovrnja*). A connective element would be placed on each square metre of the wall, connecting the outer and inner side of the wall (Figure 19).

Figure 18. The basis of rectangular houses is most commonly square (Photo: Dušan Tomanović).

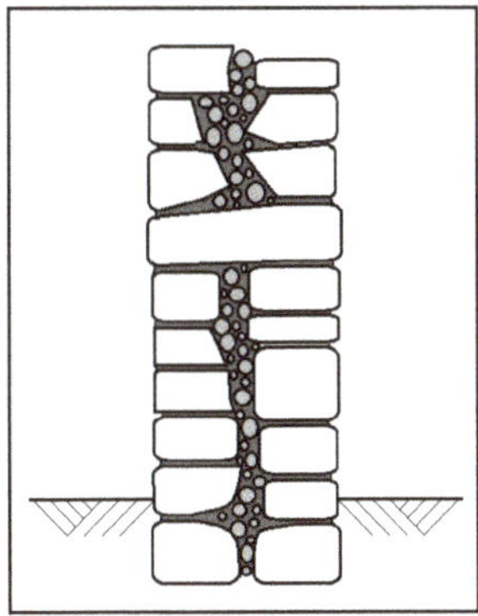

Figure 19. Stone wall without binding agent (dry construction) (Drawing: Dušan Tomanović).

In the beginning, the door was the only opening in the house. Lintels were positioned above the doors, most commonly from monolithic rock. However, there were wooden ones too (made from the hardest one—oakwood).

There were no windows on these houses in the beginning. Later, in the 19th century, windows were small in size, if existent at all, and faced the slope down towards the field and the sea (Figure 20). Windows got larger with time, so their size was about 1 m by the middle of the 20th century [13].

Figure 20. When present, the windows are narrow (Photo: Dušan Tomanović).

The kinds of openings on single-storey houses were:

- The ones which are not framed
- The ones that are framed, with "frames" made from local stone (more commonly)
- The ones with frames made from stone brought from Korčula island (less commonly)

The roof construction of these houses is very simple. It consisted of a single ridge beam stretching from the middle of one tympanum until the middle of the other one, supported by a forked pole named "soha" by the locals (Figure 21). Brotches do not lean on the ridge beam, but are simply joined at the ridge, digging into the wall at the position of the pole plate. Laths made of carved branches are thrown over the brotches and later covered with flagstones (before that, in the 18th and 19th centuries, thatch, bulrush or wooden slats and the like were used). As the brotches leaned on the inner side of the wall, the roof is slightly curved upwards at the end.

Figure 21. Roof construction is supported by a wooden forked pole (Photo: Dušan Tomanović).

2.3.3. Construction Techniques Used in Construction of Multiple-Storey Houses

The Foundation

As a rule, multiple-storey houses were built on solid ground. House foundations at the seaside were immersed in underground waters, if not in the sea itself. Nevertheless, the walls of these houses have not yet shown signs of cracking nor subsidence of any kind.

The foundations were made of flagstone or larger blocks of different sizes, and their depth did not surpass 50–60 cm. If the soil is loose, digging is done deeper than a metre, whereas foundation is not necessary if building happens on the rock. The width of the stone wall varies from 70 to 80 cm, while the binding agent used was lime mortar, sometimes also terra rossa (crljenica), known to be a good binding agent.

The Outer Wall

Walls were built using local limestone immersed in hot lime mortar or sand, clay, or sometimes even in eggs, which served as a binding agent. These walls have an outer and an inner wall 50–80 cm thick (Figure 22). Outer walls were built using the highly regarded fine stone (made from carved stone blocks) 17–25 cm in height. Side walls were built using less-regular shaped blocks than those on the front façade of the house, while their center is filled with broken and smaller stone pebbles ("sovrnja", Lat. sauornam) and lime mortar in most cases.

The mortar was prepared by mixing fine sand and lime. Lime has been known since ancient times and the Middle Ages, and was obtained from limestone in a lime kiln. The quicklime (burnt lime), when wet developed great heat, so it was cooled for three days. For the preparation of mortar, water was poured into the lime in shallow fenced plates. The slaked lime was mixed with a wooden tool consisting of a long handle with a stick vertically attached to its end. By the 1950s, it was done in the same or similar way.

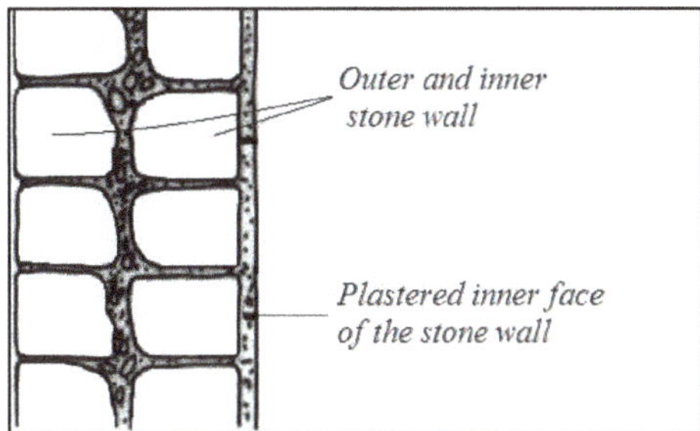

Figure 22. Stone wall with two sides (Drawing: Dušan Tomanović).

One large stone was used to connect both sides of the walls at every 2–3 m interchangeably in each row, thus fitting two walls together. This stone and procedure were called "wall binding" by stonemasons. In the corners of the house, we can see corner blocks (Figure 23). They were somewhat larger in size, more regularly processed, their colour was a darker grey, and the stability of the stone house partially depended on them. The longitudinal walls of the house were finished with wreath made of stone slabs (kotali). Gable walls are also made of stone slabs measuring 10 cm.

Figure 23. The corners of the house were made using blocks of greater length (Photo: Dušan Tomanović).

Thick stone walls provided lower inside temperatures in the summer months, and kept heat for longer in the winter time. During construction works on the main façade (in more wealthy owners), the local stone was substituted with the Korčula stone. It represented the height of the building's beauty, and at the same time its owner's wealth.

The Korčula limestone is soft and easy to process; however, due to its lightness and brittleness, it is rarely used for construction, but only for decorating the facade. The local stone is much stronger than the Korčula stone. They differ in colour—local stone is grey whereas the Korčula stone is white.

Residential buildings, aged from 100 to 160 years old, were built in straight rows up to 25 cm high. It should be noted that the width of the rows was determined only by the thickness of the layers in the quarry. Stacking the blocks into the building was done in the same position the block had in nature. This is the correct way of construction in which the joints do not come to a vertical position. In this way moisture absorption is avoided. The extraction of the blocks was done manually with wooden or steel wedges.

During the 20th century, there was a tendency for the exterior walls to be plastered to protect the house from moisture, sea and rain. In Prčanj, some houses are plastered with red plaster (most likely made from minced bricks). Plastering with cement mortar is more present in the post-World War II period when cement became more accessible to the wider population. Humidity is, however, a difficult problem during the winter period, especially in vacant homes.

Stone was not particularly processed for inner walls, except for several flagstones above the window, where they form a mild arc. Vaults were built using 5–6 cm thick flagstone taken from local quarries Verige, Strp and Lipaca.

Joints are usually filled with mortar along the wall lines and their size varies from 5 to 7 mm. Joint processing is not always the same everywhere. Those house owners who did not find their joints along the wall beautiful or efficient enough processed them in a different way: the joints were filled up until the wall line, and subsequently the horizontal and vertical ones, protruding out of the wall lines by 3–5 mm with completely straight and sharp edges (Figures 24 and 25).

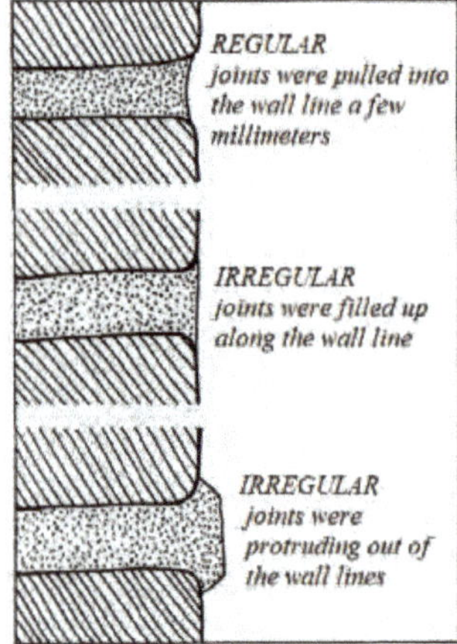

Figure 24. Types of joint processing (Drawing Dušan Tomanović).

Figure 25. An example of a joint protruding out of the wall lines (Photo: Dušan Tomanović).

Venetian feet (34.77 cm) were used to measure wall thickness. This was the unit of measurement in civil engineering until the end of the 19th century in Vrmac, in the whole Bay of Kotor area.

Some of the wall thickness measures were:

- The Verona Palace in Prčanj (built at the end of the 18th century). All outer walls are 72:72:72:65. Middle walls measure 65 cm.
- The double building Verona in Prčanj (built at the end of the 18th century). All outer walls measure 71 cm.

- The Tomanović Palace in Lepetani dating from 1846 (ground floor walls 72, upstairs 66)
- The Tomanović House in Lepetani dating from 1826 (ground floor walls 66, upstairs 60)

Walls are 72 cm (two feet) thick on some houses, while upper floors are weakened up to 66 or 60 cm.

After the 1979 earthquake, steel straps were put in place to secure the walls. They had not been used previously. Their usage began in the 19th century and was limited to tall edifices only (most commonly in the city of Kotor) [14].

Dividing Walls

In most houses in the Bay of Kotor area, dividing walls are made of wood, then plastered later (Figure 26). When property was to be partitioned, houses got divided by walls vertically, and not by floor. Dividing walls were most commonly built by vertically setting 3 cm thick planks, uneven in size, which varies between 5 and 10 cm (Figure 27). Horizontally nailed slats 3 × 3 cm in size were placed on both sides of the planks at a distance of 2–3 cm. Pre-cut planks were often used instead of slats, on the same distances. After all of this comes a 1 cm layer of lime (or later cement) mortar used to fill all cavities. The thickness of such dividing walls varies between 9 and 12 cm.

Figure 26. Dividing walls are made of wood and later plastered (Photo: Dušan Tomanović).

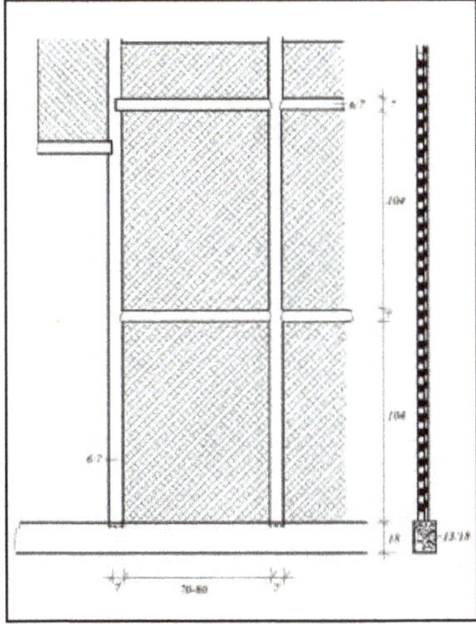

Figure 27. Cross-section through a dividing wall (Drawing: Dušan Tomanović).

Floor Joists

Mezzanine ceilings are usually made of wood. Beams are in most cases made of beech, pine or chestnut wood, or any other kind of wood available nearby. Floor joists are consisted of wooden beams 18, 20 or 22 cm in width, with significant variations in thickness: 12, 14, 16, 18, 20, 22, 28 or even 30 cm, for spans ranging from 7 to 8 m. Beams are placed at an axle distance of 70–80 cm. Floor joist beams in some houses in Prčanj (houses of the Florio and Verona families) were positioned to incline with their wider sides, which considerably reduced their load-bearing capacity.

Wooden floor joists were connected to the stone walls with tie bars. This forced them to oscillate in a harmonious way, reducing more important damages. These elements encompassed a few rows of stone on the outer side then got fastened to the joists inside and were usually set in the axle of the vertical strips of walls, where they had the most seismic effect.

Beam processing was not particularly precise. In more wealthy households, where beams were visible on the ceiling, they were finely treated, usually trimmed from both sides with slightly rounded edges.

Floor joist beams lean on constructive massive walls mainly in three ways:

- On stone tooths (Figure 28)
- On the wooden beam placed on the wall
- Inside previously prepared holes in the walls (built-in) (Figure 29).

Figure 28. The small bracket carries the floor joist beam on the inner side of the wall (Photo: Dušan Tomanović).

Figure 29. The floor joist beams are put into previously prepared holes in massive stone walls (Photo: Dušan Tomanović).

Tooths are stone brackets, sometimes made of local stone (most commonly in upper villages), sometimes of Korčula stone (on the sea), and protrude 16 cm from the wall and are 13 cm in height. They are housed in the massive outer 20–25 cm wall.

In the second case, there is a difference of 10 to 12 cm between the lower wall of the *konoba* and the floor wall, and this space is used for the wooden beam, whose function is to replace the tooths.

The third case involves floor joist beams being placed into massive walls, in previously prepared holes. These holes had the same dimensions as the beams in the axle distance about 60 cm for a 4 m span.

The ceiling heights of the *konoba*, the ground floor and the upper floor are always different; therefore, a number of heights were in active use.

In main bedrooms, the most common ceiling heights are:

$8' = 278.16$ cm
$9' = 312.93$ cm
$10' = 347.7$ cm.

In other (auxiliary) rooms, the height is somewhat lower:

$6' = 243.35$ cm
$6'1/2 = 261.12$ cm.

Roof Structure

There is a wooden roof structure on all houses. The surface is covered with tiles or flagstone. Numerous remains in the villages of the Bay of Kotor point to this conclusion. Roofs are usually double sloped (swallow style), even though there are triple-sloped, quadruple-sloped (padilion style) roofs, but also roofs with multiple slopes. The roof slopes in the Bay of Kotor range between 25–32°, most commonly 28°. This is the main type of roof structure on houses in the Bay of Kotor.

2.4. Valuable Visual Design Elements

Valuable visual design elements on stone houses are:

a. Stairs, b. Windows, c. Doors, d. Eaves, e. Roof cover, f. Balconies, g. Terraces, h. Sculptural elements (Frame, Bracket, Frieze, Baluster).

(a)   Stairways

2.4.1. External Stairways

At the time of creation of these houses, exterior stairways were always made of stone. They were mainly placed by the side wall of the house to the left or the right of the entrance, reinforced (filled out) from underneath. An extension 'solar', a space 2 to 3 m^2 in size, can be found at the entrance to the house, sheltered from the wind and rain with plenty of sunshine. It is surrounded by a *pijuo* (stone bench) with coated flagstones. Above the solar, for the main part of the 20th century, there were no eaves shielding the front door from the rain—the only protection came from the cornice. This characteristic—the lacking of eaves above the main entrance and the external stairway—is typical for the whole Kotor Bay area, and they started being added only recently.

The stairway and its elements—the tread and the front are usually made out of flagstone. The treads are uneven in size: 28 to 30 cm, while the height of the front is stable (12 + 3 = 15) or 13 + 2 = 15 cm.

The fence of this staircase is stone-made, which is almost always the case. Balusters are placed in some places, while others prefer to have their fence made from wrought iron. The stone fence is 80 to 85 cm in height, the wrought iron one is 85 + 5 for wood, while balusters come in different heights and shapes, and they are also stone-made (Figure 30).

Figure 30. The tread and fence of the external stairway are mostly made of stone (Photo: Dušan Tomanović).

2.4.2. Interior Stairways

At the beginning of the 20th century, the interior stairways were made of wood (Figure 31). They led upstairs, towards the ingle, and to the ground floor towards the *konoba*. Rev. Nakićenović describes them at the beginning of the 20th century: "The stairs reach the first floor in all houses, if they are in the house they are usually made of stone, while those that come from the yard, which is in Luština usually stone-made. On the second floor they are made exclusively out of stone. Both are spacious and nicely laid, that is to say, easy to climb. Divided houses are an exception, and there, stairs would be narrow and steep because of a lack of space [20]."

Figure 31. The interior stairs are always made of wood (Photo: Dušan Tomanović).

(b) Windows

In all houses, windows have a multiple role: illumination, solarisation, ventilation and enabling visual contact with the outside. Aside from functionality, they also have an aesthetic dimension, reflected more in the exterior, but also in the interior. They are usually positioned on the side facing the slope, towards the sea.

Single-storey houses sported windows that were small or non-existent (Figure 32). Their shape and proportions are typical for the stone architecture of the seaside in general. During the 20th century, they became somewhat larger—about 1 m.

Figure 32. Openings were small on single-storey houses (Photo: Dušan Tomanović).

Larger windows appear on multiple-storey houses. Window size and proportions of the window opening were also defined by the quality of the window frame (imported or local) (Figure 33).

Figure 33. Windows on multiple-storey houses were defined by the quality of the window frame (Photo: Dušan Tomanović).

Window frames were usually made of better quality stone, which was also better processed. Holes were dug in stone frames, on the inside, to accommodate the wings later. The span of the lintel beam is slightly larger than the breadth of human shoulders. They are somewhat higher, in order to get rectangular openings of the aspect ratio 1:1.4, which are close to 1:$\sqrt{2}$.

The shape of the casement is the same everywhere as a rule. These are double door windows, with each door divided by thin slatlets into two or three parts, for each door. Glass is single-layered and 4 mm thick. The windows were constructed so that quirks were integrated into it, making it into a receptacle for glass without putty. The glass is thus protected from the wind, but since it is movable, wind and rain can permeate through during particularly violent weather.

The outlet in the interior of the wall is slanted, so it is larger inside and smaller outside. This slant in the stone wall is done both vertically and horizontally. The vault which appears above the outlet, on the inside, follows the slant of the walls.

Cast iron lattice is an integral part of the *konoba*. Their main role was to protect the inhabitants from break-ins. Windows are smaller in size and enclosed in frames made of stone from Korčula.

Often, small brackets with holes, auricolae, could be found above windows (they were used as curtain holders in the summertime) and 'dentes' brackets (used as shutter holders or flower pots, for drying figs, etc.).

Shutters are a compulsory element of the windows. They are coloured in a characteristic green hue and protect the wood of the window frames from heavy rains and storms. Those shutters were mostly solid (made from solid wood, without any movable parts).

(c) Doors

Doors can be external and interior.

2.4.3. External Doors

External doors are different from the interior ones by their character. They are exposed to weather conditions, special functional requirements (safety, insulation), architectural and aesthetic criteria (emphasis of the entrance, singularity), which is why they differ from interior ones in terms of size, shape, materials used, construction and processing methods. External doors on the houses are the front entrance door as well as the doors on balconies and terraces.

They are sometimes also located on the stone wall fences surrounding yards of Kotor Bay houses. The openings and their number was determined by the property owner. A stone frame conditioned the dimensions of the openings, the doors were placed in different spots, and there were no irregularities.

On external walls, there are two door sizes: 5′6″ (191.235 cm) and 6′ (208.62 cm) with a minimum width of 3′ (104.31 cm), which climbs up to 5′ (173.85 cm) in half-foot intervals. The most common external door size ratios are: 6′/4′ (208.62/139.08) or 3:2 and 6′/4′6″ (208.62/156.46 cm) or the ratio of 4:3.

These doors have no fanlights, and they open to the inside. They are either enclosed in a better processed local flagstone or in frames made of Korčula stone $\frac{1}{2}$ of a Venetian foot in width, which is 17 cm. Lintels are positioned above the door, and there often is a relieving arch on top (Figure 34).

Figure 34. A door with no fanlights is framed with rectangular stone frames (Photo: Dušan Tomanović).

The door consisted of a frame—lintel and the movable part of the wing. There are single-winged and double-winged doors, divided according to the number of wings. In the beginning, they were single-winged, later developing into double-winged doors. In arch doors, we can see a separate semi-circular part above the wings, but the wings themselves sometimes end up with an arched shape. They were manufactured using wood, and mostly coloured in green and white (Figure 35).

Figure 35. A door that does have fanlights is framed with arched stone frames (Photo: Dušan Tomanović).

The main entrance to the house and yard gates or fence gates were often secured from break-ins in multiple ways. On the inside of the door, in some cases, even 3–4 closing systems can be found: locks, hasps, klav, baštun (Ital. bastone), and in some cases a "pat" is often placed. A "pat" (Ven. pato) is a strong square beam that can be inserted into special holes inside walls. When being placed, it is positioned behind a closed door, thus securing the whole opening.

The hardware is made of high-quality stainless steel, the so-called arcal, arcao (from Ital. accialio-steel). This hardware's special feature is its long durability right next to the sea shore, stretching over multiple centuries without major signs of corrosion.

Decorative elements of inner doors can be found on iron doorknobs.

(d)   Eaves

Eaves form an integral part of the roof structure and their main function is precipitation protection. The ones with kotals are the most common, while grondals are far less frequent.

Kotals are flagstone blocks, mainly rectangular in shape, protruding 15 to 40 cm out of the wall, they are 2 to 3 cm thick, embedded into the wall with one side being 5 to 10 cm. They are joined with lime mortar and act as a monolith with the façade (Figure 36). The stone eaves were a very practical solution since stone was so ubiquitous. It was the only way the house could be protected from *sijavicas* (tiny rain droplets brought by the wind during heavy wind gusts such as the *bura*, the *jugo* and other types of winds). When the time came to collect rainwater from the roof and place it in special tanks, "grondals" were used instead of kotals (Figure 37). Grondals were thick flagstone stones from Korčula with a built-in gutter for draining atmospheric waters. In some cases, grondals were supported from the bottom by small stone-made brackets (dentes) made from Korčula stone or the local kind.

Figure 36. A detail from the loft and roof structures [11].

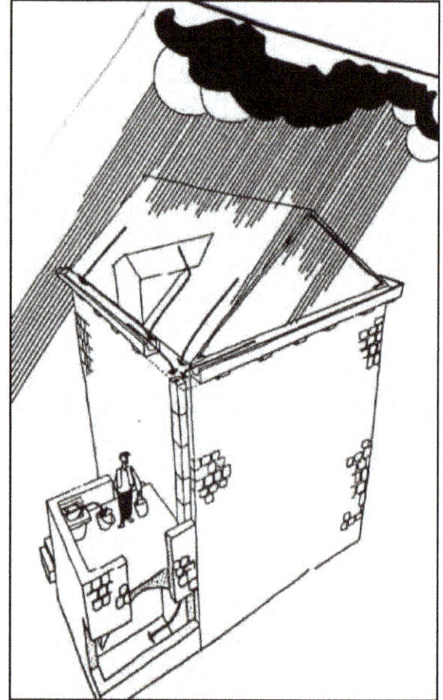

Figure 37. Grondals (stone gutter for draining atmospheric waters) [13].

(e) Roof Cover

Roofs on village houses are mainly double-sloped. Roof slope is anywhere from 28 to 32 degrees. Since the Bay of Kotor is known for its abundant precipitation, the roof protrudes slightly more with flagstone cornices (kotals). In the past, roofs were covered with flagstone blocks, found ubiquitously in the Bay of Kotor area (St. Vid, Verige, Kamenari, Strp) (Figure 38). This can be concluded after revising numerous remains in the upper villages of Prčanj, Stoliv, Veće Brdo, Gornja Lastva, Bogdašići. Before this, roofs were sometimes covered in thatch.

Towards the end of the 19th century, handmade tiles took over, and after the tile and brick factory in Krtole opened in 1907, higher quality and better processed clinker bricks were produced. They were placed between the slats, laid dry on the bottom, with a thin mortar bed being made on the upper side to adhere better in order to become impervious for any sijavica (small rain droplets brought by the rain) that may come in. An amount of effort was put into preventing the mortar from touching the wood so as not to moisten it. Larger, much wider tiles, were used for ridges and hips. The end row is covered in tiles and forms a channel (Figure 39).

Figure 38. Flagstone roofs were dominant in the 19th century—Veće brdo (Photo: Dušan Tomanović).

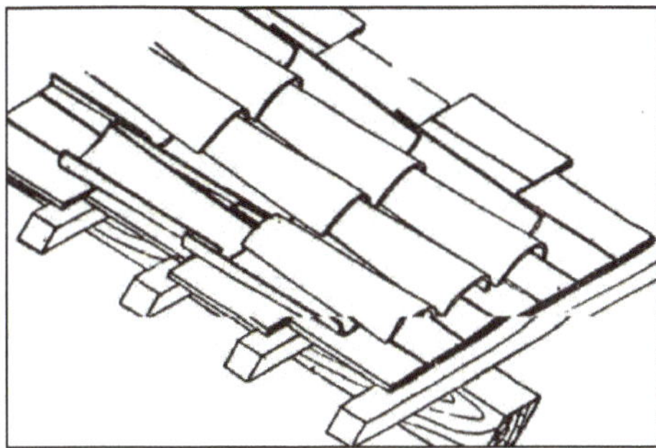

Figure 39. At the beginning of the 20th century, factory-made corrugated tiles took over [11].

(f)   Balconies

The urbanistic transformation of the city during the baroque era, when the renaissance got an urban look, remained clearly recorded in the form and function of balconies in the Bay of Kotor. The majority of houses did not have balconies. Where present, they were still in good shape and being used at the beginning of the 20th century. They were built on houses by wealthy owners (captains, ship owners, merchants), which is why we almost always see them in the 17th, 18th and 19th centuries in economically prosperous places such as Prčanj, Muo, Donji Stoliv, places that also had more of these kinds of people [12]. Architect Milan Zloković says: "I think that, when it comes to special balcony deliveries, contracts were made directly between the future house owner or their representative and the main master at one of the stone workshops, determining the size of the balcony and its final processing on that occasion, on-the-spot, according to certain reputable bracket, small pillar and baluster types [12]".

The most beautiful palaces for the noblesse were built in Prčanj in this period.

I shall cite some of them:

-   The 18th century Florio Luković Palace in Prčanj, dominated by a baroque balcony on the first floor of the front façade, facing the sea
-   The mid-18th century Verona Palace in Prčanj. It has a centrally placed balcony with four stone brackets and baroque balusters on the second floor

- The Beskuća Palace, in the centre of Prčanj. On the first floor of the main façade, we see a balcony with a stone fence (balustrade) supported by three stylised brackets. Two portals lead to the balcony, with arch-shaped fanlights decorated with a metal net
- The Luković Palace in Prčanj, 11 baroque urban houses in a succession with characteristic balconies

The balconies are not large in size and the question remains if they had a practical use, aside from the purely aesthetical belvedere (nice view) function. The balcony width ranged from 90 to 110 cm, with varying lengths (Figure 40). They are usually placed in the centre of the main façade, while the symmetry in relation to the façade (doors and windows) is strictly respected. All balconies are constructively supported by large stone brackets "regule", richly stylistically decorated and processed.

Figure 40. Typical stone balconies on houses in Prčanj (Photo: Dušan Tomanović).

In most cases, balconies are surrounded by balustrades of different shapes and heights. Balustrades are surrounded by cornices from the top and the bottom, and its length as well as the length of the entire fence is divided into 2, 3 or 4 fields (depending on the length of the balcony). The fence is about 75 cm high. Later, in the 19th and 20th centuries, on some houses, stone balustrade fences were replaced by wrought iron fences (Figure 41). The upper surface of the balcony (the tread) is not covered and it is exposed to atmospheric conditions. It is coated in flagstone from local quarries. A double-winged door, which is in most cases enclosed in *voltes* (arch-shaped stone frames), leads to the balcony. As a rule, the doors open inwards. Initially made of solid wood, they were later combined with glass. Sometimes, the part where the arch starts and ends is being glazed, while the lower part is made from wood and glass, or solid wood.

Figure 41. At the beginning of the 20th century, balcony fences were made from wrought iron (Photo: Dušan Tomanović).

Influences from the whole Bay of Kotor area can be recognised in the shapes of the balconies, which are adapted to the measures and needs of the local environment in an original way.

(g)   Terraces

Terraces are almost always present in upper villages, and significantly less in the lower parts. Depending on the slope, they are usually located in front of the house at ground floor level or at the first-floor level (Figure 42).

Figure 42. Terrace on the ground floor (Photo: Dušan Tomanović).

When placed on the first-floor level, a stone *volte* bears its load. The terrace is plate paved with rectangular stones brought from local quarries on both the ground floor and the first floor. It is surrounded by a "pijuo", a stone bench with a seat and a backrest (Figure 43).

Figure 43. Terrace on the first floor (Photo: Dušan Tomanović).

Wealthier families would make *bistjerna*s on the ground floor of their terrace. These bistjernas were used to collect rainwater with the help of a stone gutter (grondals, gurla). Poorer families would keep livestock in that same area (under the terrace), and in that case the bistjerna was located in front of the house or behind it, in the yard.

(h)   Sculptural Elements

Frames around doors and windows, frieze beams and stylised eaves above the openings, balcony components (brackets, blocks, little pillars and balusters), external stairways and their fences are all made of limestone brought over by sea from famous quarries on the Vrnik island next to Korčula known by the name *korčulanac*. This is probably due to the fact that it was not possible to find a stone easy to process in a precise way anywhere around. Thus, the inhabitants of the Bay of Kotor ordered stones processed in the finest way through their merchants and sailors and built it into their new homes.

Sculptural elements on stone houses are: the frame, the bracket, the frieze and the baluster.

2.4.4. The Frame

While houses in the Bay of Kotor were built using local stones, frames for openings were made from stones that were easier to process, brought over from the Croatian islands of Vrnik and Korčula. These frames had standardized measures, which in turn determined the dimensions of the openings themselves. Feet were used as a measuring unit up until the mid-19th century (the Venetian 34.8 cm foot was divided into 12 ounces (1″ = 2.9 cm)) [21].

On external doors, the stone frames also served as door jambs onto which wings could be attached. The width of the frames was 17 cm or $\frac{1}{2}$ of a Venetian foot, while the size of the opening dictated its height and length. In single-storey houses and in most houses in upper villages on Vrmac, door jambs were made of home-made monolithic stone whereas they used to be made of wood in older times.

On the Vrmac, openings (frames) on multi-storey houses can be:

a. With enclosures, i.e. "frames" made of local stone (Figure 44)
b. With frames made of Korčula stones and without stylised lintel cornices
c. With frames made of Korčula stones and with stylised lintel cornices (Figure 45)

Frame faces were well processed, while inner parts were done more roughly. Stone enclosures around openings (frames) play a constructive role, aside from being aesthetically pleasant. Door openings influence the door shape and determine its dimensions. Such openings could be finished in a straight or arch-shaped way. With straight frames over larger spans, a relieving arch is added to avoid cracking.

Figure 44. Window and door frames in the hinterland are mainly composed of monolithic stone with no stylised cornice (Photo: Dušan Tomanović).

Figure 45. Frames made from Korčula stones with stylised cornices (Photo: Dušan Tomanović).

Frames over arch-shaped doors and windows (volates) are somewhat more specific and wide. They are manufactured in two parts, and connected by a decorative bracket with a frieze on top (Figure 46).

Figure 46. A stone frame above an arch-shaped door, connected by a decorative bracket at the top (Photo: Dušan Tomanović).

In external doors and windows, the frames serve as door jambs or window frames, respectively, with the possibility to accommodate wings.

2.4.5. Brackets

Stone brackets find multiple usage on houses in the Bay of Kotor area. A few types were used, and those are:

- brackets with a hole (auriculae) (Figure 47)
- brackets underneath the lower window frame (dentes) (Figure 47)
- brackets bearing heavy stone balconies (regulae) (Figure 48)

Figure 47. Brackets with a single hole "auriculae" (above) and "dentes" brackets (below).

Figure 48. The regulae support heavy balconies (Photo: Dušan Tomanović).

Brackets with holes (auriculae) supported the curtains from the outer side and protected the inhabitants of the house from sunlight during the summer. Their dimensions are 30/25 cm and they are placed under the lower line of the upper lintel frame. Their width ranged from 5 to 7 cm. They exist in variations with one and multiple holes.

In the second half of the 20th century, brackets with holes (auriculae) slowly lost their primary function—to protect house inhabitants from the sun in the summer. On some houses, they were entirely removed, while some others saw them decay over centuries of use.

Brackets (dentes) come in multiple variants. They are built below the lower window frame or at the same level, to the left and to the right, in pairs. Their role differs, in some places they are used as shutter holders, while some dry their figs on them. We can also find them on the façade below the roof cornice, bearing *kotals* (stone blocks), but they also carry floor joist beams on some houses. Their dimensions are 40/20 cm, and they are mainly encountered on houses in Prčanj, Donji Stoliv and Donja Lastva.

Regulae are brackets bearing heavy stone balconies. The largest load they receive comes from the balusters. They are being placed in rows of two, three or four, depending on the balcony length. The more luxurious regulae were imported from Italy because those made on the island of Korčula were much more modest.

2.4.6. Frieze

At the beginning of the 20th century, friezes have been subsisting for more than two centuries on some houses. Despite this fact, they did not show any signs of weakness or damage because of the compact way they were manufactured in. They are mainly found on houses at the sea. They do not exist in villages in the hinterland.

They are placed above the window, the door, or gates (Figure 49). Their dimensions vary and depend on the size of the opening above which they are placed. Stonemasons from Korčula standardised them, and the largest number was bought in their workshops. Their lower part leans on the upper window frame and the door. Aside from the aesthetical function, they had a practical one, which was to prevent rainwater from entering through the openings, since rain would be falling from the roof directly onto the floor.

Figure 49. The frieze leans on the upper door and window frame with it's lower part (Photo: Dušan Tomanović).

A rich frieze means a rich house owner. Special attention was given to friezes above arch-shaped doors with a small decorative bracket at the top (Figure 50).

Figure 50. A frieze above an arch-shaped door with a small decorative bracket at the top (Photo: Dušan Tomanović).

2.4.7. Balusters

The appearance of balusters is not consistent with the tradition of folk architecture, but they did announce the gradual formation of the local architecture language and its dialects. Balusters, or stone pillars, are used to fence not only balconies, but inner yards and atria as well. They appear sporadically on the coast, while they are not present in upper villages (Figure 51). This shows us that they started being used in later stages, since all lower settlements were created afterwards. They are most common in the following places: Perast, Orahovac, Mula, Prčanj, Donji Stoliv, Donja Lastva. It was probably sea men who brought balusters from their numerous trips around the world, into the cities at first, around the 17th and 18th centuries (for example, balconies on the 17th-century Gregorina Palace in Kotor or the mid-18th century Dabinović Palace in Dobrota), and later into nearby cities sometime in the second half of the 19th century.

Figure 51. Balusters are present on the coast, and absent in upper villages (Photo: Dušan Tomanović).

They vary in shape, height and width. Framed with frieze beams and stone pillars, they were manufactured from stone and processed with high precision in this period. The baluster type with three highly convex-concave parts with sharply protruding edges towards the outer borders, with a square base and a decoration made of concentric squares (which differ from the usual baluster division in the shape of a single or double "pear") is the most common type on the Vrmac (Figure 52). It was also present in earlier architecture in the Bay of Kotor in the 17th and 18th centuries.

Figure 52. A type of baluster with three convex-concave parts with sharply protruding edges on the outer borders and with a square base (Photo: Dušan Tomanović).

3. Discussion

Representing local materials and construction techniques, the Bay of Kotor stone architecture, accumulated over centuries, is seen as one of the most important physical indicators of the culture and lifestyle of people in this region.

Up until the 1970s, all important characteristics of traditional construction were part of the forming of villages in the Bay of Kotor. Traditional materials such as stone and lime mortar were used. Dry wall construction (without a binding agent) was completely replaced by the usage of lime mortar in construction. Different types of stone suitable for construction was being extracted from local quarries.

In the 19th and 20th centuries, there were two types of stone houses in the Bay of Kotor area, and those are single-storey and multiple-storey houses. Single-storey houses ceased to serve a dwelling purpose in the second half of the 20th century. They were instead transformed into auxiliary buildings or buildings for livestock (stables). Multiple-storey houses experienced a comprehensive transformation with numerous variations in size and complexity levels. They thus became the extended type of civic house with an additional number of rooms for residential purposes.

Unfortunately, the contemporary reconstruction of these buildings all along the Bay of Kotor does not take the balanced and harmonious expression of this architecture into account. What encourages us, however, is that old stone buildings were not torn down, even if they were abandoned and ravaged by previous earthquakes that made them entirely unusable.

The most important changes occurred after the Second World War. In this period, the character itself of traditional houses gradually disappeared, just like the character of villages and houses were destroyed with the irrational and incompetent use of concrete, or at least those villages and houses worthy of mentioning. Some of the construction methods in traditional architecture also vanished, and they included collecting rainwater, usage of easily accessible materials, controlling heat input and exchange by balancing thermal characteristics of the wall mass and shadow size of openings.

Changes became even more apparent after the catastrophic 1979 earthquake. First construction laws appeared in 1981 [22]. The traditional way of building was completely lost in this period.

The development of tourism brought about the need to repurpose the existing houses and build new ones. Cost-effectiveness was not a priority because of the pressure that awareness about quick and easy earnings though building exploitation can bring. Building exploitation was most commonly the main motive behind the construction thereof, while the development of commercial activities was secondary in the coastal zone.

The biggest challenge in the revitalisation of stone houses in the second half of the 20th century was the layout of ground floor rooms and determining the position of the kitchen. In order to get as much space as possible, height was often increased using inadequate stone blocks. The roof slope was also changed, and the creation of belvederes rendered the façade asymmetrical and disproportionate.

The openings (windows and doors) on buildings were put randomly (according to need). They were enlarged and disproportionately big in relation to the surface of the facade, while the stone frame was replaced by a concrete one, as a rule. In several instances, a balcony was added onto the first floor while windows got transformed into doors.

Stone balusters made from alabaster or cement fueled the spread of kitsch in bad taste by investors or architects.

Plastering stone facades and then painting them is another example of how things should not be done. The usage of autochthonous types of stone is in regression today with regards to imported types. This is also how the painstaking way of extracting and processing stone almost disappeared. The art of building with stone is lost and stone has become mere decoration or coating, while carefully and lovingly crafted blocks of old stone houses turn into building material for new edifices. Quarries are abandoned, and do not differ from their natural environment. They can tell us a lot about the old builders' attitude towards their own living space, because they only took from nature what was really necessary.

Nothing prevents us from using the latest materials and construction styles parallelly with the building achievements of previous generations. The two will not be mutually exclusive, but can rather complement each other and leave a sign of their time, just like the predecessors left traces and know-how through the centuries together with existing ones, without disturbing or damaging their heritage.

4. Conclusions

The Bay of Kotor and its stone settlements are a valuable and unique example of traditional construction. Building stone was the natural choice for builders in this area for centuries. It was adapted to the needs of the region, easily accessible and extracted according to purpose.

Modifications in traditional lifestyles and conducting business in the 20th century permanently changed villages and influenced the moving of inhabitants from upper villages into the lower ones. The degradation of villages as a permanent process in the 20th century, particularly intense in its second half, had a twofold impact in its spatial incidence: a complete abandonment of villages or their transformation in which traditional values got lost. The complete abandonment of upper villages caused the exposure of their building structure to decay, even though it was preserved in a spatial organization and unencumbered by new construction.

Today's urbanism did not succeed in the creation of anything better than what the people created for centuries according to their needs. Alongside the coastal belt itself, new villages, completely different from the old ones, developed, even though they were "constructed with a plan". They jeopardised the old settlements both qualitatively and quantitatively, almost to their complete destruction.

When the environment proves to be historically, artistically and culturally worthy of respect, as is the case here, new buildings should have followed that logically, and fit into it with visual characteristics, often copying certain elements, naturally, adapted to today's technology and way of construction.

The authors suggest the need to better preserve these houses and settlements with the global issue of sustainability. We need to define what constitutes a threat to heritage. The new type of house should emanate from the old stone house. Mass, proportions and materials, ornaments around the doors and windows, bordures and cornices under the roof should be elegantly adapted to the olive groves, vineyards and other traditional lifestyle elements.

Another aspect of raising awareness of appropriate stone for repair, maintenance and restoration of historic structures concerns the availability of reliable technical data. Replacement stone should ideally be of the same type as the original, or the nearest possible equivalent. Use of inappropriate stone can result in significant damage to the heritage. In order to choose the correct stone, both the original stone and the potential replacements should be fully characterized. The stone industry has an important role in providing accurate technical and aesthetic information on their products [17].

Abandoned stone houses and settlements can be revived only by the return of inhabitants and commercial incentives that would motivate them to live there. One way to save traditional building as an important part of the Montenegrin cultural heritage would be to recognise such settlements as essential tourist resources.

What should be avoided in upper extinct villages are situations similar to those that arose in lower settlements, where settlements sprouted without proper spatial planning documentation or expert assistance from architects and other specialists and resemble neither a village nor a city.

A displaced and neglected village rehabilitation program should be established by the state and the local authorities, turning the settlements into tourist destinations, thus becoming a source of income for the local population in order to cease the selling out and devastation of real estate.

The most important contribution of the program in the revitalization of old heritage would be a planned approach. Tourism and agriculture should be the main driving force behind the development of these regions. Settlement reconstruction is a long and financially demanding task. High levels of protection of historical heritage should be linked with the interests of its owner/representative, as well as all those who will invest into its revitalization and quality development.

Investing in the reconstruction of such settlements should be treated as active or lasting protection of cultural property, which deserves to be exempt of certain taxes. Existing legal reliefs are insufficient to achieve lasting self-sustainability of these settlements and individual structures in them. They need to be lasting and give an incentive to everyone involved (owners, businesses, etc.). In Austria, in areas where the threat of abandonment is real, produce made by farmers in their homes are exempt from taxation. Otherwise, if such incentives are absent, these environments will require permanent subsidies.

Traditional stone architecture in the Bay of Kotor represents the compatibility of parts and the whole, that is to say a balanced and harmonious architectural expression. It is the base from which new models should be built and traditional shapes transformed according to the contemporary style of life and modern tendencies in architecture. That is how we will bring traditional architecture back to life, and return life into it.

Author Contributions: Each of the six authors, D.T., I.R., M.G., J.A., N.G. and J.L. contributed equally to designing the study; collecting the empirical material; writing the paper and substantially revising the work. T.T. translated the paper. All authors have read and approved the final manuscript and they all agree to be personally accountable for their contributions.

Funding: This research received no external funding.

Conflicts of Interest: The authors declare no conflict of interest.

References

1. Tomanović, D. Stambena Arhitektura na Poluostrvu Vrmac XVIII I XIX Veka. Master's Thesis, Arhitektonski Fakultet u Beogradu, Beograd, Serbia, 2004.
2. Tomanović, D. Transformacija Narodne Arhitekture na Poluostrvu Vrmac—Boka Kotorska u 20. Doctoral Disertation, Arhitektonski Fakultet u Beogradu, Beograd, Serbia, 2014.
3. Antović, J. *Statvta Civitatis Cathari—Statut Grada Kotora: Knjiga I, II*; Državni Arhiv Crne Gore: Cetinje, Crna Gora, 2009.
4. Available online: http://gotravelaz.com/wp-content/uploads/images/Montenegro.png (accessed on 17 May 2019).
5. Available online: https://wearetravelgirls.com/wp-content/uploads/2018/04/Montenegro-Bay-of-Kotor-Lovcen-774x516.jpg (accessed on 17 May 2019).
6. Božović, G. *Naselja i kuće Tivatskog Zaliva, Urbanistička Studija*; Centar za urbano planiranje Beograd: Beograd, Srbija, 1980.
7. Kojić, B. *Izumiranje sela u Kotorskomom Zalivu i na Poluostrvu Vrmac, Spomenik S.A.N.*; Srpska Akademija Nauka: Beograd, Srbija, 1956; pp. 165–170.
8. Korać, V. *Spomenici Srednjovekovne Arhitekture u Boki Kotorskoj, Spomenik S.A.N.*; Srpska Akademija Nauka: Beograd, Srbija, 1953; pp. 115–129.
9. Lalošević, I. *Kostanjica—Prirodno i Kulturno Nasljeđe*; EXPEDITIO—Arhitektonski Fakultet u Podgorici: Podgorica, Crna Gora, 2009.
10. Kojić, B. *Seoska arhitektura u Boki Kotorskoj, Spomenik S.A.N.*; Srpska Akademija Nauka: Beograd, Srbija, 1953; pp. 169–186.
11. Đurović, V. *O Konstrukcijama Kuća od XVI Do Konca XIX veka u Kotorskom Zalivu i Njihovim Graditeljima, Spomenik S.A.N.*; Srpska Akademija Nauka: Beograd, Srbija, 1953; pp. 147–164.
12. Zloković, M. *Građanska arhitektura u Boki Kotorskoj u Doba Mletačke vlasti, Spomenik S.A.N.*; Srpska Akademija Nauka: Beograd, Srbija, 1953; pp. 131–146.
13. Petrović, Z. Selo i seoska kuća u Boki Kotorskoj. In *III Knjiga*; Zbornik arhitektonskog fakulteta u Beogradu: Beograd, Srbija, 1956.
14. Lalošević, I. *Svojstva Istorijskih Zidanih Konstrukcija u Seizmičkim Zonama (Primjer Boke Kotorske)*; Izgradnja: Podgorica, Montenegro, 2011; pp. 11–12.
15. Agnoletti, M.; Conti, L.; Frezza, L.; Monti, M.; Santoro, A. Features analysis of dry stone walls of Tuscany (Italy). *Sustainability* **2015**, *7*, 13887–13903. [CrossRef]
16. Petrović, Z. *Tragajući za Arhitekturom*; Građevinska Knjiga: Beograd, Srbija, 1991.

17. Pereira, D.; Marker, B. The Value of Original Natural Stone in the Context of Architectural Heritage. *Geosciencies* **2016**, *6*, 13. [CrossRef]

18. Pollak, D. Ovisnost Inženjerskogeoloških Svojstava Karbonatnih Stijena o Njihovim Sedimentno-Petrološkim Značajkama (Trasa Jadranske Autoceste: «Tunel sv. Rok–Maslenica»). Master's Thesis, Sveučilište u Zagrebu-Rudars-Geološko-Naftni Fakultet, Zagreb, Croatia, 2002.

19. Vuksanović, D. Sanacija Zgrada Graditeljskog Nasljeđa uz Primjenu Energetske Efikasnosti—Ograničenja i Specifičnosti, Naučni Skup "Obnovljivi Izvori Energije i Energetska Efikasnost". In *Naučni Skupovi—Knjiga 112*; CANU: Podgorica, Montenegro, 2012; pp. 267–273.

20. Nakićenović, S. *Boka: Antropogeografska Studija*; Srpska Kraljevska Akademija: Belgrade, Serbia, 1913.

21. Archivio generale Veneto. *Principi d'Aritmetica*; L'archivio di stato di Venezia: Venezia VE, Italy, 1823; Volume 2, p. 78.

22. Službeni list SFRJ broj 22. *Pravilnik o tehničkim Normativima za Izgradnju Objekata Visokogradnje u Seizmičkim Područjima*; Arhiv SRJ: Beograd, SRIJ, 1981; p. 31.

Article

Studenica Marble: Significance, Use, Conservation

Nevena Debljović Ristić [1,*], Nenad Šekularac [2], Dušan Mijović [3] and Jelena Ivanović Šekularac [2]

[1] Institute for the Protection of Cultural Monuments of Serbia, Radoslava Grujića 11, 11000 Belgrade, Serbia
[2] Department of Architectural Technology, University of Belgrade-Faculty of Architecture, 11000 Belgrade, Bulevar kralja Aleksandra 73/II, Serbia
[3] Faculty of Applied Ecology FUTURA, Metropolitan University in Belgrade, Požeška 83 a, 11000 Belgrade, Serbia
* Correspondence: nevena.debljovic@heritage.gov.rs; Tel.: +38163432782

Received: 20 June 2019; Accepted: 12 July 2019; Published: 18 July 2019

Abstract: Studenica marble is the stone used in creating the mediaeval Serbian cultural heritage. This is a historical overview of the importance and use of stone from prehistoric times to the Middle Ages, when the most imposing religious architectural structures were built. The significance of Studenica marble is particularly manifested in the Virgin's Church at the Studenica Monastery. For its marble façades and artistic architectural elements, among other things, the church was inscribed in the UNESCO World Cultural and Natural Heritage List in 1986. Through centuries, the Virgin's Church suffered multiple destructions. In order to restore the marble façades properly and its architectural elements, the marble deposits where the stone was once quarried had to be located anew. When the stone material characterisation had been performed, the right stone was selected for the complex conservation works on the churches in Studenica and Sopocani. A practical implementation of the research results raised the awareness of the marble deposits in the Studenica vicinity, being also part of the national heritage. The Studenica Monastery Cultural Landscape Management Plan envisions preservation of the deposits and their sustainable use for restoration purposes in the future.

Keywords: UNESCO cultural heritage; Studenica; Sopoćani; marble deposits; quarry characterisation; restoration; sustainability; management

---

1. Introduction

This paper starts with the perspective of the importance and role of the natural stone—the marble as a particular type of material—whose beauty and natural properties, from the time immemorial, has always symbolically indicated the presence of the divine, and through its various forms, have emanated its artistic and architectural characteristics.

In the Serbian mediaeval lands, there was one authentic, religious and architectural work built of marble in the 12th century. It was first envisioned as a burial church of Stefan, the founder of the Nemanjic Dynasty, and it was dedicated to the Virgin Evergetis [1]. So, even today, the Studenica Monastery Virgin's Church stands as a testimony to the beauty, the varieties and durability of marble it was made of. For its marble façades and shaped architectural elements of high stonecutting, artistic and iconographic mastery, among other necessary criteria, the church was inscribed in the UNESCO World Cultural and Natural Heritage List in 1986 [2].

So far, the domain of our scholarly research has remained short of considerations about this fine material and its significance in building the Serbian religious identity and about its artistic shapes and sites it was extracted from. Studying Serbian mediaeval art and architecture for decades now, a substantial body of scholarly material has accumulated along with an extensive bibliography on the Virgin's Church of Studenica [3].

The architecture of the Nemanjic Dynasty's first endowments, including Studenica, has been extensively researched in the works of Canak Medic, M. and Bosković Dj., from 1990 [4]. The first major conservation and restoration works on the monument were conducted in the mid-1950s, as well as the works on renovating some parts of the architectural marble decorations, published by Nenadovic. S. in 1957 [5]. One of the most important studies of the mediaeval architectural sculpture on the mediaeval monuments was conducted by Maksimovic, J in 1971 [6]. Subsequently, she turned her focus on the Studenica sculpture in 1986 [6]. As for some more recent investigations, there is a special contribution by Erdeljan, J., pointing to the Constantinople models of using the marble in building the ruling dynasty identity and the Virgin's Church [1]. Besides some serious studies of style, sculpture and iconography and some analyses of architectural marble ornamentation and its individual elements, no complex research has been conducted on the origin, types and properties of the marble material. Also, there is an absence of any serious studies of an extraordinary importance of the geological area where the marble was once quarried and its definite impact on the choice of the place for building the principal endowment of mediaeval Serbia.

The only relevant studies were published back in 1956 and 1957, by an author Simic, V., which today provide a broader view of the area, along with some particular results pertaining to investigation, based above all on geological and anthropological aspects. In addition, they include references to the historic quarries where the marble for the Virgin's Church had been extracted from [7,8]. Also, general regional geological surveys give description of the area [9,10]. Furthermore, petrologist Bilbija N. initiated petrographic and chemical tests conducted on five samples taken directly from the marble façades, which is an important contribution to understanding the marble built in the Virgin's Church. The research was started due to extensive remedial and restoration works on the marble façades, that lasted for almost two decades. The investigations are recorded in a report made in 1996, being a part of the Studenica Virgin's Church Marble Façades Restoration Project of 2003 [11].

As a starting point, the research takes that Studenica marble is of special significance in terms of ideology and artistic concept in building the ruling Nemanjic Dynasty's endowments. In its preserved form, the origin of the marble deposits and the exact location can be traced to the quarries it was extracted from and then built particularly in the mediaeval religious structures. Throughout their turbulent history, they were often damaged, and the most vulnerable spots even lost their architectural sculptural elements. So, an important aspect of their restoration would be a proper choice of the original material, traditional quarrying techniques and the art of stonecutting. When the marble is applied in such a way, it facilitates its sustainable use and management of the much broader area—the cultural landscape around the Studenica Monastery. The area itself stands out even more for its quarries as a testimony to an interaction between man and nature. Since this is about a complex explanatory research process, the following goals are set: To establish a special significance of Studenica marble, which will enable a more profound understanding and interpretation of the aspects of its use that are manifold in meaning; To give a description of its use in the construction of some of the most significant monuments of the Serbian Middle Ages; To identify and confirm all the historic quarries that were opened and mined in the Middle Ages; To establish the classification of the marble from the sites and draw a comparison with the previously obtained results from petrological analyses and chemical composition tests in order to define its use in restoring the architectural heritage; Upon obtaining all the results, to conduct research and start with the restoration of the missing elements of architectural sculptural ornamentation on the most important monuments inscribed in the UNESCO World Heritage List; To draw up protection and conservation plans for a broader area around the Studenica Monastery in order to protect the marble quarries, as a special type of cultural and natural heritage, and to make it possible for the local population to participate, in a sustainable way, in managing the World Heritage Site—thus clearing a way for a sustainable development through utilisation of the cultural asset provided by the heritage itself, which would largely contribute to preserving its values for future generations.

The paper consists of a short overview about the use of stone from the prehistoric times to the Middle Ages when the most important ruler's endowment, the Studenica Monastery, was built in mediaeval Serbia. Analysing the use of marble on the most significant mediaeval church structures, we give an overview of the style features, but also of the ways and techniques of manifesting the architectural sculptural ornamentation. Surveying the area around the Studenica Monastery, quite a number of marble quarries have been discovered, with several that have been in use since the Middle Ages. Also, a poll has been conducted to establish a number of families in the surrounding villages, who have been traditionally in the business of marble quarrying since the late 19th century up to the present day.

Descriptions of a geological character of the area and the marble sample testing have determined the choice to have determined the choice of the material for the architectural decorations restoration project in both the Sopocani and Studenica.

The most significant results have been provided in two case studies, when after the works were completed, numerous hitherto lost values have been augmented. In the end, the Discussion gives a reflexive process leading to recognising and valorising the Studenica historic marble quarries as an important cultural resource, which allows for a sustainable development by raising and spreading awareness of the significance, use and preservation of Studenica marble as an integral part of the cultural and historical architectural stone heritage.

2. Historical Background of The Marble Stone Works and Use from Prehistoric Times to The Middle Ages

The use of stone and stonemasonry are as old as humanity itself. The first traces of human spirit and of the ability to act creatively were imprinted in stone; marking the beginnings of the cultural history of mankind, testifying to the intellectual strivings and endeavours of the first peoples in shaping their environment. A tactile relationship of man with stone has been identified in eastern Africa some two million years ago, where the fossilised remains of the first genus *Homo*, the *Homo habilis*, were found along with numerous stone tools. More improved stone objects, in terms of technology, design and variety, have been discovered in south-eastern Asia and India some 1,600,000 years ago, whereas in Europe they started to occur about 1,500,000 years ago. The genus *Homo erectus* learnt how to "coax" fire from stone around 600,000 years ago, leaving this precious craft to posterity, so that some 300,000 years ago, his descendant, *Homo sapiens*, could make a variety of specialised stone tools, and to use the stone as his first consecrated object—a tombstone [12]. Anthropological and archaeological investigations record that 35,000 years ago a man called *Homo sapiens* inherited all the hitherto acquired skill of shaping and working the stone, already developed as a coherent technological system. Besides their utilitarian value, the produced stone artefacts also had a powerful symbolic and religious role. Durability, strength and power of the material initiated an idea of its sacredness [13]. Sculptural imagination in a shape of a woman's figurine known as the "Venus of Willendorf", dated to the Upper Palaeolithic (c 25,000), marked the birth of the first stone art.

However, one of the most amazing prehistoric monuments of stone architecture is the artistic stone works of Göbekli Tepe, discovered in the Upper Mesopotamia region, present day Anatolia. It represents a hierarchical architectural structure made of limestone rings. They are of high artistic and craft value, a series of massive pillars topped with heavy stone blocks in the shape of the letter T, presenting a narrative, mythological, zoomorphic and anthropomorphic stories committed to relief. The site was identified as the oldest and most exciting work of religious architecture. A highly developed system of beliefs had been a prerequisite of the civilisation progress from the Mesolithic to the Neolithic era [14]. In its vicinity there were numerous prehistoric quarries and plateau-like workshops, where the T-shaped pillar "negatives" were found, as well as numerous remains of the pillars, resulting from the work process. The stone and bone tools are the evidence of the level of technology implemented in quarrying and building this magnificent architectural complex that is a real masterpiece of the human creative genius [15] (Figure 1).

Figure 1. Göbekli Tepe © DAI whc.unesco.org/en/documents/165845.

Later on (6800-5800 BCE), the Lepenski Vir culture brought individual stone figures of human and fishlike attributes, in various sizes and of specific artistic workmanship and symbolic function [16]. Similarly, on the territory of present day Turkey (6000 BCE), the use of marble as an aesthetically valued material was recorded in the shape of human figures. Then, in the fifth millennium on the Balkans, we have anthropomorphic marble figurines discovered in present day Bulgaria. Also, some marble artefacts have been discovered in present day Romania [17]. In the late fourth millennium, at the very end of the period that we call the *Stone* Age, we frequently find *menhirs*, the freestanding upright stones that were sanctifying the open space area—forever marking the human creativity in service of the divine presence [18].

The establishment of the first historical civilisations, that lasted from 4000 to 3000 BCE, relates to the end of a millennium long process of stone conceptualisation, discovery of its special properties and the possibilities of its use. Marble, as a material of special characteristics that enabled its fine processing, was used in establishing the magnificent Cycladic art. It goes back to 4500 BCE, but from 2600 to 1000 BCE, we find the real pieces of art made in marble of a highly sophisticated minimalistic form [19]. Besides the common interpretations, the prehistoric cultures art aesthetics should also be understood in the frame of religious meanings and social power within the given time and space [20]. This is evident in different and territorially remote sites where various stone ritual figures were found [21]. However, more intensive and developed stone quarrying and working for building purposes occurred in the Iron Age when the application of iron and tool making enabled, although primitive, a more organised exploitation of hard stone masses for imposing stone structures of the periods yet to come [22]. All the subsequent periods of human culture inherited the millennia long experience and knowledge of the stone complex properties and its symbolic meanings. Also, in a technological sense, certain stone types and properties were selected over others, only to be used in creating a vast variety of forms and shapes in architecture, religion and art of different peoples and cultures.

The civilisation advances throughout history are most distinctively identified by their architectural heritage made in stone. Due to its excellent physical and mechanical properties, the stone is a long-lasting material. Monumental architecture from the Ancient Egyptian period is the most striking testimony to the power of architectural expression and the durability of stone material. In the Nile River Valley and the areas towards the Red Sea, as many as 196 ancient quarries have been identified, providing various types of igneous (magmatic) and metamorphic rock material [23]. The most intensive temple building activities have been recorded in other lands of ancient civilisations such as Phoenicia, Mesopotamia or the Aegean basin, where the structures were built of natural stone. In many cases, the technology of quarrying, large stone blocks and megalith cutting in those regions have remained a mystery until the present day.

In the lands of the Far East, the history of human civilisation has recorded some of the most striking examples of cultural and religious structures made in stone. A remarkable manifestation of human artistic creation consisting of almost 110,000 Buddha stone statues, more than 60 stupas and 2800 inscriptions carved into the stelae have been preserved in the unparalleled grandiose sculptures of the Longmen Grottoes in China. In a continuous evolution from the 5th to the 8th century CE, an exceptional interaction between man and nature have been recorded, consisting of giant figures carved in the Chinese cave art, containing original Buddhist culture and its spiritual and aesthetic functions [24] (Figure 2).

Figure 2. Longmen Grottoes (China) © NDRistić.

The most fruitful architectural and artistic stone creations in Europe belong to Ancient Greece and Rome, when numerous architectural and structural works were done, such as the sacred complexes and temples areas. Those range from exceptional sculptural works, sarcophagi and votive objects to the Roman grandiose ones—fora, bridges and aqueducts.

The aesthetic and technological perfection of the stone works in architecture and art was reached with the use of marble in Ancient Greece. The Parthenon (447–432 BCE) ranks among the most harmonious and artistically perfect work of architecture of the Antiquity. It was constructed from a fine grained white Pentelikon marble quarried in the famous Mount Pentelikos quarries [25]. The marble was used for temples and various public Agora buildings, whose lavish sculptural ornamentation made them distinct from other structures. Marble polishing, the texture and polychromatic effect under the sunlight gave out an impression of a superhuman nature emanating from the gods [26]. For such creations, first there had to be a special marble extracting technology at quarries where large blocks were being prepped and the marble cut on the spot, applying advanced cutting techniques. The blocks were then transported on for further use in construction where it was worked into final shapes. Carving sculptural forms in ancient architecture was a special architectural and artistic process with an implementation of various methods—from shaping creative ideas, making the drawings and then doing the practical work, to the final mounting and fixing in position in the already designated place in the structure. The most magnificent marble sculptural works in human history on the European Continent were made in the 5th century BCE (Figure 3).

Figure 3. Acropolis, Athens (Greece) © Ko Hon Chiu Vincent. whc.unesco.org/en/documents/156667.

The use of marble flourished in Rome. Besides its extensive use in architecture, bringing new building systems and materials, it was used to make portraits, carve in epigraphs and in funerary works of art, where the sarcophagi, as permanent repositories for the remains of the dead, were usually made of marble as a symbolic material of eternity [27]. The meaning of the ancient world art made in stone is to be understood as a pinnacle and the final goal of the ancient principles of beauty made in stone.

Following the tradition from Antiquity, the mediaeval period transforms the stone into some specific inherited forms, introducing the complex Christian symbolism. Since the Antiquity sculpture was considered pagan, new messages started to be imprinted into the stone of religious buildings; accomplishing impressive artistic architectural advances, so much recognisable in Western Christianity. The pre-Romanesque and Romanesque architecture manifested a somewhat reduced expression, whereas the Gothic structures brought it to its fullest potential, often called the "books of stone" [28]. That said, the primary principles and the knowledge were drawn from the world of Antiquity where the septem artes liberales and quadrivium in particular enabled an integral development of religious architecture where the sculptural ornament was just a part of a whole that symbolised the entire Divine Universe.

Ever since Justinian, in Byzantine architecture marble was used by the Constantinople master builders, constructing the temples of the Capital City, which is manifested in many pieces of preserved architecture as well as in numerous texts—the *ekphrases*. Marble as a material and its use in creating consecrated church areas unmistakeably bring forth the presence of a divine power in generating the beauty of the temples [29]. A marvellous skill of the craftsman to create sculptural church ornaments is described in "surprising, supernatural and noble properties of stone", possessing and evoking power and mystical meanings with an aim to leave a profound and long-lasting impression on a spectator. In their theological and ideological sense, the architectural and artistic creative works of Byzantium was quite profoundly reflected in the mediaeval Serbian lands.

The mediaeval Serbian religious architecture and its artistic stonework were drawn from two different worlds. A long-lasting presence of the Roman civilisation in the region and the vicinity of the Western world in the Adriatic region on the one hand, and the artistic stonework crafts brought from the vast influence of the Byzantine cultural tradition generated some original architectural, artistic and iconographic masterpieces. Although the mediaeval Serbian art of stone carving and sculpting cannot match the high artistic spheres of the Western and/or Byzantine religious architecture, their presence on the endowment monumental temples testifies to the impact, understanding and historical developments of the Serbian culture and religious aspirations expressed in artistic stone formations [30].

This short overview of the heritage use of stone and the origins of architecture within the world culture development does not aim to provide a chronological systematisation of this immense subject, instead it aims to give a short and perhaps subjective glance at —and once again remind people of—the significant achievements of civilisation with regard to the architectural works in the time when Studenica Virgin's Church was erected. It was valorised as the World Heritage owing to its marble attire. Today, our goal is not only to protect the monument itself, but also to preserve the cultural landscape, as it includes the marble deposits whence the material for the Virgin's Church comes from.

3. Research Process Phases

3.1. The Use of Studenica Marble on The Mediaeval Serbian Churches

Artistic stone work in the lands populated by the Serbs is well-known from as far as the Lepenski Vir and Vinca cultures, through the times of Antiquity that left us skilful stone work in architecture and sculpture, all being direct sources and models in the times yet to come. A long-lasting presence of the Roman civilisation, an easy access to the Western world via the Adriatic region and the artistic stonework crafts coming as an influence from the vast Byzantine cultural tradition generated in these lands some original pieces of church architecture built in natural stone.

The architectural exterior forms of the mediaeval Serbian churches built in the period of the rise of the Serbian state and Church in the 12th and 13th centuries are reflected in an articulated three-section interior plan with a central dome, coming as a result of Byzantine influence. On the other hand, the Romanesque influence dominates the façades, featuring compositions made of carved stone decorations on the blind arcade friezes, marble portals, corbels, windows and pedestals, columns and capitals. The basic church building structure was constructed from various stone materials, whereas the exterior architectural and sculptural elements were mostly carved in marble, shaped from clearly profiled architectural elements: from modest ornaments to floral, zoomorphic and anthropomorphic images with figural compositions of highly complex symbolic content. They were carved by experienced master carvers, most probably coming from the Adriatic region. Visible impact of the Romanesque architecture came from the Apennine Peninsula, while the Byzantine influence is present in the symbolism, iconography, themes and in certain stylistic details ([6], pp. 63–82).

One of the most significant mediaeval Serbian endowments, the Studenica Monastery and the Virgin's Church (Holy Theotokos), was built by the Great Prince Stefan Nemanja. The church represents the greatest feat of an authentic architectural style called the "Rascia School" [31]. This original piece of architecture came about as a reflection of a specific position and political orientation of the mediaeval Serbian state. By making a synthesis of the two architectural models—one coming from Byzantium and the other from the West—the marble-wrapped Virgin's Church was materialised. The use of the polychromatic marble on the church exterior façades may have been drawn from the extensively applied material in the church architecture of Constantinople [1]. Its "Romanesque" portals and windows are shaped according to an already set geometrical scheme, completed with various figural images, lavish floral ornamentation and interlaced combinations of both (Figure 4).

On the one hand, the entire aesthetic of the church structure is visible in a synchronised materialisation of the portal and the windows, while on the other, the stone carved decorations on certain architectural marble elements were formulated in a variety of ways. There is a special classic feeling in shaping the capitals of the free-standing colonnettes, showing visible differences in the manner and the extent of their construction. On the south portal there is a luxuriously carved frieze made of Corinthian capitals with human heads, whereas the north ones feature only an abacus. It is quite normal for the Romanesque portals to have differently carved capitals that flank the portal, as well as the entire door frame [32].

Figure 4. (**a**) The Church of the Holy Virgin 12th century view from the northeastern side. (**b**) Southern portal © LiveViewStudio.

The bifora windows feature a variety of carved capitals unevenly decorated but evenly shaped at one and the same workshop. Besides sculptural carvings with the use of chisels and picks in order to generate classical motifs of a stylised Ionian volute with a Romanesque-like leaves or floral pattern on the calathos, here we also have a specific use of drills for making numerous holes in order to free the carvings from the rest of the material, but also for decorative purposes, for creating a sense of inner dark shades. In order to make a better visual impression, the holes were filled with molten lead. Such a way of doing the stone capitals was going to be used as a model for carving capitals on other church structures in the 13th century [33] (Figure 5).

Figure 5. (**a**) Sopocani Monastery 13th century (**b**) Gradac Monastery 13th century © NDRistić.

By the end of the 13th century, sculptural decorations were applied on the churches from the Nemanjic Dynasty period. In somewhat simpler architectural shapes, the stone sculptural ornamentation was accentuated with the use of polychromatic marble, quite characteristic for the Holy Trinity Church of Sopocani, an endowment of the King Uros I and for the Holy Virgin's Church in the Gradac Monastery, an endowment of King Uros's wife, the Queen consort, Helen of Anjou. Wishing to reach the architectural and aesthetic harmony of the temple built by the founder of the Nemanjic Dynasty, the ktetors of the subsequent churches brought in the carved ornamentation as a heritage of dynastic architecture, trying to emulate Studenica. The marble for the ornaments was being transported from the Studenica deposits, from the same quarries the Studenica Virgin's Church was made eternal [34].

However, the mediaeval Serbian church architecture saw some rather turbulent time. Some of the most significant buildings endured both destructions and renewals. The most sensitive spots of destruction were the elements of architectural sculptural ornamentation, particularly at times of tectonic movements and earthquakes. However, the material traces that remained, made it possible to restore what had been demolished.

The 20th century was focused on renewing the most significant monuments of mediaeval Serbia. If the level of damage or the total level of authentic data and architectural elements that were preserved made it possible to restore the original shapes without any arbitrary inputs or hypothesising, then either partial or total restoration was to be undertaken. Such restoration comprised all the primary steps and procedures based on detailed and reliable analysis of all the available data and knowledge gathered through archaeological, architectural, historiographical, archival and other investigations [35].

The full awareness of the importance of the artistic and aesthetic valorisation in interpreting the historic monument's architectural characteristics was a prerequisite for their conservation within their twofold nature of both tangible and spiritual artefact [36]. The process of defining the steps in conservation and restoration of architectural elements, with regard to an overall architectural structure, was always an important starting point for each subsequent activity. Depending on the material data and the number of archaeological finds of architectural sculptural ornamentation, different steps were taken in their recording, processing, further use, inset into a building and presentation. An order of possible interventions was changing from case to case. At first, materials used in restoration, particularly in carving new marble elements, were not considered much, save their type and colour.

Therefore, in recent decades, while working on remedying the marble façades on the Virgin's Church of Studenica and in the restoration works on the churches in both Sopocani and Studenica, there was a special focus directed to selecting the material from the original marble quarries that had been used in the Middle Ages to obtain the marble blocks. Bearing in mind that both sites are on the UNESCO Heritage List, special care was taken in analysing the available data for the restoration and conservation, as well as in selecting the material for the reconstruction of the missing architectural elements.

3.2. Identification of the Mediaeval Quarries in The Vicinity of The Studenica Monastery

The conservation and restoration works on an architectural heritage built in natural stone require investigations and tests to be done previously to determine the overall condition of the monument, the nature and forms of periods alternations on the structure itself, a macroscopic examination, sampling the used materials, then certain petrographic and chemical tests, and also going back to the places where the stone material was quarried for that particular structure because of its potential use today, as well as the cultural, historical and natural significance of preserving the broader area of the monument site.

The Studenica Virgin's Church is a unique monument of the Serbian national heritage where the entire structure, save the narthex that was built on a later date, was erected in one single type of marble, the varieties of the Studenica white marble that come from the same petrographic structure but from different quarry sites (Figure 6).

Marble is a compact crystalline rock consisting of minerals and is used in shaping architectural elements, since it possesses properties necessary for ornamentation, so it is classified as a decorative stone [37]. This fine material of quite specific properties is suitable for building structures and here it was most probably used for the first time for erecting the Studenica Virgin's Church.

Figure 6. Map of Studenica river basin—Monastery, Church, Church ruin, Anachorete cell, Fortress and Quarries © A. Stanojlović (Management Plan of Studenica Monastery 2018).

About 4000 marble elements of various sizes and shapes were used to build the exterior church walls, which points to substantial amounts of the rock that had to be extracted from the local quarries. The largest elements were used for the portals, but a macroscopic survey of the façades pinpointed a certain number of marble blocks in the architectural structure that were even 2.25 m long [11]. The marble elements in the façade walls possess significant properties of the rock mass where they were quarried from: heterogeneous appearance coming from a pronounced lithological discontinuity in the form of thin layers with exokinetic fractures of various prominence. The structure was built in an alternating horizontal row of perfectly dressed stone elements of uneven dimensions and heights, which may be attributed to the rock mass they were extracted from [11]. The whole process of extracting, selecting, shaping and transporting the blocks may be reconstructed to a certain extent. As the quarry sites are distant from one another, on average about 4 km between them, the church marble is pretty heterogeneous. Land trails have been identified, still not archaeologically investigated, though, which indicate a possible path the marble was being brought to the construction site. The abundance and a fine quality of the stone, as well as the particular features of the natural surroundings may have been a deciding factor in the choosing of the site for erecting the most significant endowment of the Nemanjic dynasty.

According to his biographer, the selected site for building the Studenica Monastery, as an endowment and a burial place of Stefan Nemanja, the place had once been "an empty hunting ground" [38]. The place where the monastery was built is a naturally flat area above the Studenica River right bank, lying under the north slope of the Radocelo massif. The existing shape of the terrain, the surrounding rock massifs and the composition of the soil were obviously primary factors in selecting the area and the location for building the monastery complex, as well as in deciding on an architectural concept of shaping and constructing the monasterial settlement, the Virgin's Church in particular [39] (Figure 6).

The Radocelo mountain massif belongs to a group of older geological formations with areas of white and polychromatic marble, while the flat area itself consists of a hard rock bed of green-brown schist serpentine rock occurring at various depths in regard to the land surface [40]. A particular natural resource was a productive marble belt in the eastern and north-eastern slopes of Radocelo, with quarry sites between three villages, today known as Vrh, Dolac and Brezova. In that area, where marble has been quarried until the present day, numerous deposit sites have been identified, which we know have been exploited since the Middle Ages. With time, along the rocky massif, in places that were almost inaccessible, stonecutting settlement communities were formed, which today help us in orienting and locating the marble quarries. Each of the said villages extend to several hamlets—Vrh with its Gobelja hamlet, Dolac with the hamlets of Vranjevo, Backulje and Bozici and Brezova with its hamlets of Godovic and Gusterica (Figure 7).

Figure 7. Historical Quarry in Vrh (Kotline) © NDRistić.

The Vrh village quarries are situated on the south slopes of Backa Stena, at elevations of 1150 to 1050 m, known as Draskovac. One old Studenica quarry was identified also near Gobelja hamlet, in the place called Kotline. Along the quarry goes the Brevina River gorge formed by the deposits of marble material. Here, the traces of old marble quarrying techniques have been found. Inspecting the rock, the positions of the holes have been observed, where wedges and iron levers were used on the rock to extract large blocks. Also, some old tools have been found such as a massive sledgehammer of about 7 kg with one side shaped into a pick, as well as one plug made of hard solid iron [7].

Strip blocks of blue-grey marble on the Virgin's Church come from this quarry site. Although there are varieties of marble strips in other quarries in the vicinity, these fine types are of very fine grain structure, basically white or blue-grey in colour. The exquisite beauty of the marble is the section where white, unpigmented stone alternates with the pigmented ones, mostly grey to black-grey shades. Horizontal marble strips are of 2 mm to 10 mm thick, quite eye-catching at places or sometimes rather low-key. When looked at closely, it may be observed that within one pigment vein there is a strip-like structure with millimetre thick blue-grey lines of different intensity [7]. Some of the most beautiful are varieties on the church façade with 4 strips at 12 cm. Their depth of extraction could not have been more than 60 cm, which apart from the aesthetic effect, points to a functional choice of this construction variety. In its consistency, it is classified as natural stone easy to carve, cut and polish, which is best manifested in the work done on the Studenica Virgin's Church windows and portals. The marble stripe layers were cut into blocks that were then very carefully, depending on the cutting directions, fitted into the façade walls. Choosing different directions of cutting relative to the stone structure, the builder wanted to accentuate certain church elements. The portals are mostly built of this variety marble blocks, although there are details of carved architectural elements in pure white blocks, like for instance, toned portal columns in the south vestibule. A special dynamism is achieved in vertical pilaster strips cut in blue-grey veined marble, combined with white short columns.

Those quarries, at much deeper layers, there are varieties of pure white, fine grained marble deposits, however, in smaller amounts. The fact that there were small amounts of the pure white marble at the time of the church construction is confirmed in only a few architectural elements carved in white stone, such as the altar trefoil, the west façade *biforas* and 3 m tall cylindrical columns on the main portal. There is less than 10% of the white block on the exterior façades.

The finest small grain marble varieties come from the quarries in Draskovac, Kotline and Godovic, belonging to the same name hamlet of Brezova-Godovic. The Godovic quarries are situated on the north-east slopes of Krivaca, at elevations of 1280–1380 m. There are several deposit sites, but the largest and most productive ones are the Godovic quarry, the Stari Majdan (Old Quarry) and Secina. It is believed that most of the marble used in the church construction was quarried from these sites. As for the old tool finds, only one iron plug has been discovered, but no old traces of stone extraction have been observed, as marble quarrying has been going on constantly since the construction of the monastery [7]. The remains of an erstwhile settlement of dugout dwelling houses lined with broken marble set in an uneven course indicate that it was the period of the Middle Ages. Nearby, at the settlement there is a water source and quarry above. The very name of the place, Secina (from Serb. *Seći*—to cut) is a typical mediaeval toponym for places where marble blocks were extracted, cut into smaller ones and transported to the construction sites. The marble from these sites are of pleasant tones from white and blue-grey to ochre and ivory shades. In the same area, there are belts of marble deposits called Zmajevac, of monochromatic character, at 1640 m of elevation. However, there is a drawback due to the large amount of schist, so when cut, it separates along the diaclases.

The area and steep cliffs make the sites rather hard to access in modern times, as well as the application of modern stone extraction techniques, save the old ones using picks and hammers. So, until the present day, marble extraction has been done manually. Furthermore, and interestingly, in cutting and polishing processes, no water was ever used, although the area abounds with it (Figure 8).

Figure 8. Historical Quarry in Godović © NDRistić.

Godovic is a stonecutting settlement with centuries long marble quarrying tradition. It has always been far and wide famous for its marble gravestones that caravans transported to distant lands [7]. Similarly, many other hamlets and settlements in the area fostered the stonecutting craft. Almost every house earned their living by working the stone. The mastery of the craft was conveyed from generation to generation, and entire families have been known as stonecutting masters for several generations. Some old census records show all the villages and houses that have been working with marble since the late 19th century. The latest publicised information about it dates from 1957. However, a recently conducted poll have shown a rapid decline in the craft due to migrations to towns and cities.

In the mid-20th century there were some attempts to extract Studenica marble using an industrial exploitation model and on that occasion the mining and geology experts surveyed the structure and character of the stone from several quarrying sites. In addition, this form of exploitation lasted for several decades and on various sites, particularly in Kotline. However, due to land ownership structure changes and rather unfavourable exploitation conditions, this model was abandoned. Nevertheless, at that time, the commercial names for Studenica marble came about: fine grained "Studenica White" and large grained "Studenica Crystalline" that are dominant varieties in the north slopes of Radocelo, from Brezova towards the mid-section of the Cemerno mountain range where new quarries have been opened up [41].

Surveys of the geographical area of the Studenica Monastery and the surrounding quarries once again highlighted the primary significance of this precious resource that is manifested in an interaction between the features of the area, architecture and local traditions. A special aspect here focuses on traditional craft mastery, techniques and technologies that are still being conveyed to future generations. Also, the information gathered in the field show that, for instance, a proprietor of the Godovic quarries, a resident of Godovic, is at the same time the quarry manager, and that his family have been in the stonecutting business for more than a century.

Owing to the preserved stonecutting tradition, this was the quarry where the original marble blocks for restoring the bifora on the Sopocani Monastery St Trinity Church came from. In addition, this was the same source of marble for conducting one successful restoration on the Studenica Virgin's Church.

All the anew collected information and knowledge point to a significant resource in the form of a natural stone—Studenica marble. The natural landscape has been shaped for millennia, now featuring rural scenery marked by the quarries that are an integral part of the present, so recognisable, Studenica Monastery cultural landscape. Raising awareness of the tangible and intangible characteristics and values of this natural material facilitates promoting the area sustainable preservation and transference of knowledge and cultural meaning to future generations. This section may be divided by subheadings. It should provide a concise and precise description of the experimental results, their interpretation, as well as the experimental conclusions that can be drawn.

4. Studenica Marble Geology and Petrographic Description

4.1. Geological Fabric of Studenica Olistoplaque and Its Immediate Surroundings

According to its position, the Studenica olistoplaque as a geotectonic unit of the lower level is bounded to the west by ophiolitic mélange of the Senonian age at the point of contact with the Dinaric flysch of the same age, in the east its border is almost everywhere covered with allochthonous ultramafic rock, and only in the northeast clearly follows its tectonic contact with the Triassic metamorphic rock of the Kopaonik Mt. "In the south, this unit spreads over Cemerno, with characteristic metamorphic rock of the 'Studenica series' as a substrate [10] (75)" (Figure 9).

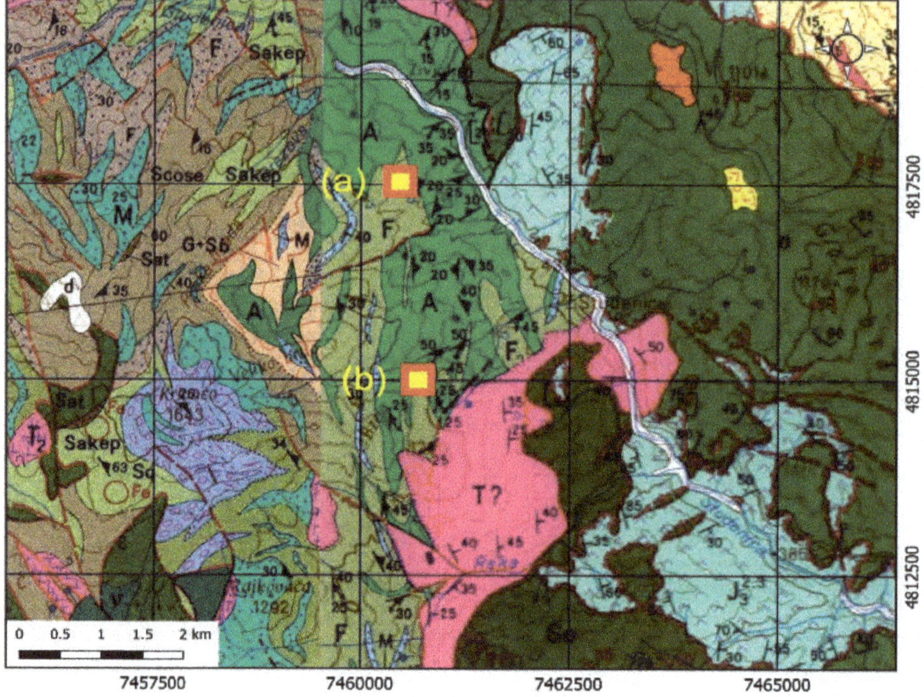

Figure 9. The geological map of Studenica olistoplaque, according to Basic Geological Map. Ivanjica and Vrnjci sheets [42] (**a**) Quarry in Vrh. (**b**) Quarry in Godović.

According to its position, the Studenica olistoplaque as a geotectonic unit of the lower level is bounded to the west by ophiolitic mélange of the Senonian age at the point of contact with the Dinaric flysch of the same age, in the east its border is almost everywhere covered with allochthonous ultramafic rock, and only in the northeast clearly follows its tectonic contact with the Triassic metamorphic rock of the Kopaonik Mt. In the south, this unit spreads over Cemerno, with characteristic metamorphic rock of the 'Studenica series' as a substrate [43]. Due to the complexity of this geological mass, the mentioned authors divide it into two parts. In the upper part of this series there are limestone and dolomite, that is, marble and dolomite marble ([44], p. 200), depending on the degree of metamorphosis. The floor of this carbonate series includes phyllites, green rocks and amphibolite. According to Rampnoux J. ([44], p. 202), in this marble floor, crystalline schist alternates with marbles, which is a result of a very complex tectonic structure. Dimitrijevic, M.D. assumes that the Studenica series can be an original part of the Drina-Ivanjica element that, during geological history, has separated and merged with the Vardar zone, " ... , a part of which it can be (but not necessarily!) considered from the Senonian age onwards ([10], p. 81)."

4.2. Petrographic Description

Marble deposits in the Studenica olistoplaque have been used for a very long period as a technical-construction and architectural-construction stone [10] Studenica marble quarries which are even today exploited in a traditional way, are located on the right bank of the Studenica River, between the settlements of Gušterica and Vrh.

As a universal medium of multiple architectural significance, Studenica marble is characterised by durability in terms of time, resistance to the adverse effects of atmospheric agents, abrasive influences and chemical-biological agents. These features were, in addition to old master, recognised in modern

Serbia by the pioneers of geology and mining as peculiar, thus in the Review of Minerals in the Kingdom of Serbia for the Paris Exhibition in 1900 the following was stated:

"*Studenica marble.* At Radočelo, near the Studenica Monastery, and especially in Brezova, there are still famous marble quarries from ancient times. This Studenica marble was used for the embellishment of many endowments of the Serbian kings from the Middle Ages. Marble intercalations are found here in amphibolite, actinolite and various other Archaean schists. Two varieties can be distinguished: grainy and schisty; the former is white and forms thicker banks than the latter which is often bluish ([45], pp. 101–102)."

Previous studies of five samples of the Studenica marble, which were obtained by probe drilling of the built-in stone of the Church of the Virgin, show the following characteristics:

Petrographic studies have found that the marble for its mineral composition is heterogeneous and anisotropic. The Studenica marble is distinguished by an irregular grainy structure, thus small, medium and large-grained varieties can be found in the same mass ([45], pp. 101–102). The marble is permeated with millimeter-size strips, indicating the lamination of the primary carbonate rock. In some lamina, richer in clay component, the formation of non-carbonate minerals—finely-flaked sericite, quartz, and subordinate albite occurred during metamorphism. Graphite powder, anthracite, formed by transforming organic matter, is the cause for blackish-striped parts.

Based on chemical tests, a very high content of calcium carbonate (98.9%) in an average sample dominated by non-pigmented parts was identified [11]. The chemical composition of marble is shown in Table 1.

Table 1. Average chemical composition of samples of Studenica marble ([11], p. 47).

Chemical Composition	SiO_2	Al_2O_3	FeO	Fe_2O_3	MgO	CaO	Other
Presence	0.82%	1.13%	0.20%	0.17%	0.87%	55.55%	41.26%

Tests of marble fragments with schisty texture and macroscopically visible sericite content and dark pigmentation indicate an increased non-carbonate component content in the insoluble residue that varies about 9%, of which an organic component is 0.15%, on average.

Based on X-ray analysis of an insoluble residue, which is not annealed, the following minerals have been identified: quartz, mica (muscovite), dolomite and calcite. In annealed insoluble residue, using the same type of analysis, the following minerals have been identified: quartz, mica (muscovite), periclase and graphite ([11], p. 48).

The properties of the stone include a series of properties that condition its application and impose restrictions on its use.

The physical properties of Studenica marble include color, density, porosity and hardness.

No new data were found on the color of the marble, while the following descriptions refer to its visual appearance:

"The Godovic quarries marble: fine, white, bluish, rarely striped ([7], p. 252)."

"The most widespread are masses of monochromatic, fine-grained marble, the color of light ivory at the Zmajevac mines site and Preka Livada" ([7], p. 253)."

"Marble mass is fine-grained; it is white with greyish tones. There are also grayish, monochromatic and striped marble masses in the village of Vrh, Kotline quarry ([7], p. 250)."

"The Brezova marble is exclusively coarse-grained. The color of marble is white with grayish tones ([7], p. 250)."

"The Brezova marble is exclusively coarse-grained. The color of marble is white with grayish tones ([7], p. 253)."

"Marble from Radočelo-Studenica: Full crystallinity and varied grain orientation. Shiny shale gloss. Beautiful whitish and yellowish color ([9], p. 140). "White types of the Studenica marble still tend to "absorb" dust, thus if exposed to the east winds for a long time can turn yellow ([9], p. 142).

Tests on bulk density, water absorption and open porosity were performed on all five samples, and the results are presented in Table 2. Based on the obtained results of open porosity and water absorption, it has been established that this is a very compact marble.

Table 2. Physical properties of the Studenica marble ([11], p. 48).

	Sample 1	Sample 2	Sample 3	Sample 4	Sample 5
Bulk density (g/cm^3)	2.68	2.69	2.67	2.67	2.64
Water absorption (% of mass)	0.16	0.11	0.18	0.30	0.36
Open porosity (%)	0.44	0.30	0.49	0.80	0.97

According to Janjic ([9], p. 140), the Studenica marble in situ is characterised partially by stronger brokenness and sporadically large grain size. This is supported by the findings of Bilbija, on the physical defects manifested by the separation of smaller or larger fragments and the appearance of very pronounced secondary cracks, which links these phenomena to very strong stress in the construction of buildings ([11], p. 51).

Considering that the Studenica marble mass consists of small, medium and large-grained varieties, the following observation by Janjic, is very important: "In marble, medium-sized types are more favorable for dressing than small-grained ones, because in their decomposition they rarely create uneven and shell-like surfaces—as in onyx and marble limestone. On the other hand, lamellar grains of large-grained marbles are always more easily damaged, so they are more resistant to weathering ([9], p. 130)".

4.3. Results of The Study Of the Studenica Marble at The Sites of Godovic and Vrh

During the field visit at the sites from which the Studenica marble was extracted for the restoration works on architectural sculptural ornamentation on the Church of the Holy Trinity in Sopoćani and the Church of the Virgin in Studenica, a total of three samples were taken, that is, two from the quarry in Godovic, and one from the quarry in the village of Vrh.

Leica DMLSP Polarizing Light Microscope and Leica DFC290 HD digital camera were used for the tests which were carried out at the Geological Institute of Serbia in Belgrade.

Sample 1 (quarry in Godovic)—marble (Figure 10)

The rock has saccharoidal appearance, it is white with very rare dotted pigments—yellow-brown stains. It has massive texture and granoblastic structure. In the specimen analyzed with the microscope, it was noticed that it was formed by irregular to polygonal growth of grains of carbonate—calcite ± dolomite in the size of 0.2 × 0.1 to about 1.5 × 1 mm (dominated by grains of size about 1 mm).

Figure 10. Macroscopic structure of sample 1 marble: (**a**) With parallel nichols (**b**) With crossed nichols.

Very rarely, small deposits (0.1 to 0.3 mm) of fine-grained chlorite were observed in the interstices between the grains. It is dominated by calcite and CaCO3 component.

Sample 2 (quarry in the village of Vrh)—**marble** (Figure 11)

The rock is white and has saccharoidal appearance. It has massive texture and granoblastic structure.

It was formed by irregularly polygonal intergrown grains of carbonate—calcite ± dolomite in the size of 0.2 × 0.1 mm to about 1.5 × 1 (dominated by grains of size about 0.7 to 1 mm). Very rarely were fine rounded grains of quartz up to 0.1 mm in size observed, as well as occasional sericite and chlorite flakes. Rare irregular forms (0.1 to 0.2 mm) of non-transparent matter (metallic or organic?) were also observed. It is dominated by calcite and CaCO3 component. The rock has very similar properties as in sample 1.

Figure 11. Macroscopic structure of sample 2 marble: (**a**) With parallel nichols (**b**) With crossed nichols.

Sample 3 (quarry in Godovići)—**schistose marble** (Figure 12)

The rock is light grey. The texture of the rock is schistose with strips, and granoblastic structure.

It was formed by irregular polygonal inter-grown grains of carbonate—calcite ± dolomite, which, due to the effect of pressures, were elongated and with its long axis oriented parallel schistose. Grain size varies from 0.2 × 0.1 to about 1.5 × 1 mm (dominated by grains of size about 1 mm). Muscovite lamina up to 1 mm, oriented parallel schistose, which in certain parts form small stripe-like deposits were occasionally observed. Apart from these, striped and luminal deposits made of small flakes of sericite, chlorite and non-transparent matter were also observed.

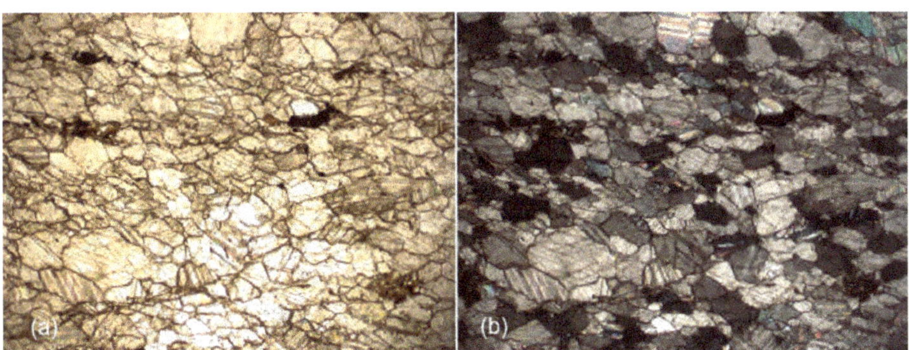

Figure 12. Macroscopic structure of sample 3 rock schistose marble: (**a**) With parallel nichols (**b**) With crossed nichols.

In addition to the aforementioned form, the deposits of the aggregates of these minerals were observed under the microscope at the contacts between the calcite grains. Rounded quartz grain (about 0.05 mm in size) was very rarely noticed in the rock.

On the basis of an analysis of the available published papers on the Studenica marble, the tests of the built-in marble on the façade of the Virgin's Church of the Studenica Monastery, as well as the tests of the samples from the quarries in Godovic and the village of Vrh, the following has been established:

The occurrence of the Studenica marble on the right bank of the Studenica River is metamorphosed rock that occurs in very complex geological conditions, which are not yet fully understood.

Petrographic tests carried out on the embedded material indicate that the Studenica marble is of very high quality and favorable technical properties. According to the three quarries specimen's microscope analysis, it is clear that it is the same type of marble, but should be also furnished with chemical and analysis of its mechanical characteristics.

Although the Studenica marble was observed as an exceptional rock for construction and architectural-construction, used for building churches in the Middle Ages, its limited extent of occurrence (about 20 km) and difficult exploitation conditions contributed to the lack of interest in greater commercial use and hence more detailed study.

Based on the analysis of the samples from quarries, it was concluded that marble sample 1 (Godovic) and sample 2 (the village of Vrh) have very similar properties, saccharoidal appearance and white color, and that sample 1, due to its very rare yellow-brownish dotted pigments falls behind sample 2 in terms of beauty.

By comparison with the results of the examination of the built-in stone, it was found that the white marble was used from quarries in both villages, and the light gray marble from Godović.

With regard to the Studenica marble use in the monumental medieval endowments of the Rascia Style, it is necessary to carry out additional geological tests on all existing quarries, sampling and petrological testing of rock samples in order to define all the technical characteristics of this material.

After carrying out the aforementioned research and obtaining the results, a selection of the representative Studenica marble profile should be performed, which should be then protected as natural geological heritage sites.

5. Application of Studenica Marble in Restoring the Architectural Ornamentation Elements

5.1. Restoration of the Biforas on the Sopocani Monastery St Trinity Church

Defining the right procedure in architectural elements restoration and conservation relative to the entire architectural structure—the whole building, was the primary step in renewing the biforas on the Sopocani Monastery St Trinity Church. The hitherto conducted investigations comprised thorough and careful analysis of the structure and its component, diagnostics of the overall state, including all the previous interventions. Survey and assessment of the structural system, damages to the biforas and identifying all the missing elements or those that had been restored in the past with the use of inappropriate materials, enabled us to draw up the proper documentation with the evidence of the extant situation.

The primary goal in the biforas restoration was not to impair the church building authenticity but to enhance its aesthetic, artistic and functional values that were lost through centuries. Such an approach included considering the entire configuration, the integrity of the monument, its architectural and artistic values, as well as the use of original materials, their right source, quality, special characteristic and texture.

The St Trinity Church features some special properties with regard to its architectural sculptural ornamentation of the Romanesque style. The first step was to devise an intervention plan that would ensure not only protection of the monument, but would also achieve the structural and functional entirety of the monument. After the evaluation of all the previously collected results and the works on the monument had been completed, another priority emerged: the question of restoring the architectural sculptural ornamentation on the church main nave biforas and the monumental altar one.

In the second half of the 20th century, there were some decades long architectural and archaeological investigations on the Sopocani Monastery. The monument was one of the first to be inscribed in the UNESCO World Heritage List in 1979 as part of the Stari Ras and Sopocani complex. The said works were managed with care and in accordance with the World Heritage guidelines [46]. The available information on the marble ornamentation is the result of archaeological investigations that lasted from 1973 to 1985 [47]. Fragments of the architectural elements found as archaeological remains had been carved from the Studenica white marble and were considered as a highly valuable starting point for a theoretical reconstruction. Furthermore, this issue had been a subject of investigations and various considerations of earlier conservators [34]. The elements consisted of fragments of four capitals, several fragments of small columns in various dimensions, one whole base of a bifora central column and fragments of two bases. Although severely damaged, the capitals contained enough information for reconstruction, so modes could be cast in gypsum plaster. The process provided some very important information about the proportions and the spatial geometry of the capitals and bases, but also the very shape of the capitals was of a type that could be recognised in certain biforas on the Virgin's Church of Studenica.

In their proportions, the simple bifora form with a tympanum in the line of the wall and the twice recessed window frames with the frontal sickle arches gave an impression of harmony and balance. The photo documentation dating from 1962, obtained before the first extensive church restoration, clearly gives the preserved shapes of the south bifora below the dome. It is obvious that at that time there were no elements of the dividing mullion or any decorations in the shape of a capital. The extant shapes show that the north bifora below the dome was quite preserved as it possessed all the original parts, again except the mullion, the capital and the base, which are generally the most vulnerable spots in times of any tectonic movements, known to have occurred in the past. Other four biforas on the main nave had been substantially damaged, so during the first renovation works, their primary frame shape was restored. Those elements were done in concrete and sandstone, while the mullions, bases and capitals were not restored since at that time no information existed about them. All six windows on the central nave, as well as the altar window were found without the capitals, bases or mullions [47].

Measurements and proportional analysis showed that all the windows are of almost the same dimensions. Based on the available original archaeological finds, the shape and length of the mullion and the capital were determined and later on applied in all the windows. All the previous inadequate interventions in concrete were documented and then removed. The original marble fine grain structure, colour and texture were compared with the analysis results from the previously described samples from the Virgin's Church in Studenica. The structural uniformity with the original mullion elements that were discovered in archaeological investigations and were used as primary models confirmed that the original marble belonged to a group of monochromatic marbles, identical to that from the Godovic quarries in the Studenica area (Figure 13).

Stone carving and sculptural works were done by hand in a workshop, while respecting the essence of the material and the carving techniques whose traces are discernible on the archaeological remains of the original fragments. The missing elements in all the six windows were put back in a synchronised manner. Some smaller interventions, such as carving the tympanum on the nave north window was done in situ. In places where the tympanums had been made of concrete, carved marble elements were put in [48].

The altar bifora was restored according to an architectural ornament on the only one preserved bifora between the nave and the narthex. In terms of proportions and shapes, the same thing was done on the west façade and the altar. The mullion base, shaft and capital dimensions were reached through an analogy and proportions analysis conducted with regard to the opening length and width proportion, as well as to a partially preserved point where the bifora final recessed arches join [49].

Figure 13. St Trinity Church (**a–d**) Restoration and coservation process © NDRistić.

Restoration of the biforas on the Sopocani St Trinity Church required traversing a long road—from the finds of the first marble ornamentation elements, the years long study of the origins and project drafting, some concrete finds during the conservation and restoration works, to finding the original material—Studenica marble, to final shaping and producing the missing elements, solving the glass pane surfaces, to their final installation and finishing steps in such a complex task. The result is not only a re-construction and/or restoration of the architectural ornamentation in the form of the bifora, but also a materialisation of a new dimension that points to the traces of the past, to redoing something that actually cannot be repeated, to the markings in marble that strongly highlight the values of a synthesis occurring at the border between the East and the West, leaving behind works of eternal value.

5.2. Restoration of the North Bifora on the Studenica Virgin's Church West Façade

The Virgin's Church of Studenica has suffered multiple destructions throughout its history. In one of the most severe fires and devastations in 1805, the entire church structure was seriously damaged. The proper conditions required for its renewal for the first time occurred after the final liberation of Serbia from the Ottoman rule in 1833 [4].

Some extensive works on the church restoration were conducted in 1839 and they were particularly related to opening thirty-six windows, as well as making and fixing "thirteen marble columns" [50] that still today testify to a special care applied to bringing back the marble parts of architectural ornamentation on the church windows and portals. A unique restoration process was then implemented on fitting in the missing parts in recreating the altar trifora form. The newly carved elements were made clearly visible, completing the overall geometry of the trifora [4]. All the elements were made of white marble, most probably extracted from the nearby quarries. So, most of the primary and sculptural markings of the Virgin's Church were restored then.

Quite extensive investigative, conservation and restoration works were conducted in the Studenica Monastery in the period between 1952 and 1956. Besides various rather complex architectural and engineering actions done on the Virgin's Church, certain segments of architectural ornamentation were restored in that period, like the west façade south bifora [5], among other things.

Pieces of a lunette with a rosette were discovered in the Radoslav's narthex, which in their size and the place they were found indicate that they belonged to the south bifora. There was its base preserved in situ, as well as a fragment of an octagonal bifora column, which left no doubt that it could be completely restored.

All the new parts were cast in toned concrete [11]. In the way it was made and in its style characteristics, the capital that was used for restoring the south bifora mullion belonged to the Virgin's Church architectural programme. However, the two fragments with a sculptural relief of birds, forming a lunette, may have belonged to another, the west façade north bifora, were found built in on top of the pediment of one of the monastery residential buildings. Even then it was assumed that the fragments may have been parts of the west façade north bifora, since its central part was "carved similar to the tympanum trifora" [51]. Although a masterpiece of the Virgin's Church carved ornamentation, in the decades that followed, the preserved north bifora lunette fragments have never been returned to their original place.

Things changed with regard to completing the lunette when in the late 1990s, during the archaeological excavation works, another fragment was found that made it possible to define precisely the radius of the lunette outline, as well as the two arches that determine the bifora openings. In the monastery gallery of the Virgin's Church stone decoration fragments a piece of marble was hiding, whose atypical shape did not indicate any larger place it could belong to. However, it was noticed for its fine grain crystalline structure, the way its outer surface was finished, as well as for a gentle unrecognisable smooth tracery in the shape of an entwined ribbon and the slight curves on its outer and inner borders of the fragment. Although none of its sides could be fitted to the existing three fragments, the lunette's geometry indicated a position where it could fit in. When the lunette tracery was closely observed, what looked like an entwined ribbon was actually a tail of a bird. Further analysis and measures of the archivolt radius and its recession depth finally confirmed that the lunette with the birds carrying a Holy Trinity symbol belonged to the church west façade [33]. The bifora mullion base and an imprint of an octagonal column were in situ. Analysing the dimensions of the archaeological remains of the capital, it was found that a capital fragment No. P/3, now kept at the stone fragment gallery, may have belonged to the same lunette [32]. Based on the available fragments, a capital was cast in gypsum plaster and then a new marble one was carved to be placed in its original position. So, through proportions analysis, graphical method, assembling and connecting all the available fragments, the original form of the west facade north bifora was obtained.

For the restoration of the missing parts, blocks of marble were extracted manually from the Godovic quarry according to the already defined geometry and dimensions. The hitherto investigations and the experience from the restoration works of the mullioned windows on the Sopocani St Trinity Church, the previously chosen particular spots for extracting the marble, as well as the knowledge of the craftsmen who had been traditionally in the marble business for generations, all indicated one and the same quarry. After the primary extraction, the blocks were transported to a marble carving workshop where the master craftsmen first closely studied the carving methods from the lunette fragments and then, with a use of a hammer and a chisel, sculpted the missing shapes and figures. In parts where there was not enough information for a proper restoration of figural motifs, roughly carved sections were left without any improvisations.

There is visible damage on an important image of two affronted birds carrying an entwined vine scroll; a Holy Trinity symbol was restored with polymerised plaster. The material was also used in restoration works on the Studenica Virgin's Church façades. The content of the plaster was first chemically tested and then carefully composed, making it a reversible material consisting from a mixture of ground marble of various granulation, cement adhesive and additives, making it highly malleable and elastic [52]. Some major damages were observed on the lunette central trinity motif and on its borders. Plaster preparation and application required an extremely careful treatment and toning so that an aesthetic and visual, as well as structural harmony could be accomplished between the applied material and the original stone surface. A complex methodological process also required consistency and compatibility and synchronised activities between a master sculptor, a stone restorer and an architect conservator. The works on bringing back the bifora to the Virgin's Church west facade in the Studenica Monastery was of an exceptional cultural significance not only for the professionals but for the broader public as well (Figure 14).

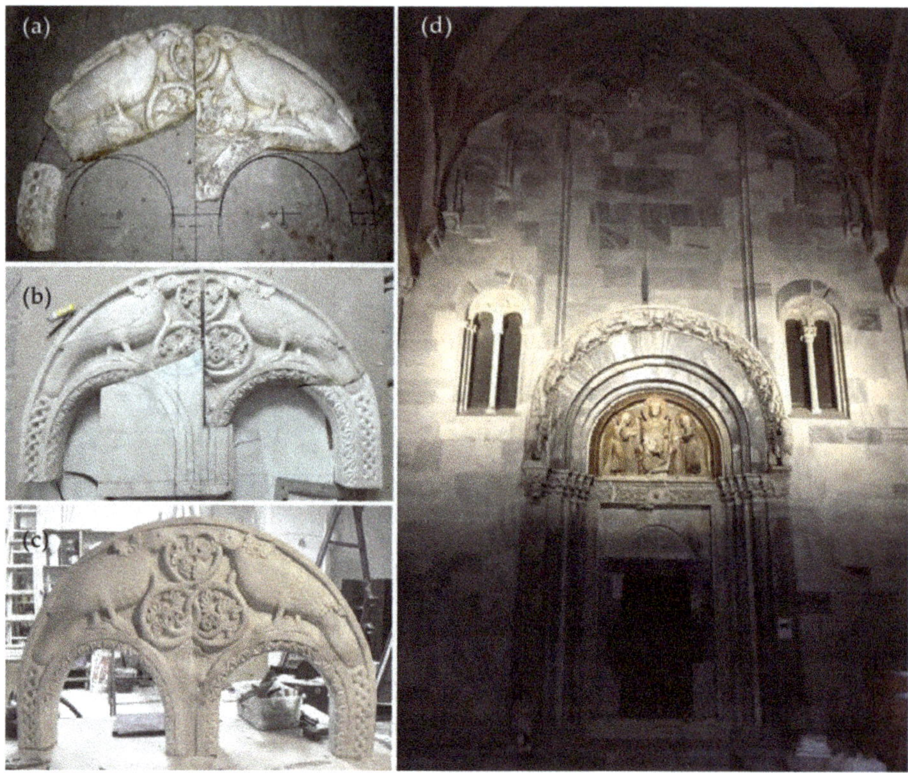

Figure 14. Virgin's Church (**a**) (**b**) (**c**) (**d**) Restoration and coservation process © NDRistić.

This is an important piece that represents a pinnacle of a subtle stonecutting craft, a fine sense and knowledge of the nature of the material—the white marble. The entire restoration was conducted in 2018, thus completing an envisaged programme of architectural sculpture embodied in the Virgin's Church exterior symbolism. The complex sign-symbolic image of the two affronted birds carrying the Holy Trinity symbol reveals and confirms its strong rooting in profound theological concept held dear by the church founder. In the conducted conservation and restoration works, the Virgin's Church west façade gained its full architectural meaning, whilst its entire sculptural ornamentation and the stone tracery substantiate its deep theological significance.

6. Discussion/Conclusion

The heritage in marble is of a special importance in the making of the mediaeval Serbian cultural and artistic identity expressed in a highly sophisticated understanding of symbolic and aesthetic properties of the material used in various forms throughout the world history of architecture. A need for a methodologically based assessment of Studenica marble values and significance has been a pivotal aspect of this research.

Primarily, its significance has been confirmed "in itself" as it was the material that was used to build some of the most outstanding Serbian mediaeval endowments. Then, by establishing a relation between the historic quarries where it was being extracted for erecting the Virgin's Church of Studenica, using it in its dressed form, the "instrumental" value of the quarries has been established, as places the material originated from, which then gave a special character to the architectural heritage, bestowing it its outstanding value. Its significance is also evident in an authentic synthesis between the local marble shaped into forms of architectural elements featuring Romanesque style combined with the Byzantine iconographic content [32].

When looking at the Studenica Monastery surroundings, the importance of the area geological structure has to be emphasised once again, as it also had to have an impact on the choice of a spot where the endowment of the ruler, the founder of the Nemanjic Dynasty, was to be built. So far, the investigations have been conducted in the Studenica area geological formations structure and the marble deposits, together with the Virgin's Church marble samples that had been tested earlier in order to establish the methodology and materials in the remedial works and the church marble façade restoration. Also, the samples taken from the two major historic quarries have been tested recently. Combining all the results obtained from those investigations confirmed that the marble was one and same type with certain high-quality varieties of favourable technical properties. Difficult exploitation conditions and a rather limited area of marble deposits made it hard to increase its commercial use, which actually proved as an advantage for a limited, sustainable exploitation and a preservation of traditional quarrying methods. Such an advantage comes as beneficial for any future restoration of the marble monument architectural and other elements that have disappeared and been damaged with time.

The quarry where the material for restoring the architectural ornamentation was extracted from was determined according to several indicators: the research results obtained from petrographic and chemical testing of the fine grained white marble of outstanding technical properties, as in the Godovic quarry; the site where the stone is still actively mined today by the locals in a traditional old way; the possibility of obtaining the right dimensions and uniform aesthetic stone characteristics.

The first major restoration project of the architectural marble ornamentation was conducted on the Holy Trinity Church of Sopocani, where the marble was mined, used and dressed, applying traditional extraction and stonecutting techniques in shaping the elements. The work process included a phase of thorough checking of all the marble components, situation diagnostics, checking the validity of all the previous interventions and subsequently, assessment of all the damages, missing elements, as well as inappropriate materials applied in previous interventions.

Preliminary testing had been conducted of marble samples from both the quarry and the archaeological material, giving positive results. Subsequently, based on the collected documentation, the stone was cut and all the parts made from inadequate material from previous interventions were first identified and then replaced, as well as the missing elements restored in the original material.

On the Virgin's Church of Studenica, a bifora was restored, an important one, as the marble remains of its lunette held a masterpiece of the stonecutting craft, expressed in a symbolic form of two birds carrying a Holy Trinity symbol. Carefully selected white marble from the historic quarry, carving the missing elements and a meticulous restoration of the relief motifs manifested a high level of technical and technological skills of the artists and conservators working on it. Understanding the nature of the material and reading the hidden signs preserved in the original remnants brought back the full character of the missing image.

The choice of applied interventions and steps in the original marble material enhanced the cultural significance of both monuments, thus re-establishing their aesthetic, symbolic, technological, functional and scientific values.

The entire investigative and research process actually started with a question: Where could the original restoration material for the missing elements of architectural marble decorations on the Serbian mediaeval monuments be found? A renewed interest in sites where the marble was quarried in the Middle Ages, visits to the quarries in the vicinity of the Studenica Monastery, talking to the locals, research of the abundant literature on the mediaeval monuments, as well as a rather sparse literature on geological, anthropological, technical and technological investigations of the area, all pointed to the precious natural resource, still hiding unveiled meanings of the sites and the material built in one of the most beautiful edifices of Mediaeval Serbia (Figure 15).

Figure 15. Area of Studenica Monastery © LiveViewStudio.

After completing the restoration and conservation interventions on the architectural marble decorations on two very important cultural monuments, inscribed in the UNESCO World Heritage List, Studenica marble and its still active quarries in the natural landscape have been recognised as a significant resource of sustainable development in the Studenica Monastery entire area. All the marble quarries have been surveyed and valorised as an integral part of the cultural landscape. The cultural landscape protection and conservation plan as part of the Studenica Monastery Special Purpose Area Spatial Plan has established the historic marble quarries as scheduled cultural/natural monuments. They have been identified as geological, natural and cultural and historic monuments.

The cultural landscape boundaries hold three scheduled sites: Korita quarry 43°27′13.73″ N, 20°30′31.26″ E, Godovic quarry 43°28′22.83″ N, 20°30′15.51″ E and Brevina quarry 43°27′12.97″ N, 20°30′42.75″ E [53].

While drawing up a management plan for the UNESCO World Heritage site—the Studenica Monastery, a unique cultural landscape has been identified. The previous evaluation of the Studenica quarries value and significance resulted in their integration into the Plan, as valuable monuments of interaction between man and nature. An important starting point in making the Plan was to broaden the concept of heritage and its special role in sustainable development, which can enhance the contribution to an environmental, social, as well as economic aspect of sustainable development [54]. In those terms, the Studenica quarries were recognised as a cultural resource not only in the context of their utilisation in the cultural monuments restoration and conservation but also as a witness of the past, something that had become a part of the cultural routes system, which would improve a wide spectre of benefits: social, economic and cultural—directly linked to the heritage [55].

Therefore, grasping a cultural heritage wider context occupies a fundamental role in understanding and empowering the identity of a social community. Putting an accent on managing the historic environment based on a participative approach, allowed for the local community to be involved in the cultural property management in a much broader sense. In this particular case, the management of the quarries has been entrusted to the local residents who are also the proprietors of certain sites, which actually means giving support to the locals to provide for themselves rendering various services and goods. The area is identified a one that has been traditionally utilised. Here, education and research programmes are envisaged about recording, preserving and renewing traditional knowledge and skills applied in building and marble art works of architectural heritage imbedded in the Virgin's Church of Studenica.

All the presented results mirror an aspiration to cast light, as much as possible, on the significance of Studenica marble that has been a building material in erecting an identity of the Serbian most important mediaeval monument. Furthermore, they emphasise the extraordinary role of the marble in any future approach to restoration work, but also direct the focus of the public, the scholars, the experts and anyone else, to some special values of the Studenica Monastery surroundings, where the marble quarrying tradition has been going on from generation to generation for centuries, which has unfairly been neglected and underexplored. The paper also gives a list of the various activities that have, almost instinctively, formed a research process whose results are actively included in the cultural heritage development strategic documents, anywhere with stone as a key material in conserving the outstanding values for both the present and the future generations.

Author Contributions: All authors contributed equally to this paper. Conceptualization and methodology, N.D.R., N.Š. and J.I.Š.; validation, investigation and resources, N.D.R., N.Š., D.M., and J.I.Š.; writing-original draft preparation N.D.R., N.Š., D.M., and J.I.Š.; writing-review & editing N.Š., N.D.R., and J.I.Š. All authors read and approved the final manuscript.

Funding: This research received no external funding.

Acknowledgments: We would like to thank the Geological Institute of Serbia for the analyzes of the samples, and Brkic Milan, B.S. in geology from Geoprofesional in Belgrade for technical assistance and kindness

Conflicts of Interest: The authors declare no conflict of interest.

References

1. Erdeljan, J. Studenica. An identity in marble. *Зозраф* **2011**, *35*, 93–100. Available online: http://www.doiserbia.nb.rs/img/doi/0350-1361/2011/0350-13611135093E.pdf (accessed on 16 July 2019). [CrossRef]

2. Studenica Monastery, WHS. Available online: https://whc.unesco.org/en/list/389 (accessed on 16 April 2019).

3. Malcer, B. Bibliography.Ed.V.Djuric. In *The Studenica Monastery Treasure*; SASA Gallery: Belgrade, SFR Yugoslavia, 1988; pp. 325–371.

4. Čanak-Medić, M.; Bosković, D.J. *Architecture of the Nemanja's Period*; RIPCM: Belgrade, SFR Yugoslavia, 1986; pp. 77–117.

5. Nenadović, S. *The Studenica Issues*; Bulletin III. RIPCM: Belgrade, SFR Yugoslavia, 1957.

6. Maksimović, J. *Serbian Mediaeval Sculpture*; Matica Srpska: Novi Sad, SFR Yugoslavia, 1971.

7. Simić, V. Marble Stonecutting in Studenica and Cemerno. *Ethnogr. Mus. Her.* **1956**, *19*, 273–289.

8. Simić, V. Ornamental and Structural Stone in the Studenica Wider Area. In *The Herald*; Geological Institute of Serbia Geological Institute of Serbia: Belgrade, SFR Yugoslavia, 1957; Volume XIV, pp. 245–278.

9. Janjić, M. *The PR of Serbia Geological and Engineering Features of the Terrain, Chapter: Carbonate Rocks*; special editions; The Geological Institute of Serbia: Belgrade, SFR Yugoslavia, 1962; pp. 123–143.

10. Dimitrijević, D.M. *Geology of Yugoslavia, Chapter: Studenica Formations*; GeoInstitute and Barex: Belgrade, FR Yugoslavia, 1995; Volume 79–81, p. 205.

11. Bilbija, N. *The Stone on the Studenica Virgin's Church Façade Walls. The Stone Condition Research Results. The Studenica Virgin's Church Marble Façades Restoration Project*; RIPCM: Belgrade, Serbia and Montenegro, 2003; pp. 43–56.

12. Hawkes, J.; Woolley, L. *History of Mankind: Cultural and Scientific Development. Prehistory and the Beginnings of Civilization*; Allen and Unwin: London, UK, 1963; Volume 1, pp. 63–103.

13. Srejović, D. *Dialogue between Stone and Man in The Stone and Man*; The SASA Gallery: Belgrade, SFR Yugoslavia, 1990; pp. 92–119.

14. Schmidt, K. Göbekli Tepe–the Stone Age Sanctuaries. New results of ongoing excavations with a special focus on sculptures and high reliefs. *Doc. Praehist.* **2010**, *37*, 239–256. [CrossRef]

15. Göbekli Tepe, WHS. Available online: https://whc.unesco.org/en/list/1572/documents/ (accessed on 16 April 2018).

16. Borić, D. Body metamorphosis and animality: Volatile bodies and boulder artworks from Lepenski Vir. *Camb. Archaeol. J.* **2005**, *15*, 35–69. [CrossRef]

17. Bailey, D.W. *Balkan Prehistory: Exclusion, Incorporation and Identity*; Routledge: London, UK; New York, NY, USA, 2002; pp. 230–233.
18. Renfrew, C. *Before Civilization, Alfred A*; Knopf: London, UK, 1973; pp. 120–122.
19. Hood, S. *The Arts in Prehistoric Greece*; Penguin Books: London, UK, 1978; pp. 23–26.
20. Gosden, C. Making sense: Archaeology and aesthetics. In *World Archaeology*; Godsen, C., Ed.; Routlege: London, UK, 2001; pp. 163–167.
21. Srejović, D. *Archaeological Lexicon: Prehistory of Europe, Africa and Middle East, the Greek, Etruscan and Roman Civilisations*; Savremena Administracija: Belgrade, FR Yugoslavia, 1997; p. 237.
22. Scarre, C. Monuments and miniatures: Representing humans in Neolithic Europe 5000–2000 BC. In *Image and Imagination: A Global Prehistory of Figurative Representation*; Renfrew, C., Morley, I., Eds.; McDonald Institute for Archaeological Research: Cambridge, UK, 2007; pp. 17–29.
23. Fiora, L.; Alciati, L. Marmi e pietre antiche nel bacino mediterraneo=Ancient marbles and stones in the Mediterranean Basin. *L'informatore del Marmista* **2005**, *44*, 20–27.
24. Longmen Grottoes, WHS. Available online: https://whc.unesco.org/en/list/1003 (accessed on 16 April 2018).
25. Fischer, M. Marble from Pentelicon, Paros, Thasos and Proconnesus in ancient Israel: An attempt at a chronological distinction. In *ASMOSIA VII, Proceedings of the 7th International Conference of Association for the Study of Marble and Other Stones in Antiquity, Thassos, the Aegean Sea, 15–20 September 2003*; École française d'Athènes: Thassos, Greece, 2003.
26. Camp, J.M. *The Athenian Agora. Excavations in the Heart of Classical Athens*; Thames and Hudson: London, UK, 1986; p. 38.
27. Greenhalgh, M. *Marble Past, Monumental Present: Building with Antiquities in the Mediaeval Mediterranean*; Brill NV: Leiden, The Netherlands; Boston, MA, USA, 2009; Volume 80, pp. 20–26.
28. Kergall, H.; Mine Sève, V. *La France Romane et Gothique*; La Martinière: Paris, France, 2001; p. 19.
29. Macrides, R.; Magdalino, P. The architecture of ekphrasis: Construction and context of Paul the Silentiary's poem on Hagia Sophia. *Byz. Mod. Greek Stud.* **1988**, *12*, 47–82. [CrossRef]
30. Debljović Ristić, N. The Utilization of Traditional Technique of Architectural Stone-Carving in Serbian 12th and 13th Century Medieval Churches. *Mod. Conserv.* **2016**, *4*, 117–126.
31. Millet, G. *L'Ancien art Serbe. Les églises*; Picard: Paris, France, 1919; p. 10.
32. Djurić, S. Discovering the Studenica Ornamentation. In *The Studenica Monastery Treasure*; The SASA Gallery: Belgrade, SFR Yugoslavia, 1988; pp. 100–102.
33. Nenadović, S. *Construction Techniques in Mediaeval Serbia*; Prosveta: Belgrade, Serbia and Montenegro, 2003; pp. 236–250.
34. Kandić, O. Sopoćani. In *History and Architecture of Monasteries*; The SCS Herald: Belgrade, Republic of Serbia, 2016; pp. 66–74.
35. Debljović Ristić, N. Mediaeval Religious Heritage—Principles and Procedures in Architectural Conservation. *SCS Her.* **2015**, *39*, 290–301.
36. Brandi, C. *Theory of Restoration*; Publicum: Belgrade, Serbia, 2007.
37. Primavori, P. *Il Primavori: Lessico Del Settore Lapideo:[2500 termini]*; Zusi, G., Ed.; Unlibrio: Verona, Italy, 2004; p. 415.
38. *St Simeon's Hagiography*; Collected Writings, translation: Mirković L., Bogdanović D.; Prosveta: Belgrade, SFR Yugoslavia, 1986; p. 97.
39. Popović, M. *The Monastery of Studenica Archaeological Discoveries*; RIPCM: Belgrade, Republic of Serbia, 2015; p. 31.
40. Janković, M. *Archaeological Data on the Construction of the Virgin's Church, Collection Dedicated to Boško Babić*; Museum of Prilep: Prilep, SFR Yugoslavia, 1986; pp. 109–111.
41. Pajković, V. *Studenica Marble Basic Types with a View on Stone Slabs; Novopazarski zbornik*; Museum"Ras: Novi Pazar, Republic of Serbia, 2010.
42. Urosević, M.; Pavlović, Z.; Klisić, M.; Brković, T.; Malesević, M.; Trifunović, S. *Basic Geological Map Vrnjci Sheet K34–18*; The Federal Geological Institute: Belgrade, Serbia, 1964.
43. Brković, T.; Malešević, M.; Urošević, M.; Trifunović, S.; Radovanović, Z.; Dimitrijević, M.; Dimitrijević, M.N. *Guidebook of Basic Geological Map Ivanjica Sheet K34–17*; The Federal Geological Institute: Belgrade, Serbia, 1977; p. 61.

44. Aleksić, V.; Kalezić, M.K. *Geology of Serbia, III-2. The Metamorphism of the Transitional Oceanic Lithospere—The Studenica Region*; Aleksić, V.M., Petković, K., Eds.; Faculty of Mining and Geology, University of Belgrade: Belgrade, SFR Yugoslavia, 1977; pp. 99–209.

45. Antula, D.J. *Review of the Results in the Kondom of Serbia for the Paris Exhibition*; State Printing Office of the Kingdom of Serbia: Belgrade, Kingdom of Serbia, 1900; p. 134.

46. Operational Guidelines for the Implementation of the World Heritage Convention. Available online: https://whc.unesco.org/archive/opguide12-en.pdf (accessed on 19 May 2019).

47. Kandić, O. *Architectural Investigations and Conservation Works on the Sopocani Monastery*; RIPCM: Belgrade, SFR Yugoslavia, 1984; pp. 7–29.

48. Debljović Ristić, N. Restoration of the mullioned windows on the Holy Trinity Church of Sopocani. *SCA Her.* **2011**, *35*, 89–93.

49. Debljović Ristić, N. Window restoration and conservation on the Holy Trinity Church of Sopocani. *SCA Her.* **2012**, *36*, 50–55.

50. Tomić, S. *An Unknown Document on the Studenica Monastery Restoration*; The SASA Herald: Belgrade, SFR Yugoslavia, 1953; Volume V, book1; pp. 180–181.

51. Debljović Ristić, N. Mullioned windows in the west façade of the church of the Virgine at Studenica. An analysis of the shapes and sculptural decoration. *Bulletin* **2017**, *49*, 26–42.

52. Dragicević, L.J. Research of binders for rehabilitation the Studenica Virgin's Church marble façades (1991). In *The Studenica Virgin's Church Marble Façades Restoration Project*; RIPCM: Belgrade, Serbia and Montenegro, 2003; pp. 1–16.

53. The Studenica Monastery Spatial Plan. Available online: https://www.mgsi.gov.rs/lat/projekti/prostorni-plan-podrucja-posebne-namene-manastira-studenica (accessed on 2 June 2018).

54. *Managing Cultural World Heritage Manual*; United Nations Educational; UNESCO WHC: Paris, France, 2013; pp. 19–22.

55. Feilden, B.M.; Jokilehto, J. *Management Guidelines for World Heritage Sites*; ICCROM: Rome, Italy, 1993; pp. 102–103.

 *sustainability* 

Article

The Use of Natural Stone as an Authentic Building Material for the Restoration of Historic Buildings in Order to Test Sustainable Refurbishment: Case Study

Nenad Šekularac [1], Nevena Debljović Ristić [2], Dušan Mijović [3], Vladica Cvetković [4], Slobodan Barišić [2] and Jelena Ivanović-Šekularac [1,*]

[1] Faculty of Architecture, University of Belgrade, Bulevar kralja Aleksandra 73/II, 11000 Belgrade, Serbia
[2] Institute for the Protection of Cultural Monuments in Serbia, Radoslava Grujića 11, 11000 Belgrade, Serbia
[3] Faculty of Applied Ecology Futura, Metropolitan University in Belgrade, Požeška 83 a, 11000 Belgrade, Serbia
[4] Faculty of Mining and Geology, University of Belgrade, Đušina 7, 11000 Belgrade, Serbia
* Correspondence: jelenais@arh.bg.ac.rs; Tel.: +381-64-11-65-036

Received: 19 June 2019; Accepted: 19 July 2019; Published: 24 July 2019

Abstract: This study deals with the integrated process of conservation and restoration of architectural heritage and sustainability. The objective of the research was to define adequate methodologies for the structural restoration of historic buildings, their re-use, and sustainable refurbishment in accordance with modern requirements and conservation standards while maintaining the original visual character by using natural stone as an authentic building material. The main research method was the in-situ observation of the historic structures during the restoration and adaptive re-use, the analysis, and evaluation of the research findings regarding energy efficiency improvements and energy saving in the Haybarn complex within the monastery Hilandar, Mount Athos, Greece. Due to its cultural and natural values, Mount Athos has been inscribed on the UNESCO World Heritage List. The research included the damaged and abandoned agricultural structures that belong to the Haybarn complex and the analysis of the obtained results after the restoration had finished and the abandoned premises had been turned into guest rooms for the visitors of Hilandar monastery. The result section states the findings of the research arranged as recommendations for historic building restoration and re-use, emphasizing their new function in accordance with modern comfort requirements and environmental protection standards. The main contribution of this study is the analysis of the research findings and the possibilities of energy refurbishment of the restored historic buildings, through the use of natural stone as authentic local construction material, in accordance with energy efficiency measures and principles, conservation requirements and cultural heritage conservation standards.

Keywords: cultural heritage conservation; sustainability; architectural conservation; stone architecture heritage; modern principles of double-layered ventilated roofs; conservation requirements; UNESCO World Heritage List

---

1. Introduction and Premises

The cultural value of heritage buildings is our legacy that we will pass on to our future generation. However, it is evident that not all heritage buildings still function with their original use, which means for these cases a new function is required so that we could preserve the intrinsic value of heritage buildings [1].

Adaptive re-use of old, damaged and restored historic buildings ensures environmental benefits and revival of dilapidated buildings in order to follow modern trends and architectural, functional and economic values [2]. Nowadays, the process of retrofitting a great number of old buildings, such as churches, public facilities, industrial facilities or agricultural buildings, for new functions, is carried out in accordance with the current needs and adaptive re-use requirements, ensuring the preservation of these architectural structures from destruction. Restoration and revitalization turn the abandoned and demolished historic buildings into new business premises, galleries, alternative theaters, commercial buildings, hotels, tourist accommodation facilities, and so on.

In the past, construction workers would use local materials, natural light and ventilation and optimal orientation of a building in order to take advantage of natural sources of heat energy, in accordance with bioclimatic factors. Therefore, they were implementing the principles of environmental and energy efficient construction of that period. The construction of buildings used to be carried out in accordance with bioclimatic design—specific resilience of architectural structures, the durability of local construction materials, the abundance of greenery and free land space around the buildings. After hundreds of thousands of years, the environmental endurance and durability of these construction materials testified on their sustainability [3,4].

Restoration and re-use of a historic building, energy-save refurbishment measures and the use of stone as authentic local material, represents one of the main conditions of the successful revival of the existing building. Construction materials and any performed additions have to be in harmony with the original structure. When it comes to interventions, they also need to be carried out without harming the quality impartment and integrity of the building itself. The new function of the restored building must complement its space requirements [5].

This study deals with the restoration and adaptive re-use of the abandoned buildings within the Haybarn complex on Mount Athos, in Greece, in order to improve the energy performance, to contribute to environmental protection and to maintain the authenticity of these buildings. The use of stone as the main building material on Mount Athos is a consequence of its availability and easy extraction. Stone is a natural material extracted from natural deposits called quarrying. Stone needs very little maintenance, and after a simple method of stone processing, it is turned into construction material. Building stones were widely used in the Middle Ages for massive foundations, walls, domes or vaults. Several types of stone were used for decorative facade ornaments, the interior design of these buildings as well as for indoor flooring. Natural stone slabs, used as a roofing material, were almost untreated (Figure 1A). This stone roofing technique still stands in some Mount Athos buildings. Numerous structures within the monastery complex of Hilandar were covered with stone slabs: mansions, towers, churches and chapels [6].

The use of natural stone extracted from a local quarry maintains the authentic architectural value of facades. Natural stone has a direct influence on the technical and aesthetical aspects of the restored historic buildings. Appropriate extractions from historical stone quarries reoccur in remote and isolated areas, such as Mount Athos. Inaccessibility and isolation of the Athos peninsula enabled the long-term conservation of natural stone resources in these historic quarries of Mount Athos. The extraction of stone, from the same quarries from which the extraction had been carried out a few centuries ago for the purpose of the construction of the buildings, which had been included in the restoration and revitalization process described in this study, has a great positive value for heritage conservation. The exploitation of stone from these historical quarries and the use of the same type of stone contributes to the appropriate restoration of heritage buildings. It also prevents potential negative aspects regarding durability and endurance, thus enabling the conservation of architectural heritage.

The new function of the restored building should ensure a significant financial gain so that the future maintenance of the building is provided for. A successfully realized adaptive re-use increases the value of the property and might also contribute to cultural tourism. The physical survival of heritage buildings is not enough—these buildings are to be economically sustainable as well [5].

When it comes to contemporary conservation theory and practice, adaptive re-use is regarded as a significant and qualitative approach to cultural heritage conservation [7]. Cultural, social and economic benefits of a community significantly depend on architectural heritage conservation. Therefore, a new function of the building should be the most appropriate and useful one, in order to ensure the preservation of the cultural significance of the heritage building. Numerous factors contribute to the selection of a new destination for the heritage buildings. All the aspects and factors must be detailed and considered for a sustainable re-use following an adaptive project [1].

The concept of the energy refurbishment of heritage buildings is not different from the concept meant for the buildings that are not inscribed in the heritage list. Energy refurbishment methods depend on cultural heritage aspects and the selection of a new function of historic buildings. Before the beginning of any restoration, all current conditions and characteristics of a historic building should be assessed in order to choose the most appropriate method of energy efficiency improvement. These energy improvement measures should include potential energy savings as well as the preservation of materials and features of a historic property [4]. Optimal energy refurbishment of historic buildings must take into consideration the existing energy aspects of the historic construction in order to preserve them for further efficient use, along with new measures aimed to improve energy performance.

Historic buildings are heterogeneous and preserved in different ways. So far, general recommendations for a minimum quality level of energy efficiency improvement in historic buildings have not been defined on a large scale [4].

Restoration of historic buildings and energy efficiency measures should be planned and carried out carefully, avoiding potential negative impacts on the historic character and integrity of the restored buildings. Ongoing monitoring of the restored buildings detects unexpected changes or defects, as a consequence of the restoration. Instantaneous measures for problem removal prevent irreparable damage to historic materials. These measures, along with regular maintenance, ensure the long-term preservation of our historic built environment and sustainable use of construction materials resources [4,8].

Preserving the authenticity is one of the basic concepts of most conservation requirements and the authenticity maintaining is the overall goal of heritage conservation. The concept of authenticity (The Nara Document on Authenticity) [9] is fundamentally significant for cultural heritage research, conservation measures and World Heritage List criteria selection. The Nara Document on Authenticity, formulated in 1994, defines the main aspects and assessment of authenticity [9]. The preservation of architectural heritage, the authenticity of heritage buildings, their original and new functions, comfort improvements as part of restoration and energy savings ensure a sustainable model. Sustainability and restoration of historic buildings are inseparable and interdependent.

Knowing that keeping of the authenticity maintenance and the cultural heritage preservation are significant aspects of any restoration of historic buildings, the retrofitting is mostly performed in the interior of the building while maintaining all of its original exterior wall elements [10]. A building envelope is the physical separator between the conditioned and unconditioned environment of a building including the resistance to water, air, light, heat and noise transfer.

The thermal insulation of the external wall, part of the building envelope, should be avoided since the loss of the authenticity is incomparable in its importance with an insignificant energy saving. In this way, not only the visual appearance of historic buildings are protected, but also the construction system itself [11].

The cooling and heating of historic buildings without appropriate thermal insulation requires additional energy, which increases energy costs for using and maintaining that building. It also leads to increased environmental pollution. The existence of cold peripheral space of a building, such walls and attics, which are uninsulated or inadequately insulated, is responsible for: increased heat loss in the winter months, structural defects due to condensation from within the structure or overheating in the summer months. Poor condensation control affects the performance of building elements, damaging the durability and lifetime of the building itself. Condensation may cause the occurrence of mold allergens, which leads to chronic health problems and unhealthy living and working environments.

The restoration carried out in specific conditions of the isolated Mount Athos property contributes to the conservation of the listed buildings [4].

The European Energy Efficiency Directives refer to energy saving within the existing buildings and do not exclude the functional historic buildings. However, there is not a single international legislative act, in the field of architectural heritage conservation, to regulate the energy retrofit.

The present European laws do not specify the minimum level of energy performance for historical buildings. It is obvious that this issue is of great significance since the buildings are heterogeneous and have different protection levels. So far, these regulations do not include yet general suggestions about how to manage the energy efficiency of historical buildings [12]. However, there are some individual examples of regulations, adopted on a national level, related to the historic buildings treatment during restoration processes.

Taking into consideration the above mentioned, it is obvious that any country could adopt its own regulation in order to decide whether or not to apply energy efficiency requirements to both existing and heritage buildings. In order to cover the gap between heritage buildings and energy retrofit, the National Heritage Institutions should launch an initiative in finding effective ways towards the retrofit of operating heritage buildings or found in conservatory or restoration programs [13].

Since the buildings included in this study are located in the territory of Greece, i.e., the European Union, they are taken care of by both Greece and the Republic of Serbia, considering the fact that the Serbian nobles founded the monastery and that most of monastic brotherhood are Serbs as well. Restoration financing is provided by Greece, the Republic of Serbia and a great number of donors, Orthodox believers and admirers of Hilandar Monastery from all over the world.

The conducted study aimed to measure the level of energy efficiency of these buildings according to the current Serbian requirements and based on the following legislative documents and references:

1. Regulations on Energy Efficiency of Buildings [14], (Rulebook of energy efficiency of buildings),
2. Regulations on Conditions, Contents, and Methods of Certificate Issuance regarding energy efficiency of buildings [15], (Rulebook of energy efficiency of buildings), according to the European standards and norms in this field:
3. Energy Performance of Building Directive—EPBD No 2002/91/EC—Directive of the European Parliament and of the Council of 16 December 2002 on the energy performance of buildings [16],
4. Directive 2010/31/EU of the European Parliament and of the Council of 19 May 2010 on the energy performance of buildings (recast) [17].

The Regulations on Energy Efficiency of Buildings, adopted in 2011, introduced mandatory energy certification of buildings and imposed significantly lower permitted thermal transmittance values. The assessment of achieved energy efficient of a building is expressed by energy label, which corresponds to a calculated value of annual energy use, graded from the most energy efficiency energy class "A", to the lowest efficiency energy class "G" [14]. Based on the Regulations on Energy Efficiency of Buildings [15], energy efficient buildings will have grades of at least "C".

Based on the Regulations on Conditions, Contents, and Methods of Certificate Issuance regarding the energy efficiency of buildings, the heritage buildings do not comply with the mandatory energy certification of buildings, since energy efficiency requirements would be in direct conflict with the principles of conservation conditions [15].

In order to achieve optimal energy efficiency, the restoration and renovation of architectural structures should include the following operations [18]:

- Analyzing of related architectural form, orientation and location,
- Use of local construction materials, whenever it is possible,
- Applying methods for high-level thermal insulation of the entire external cladding (thermal bridges should be avoided), and solar gain and protection from excessive solar exposure,
- Implementation in the further exploitation of energy efficient cooling, heating and ventilation systems, along with renewable energy resources, as much as possible.

Mount Athos is a unique and secluded site where the use of natural resources for the purpose of construction had been limited to local use only. Seclusion and isolation of this site ensured the protection of natural construction resources from uncontrolled human activities and excessive exploitation, thus enabling sustainable exploitation and environmental protection. Careful and competent planning of natural resources exploitation and the conservation of architectural heritage is the main prerequisite for the achievement of cultural sustainability for future generations. Cultural heritage protection refers to the maintaining of the authenticity, architectural values, and composition [9]. The use of natural stone in historic buildings reflects local geology, culture, historical and economic features.

The hypotheses underlying the study were:

1. For this secluded and protected site, the use of natural stone, extracted from a local quarry, for the restoration of façade walls and the use of slate for roofing, are the only adequate choices for the building materials,

2. Adequate construction methods, in accordance with the conservation requirements, and the use of stone as an authentic building material enabled the restoration of the damaged buildings and provided them with a new function, maintaining the authenticity and harmony with the listed property of Mount Athos;

3. The restoration of the damaged buildings shall be based on the energy saving and energy efficiency criteria.

The goals of the study were:

- Taking advantage of property conditions for the purpose of historic preservation, in accordance with environmental protection requirements,

- Increasing the range of knowledge about the purpose and significance of historic buildings and their re-use, restoration methods and use of natural stone as authentic building material, in accordance with the conservation requirements,

- Using renewable energy sources and tailored construction methods to restore historic buildings, in order to improve energy performance and preserve environmental values of this secluded property,

- Defining the general key factors of preservation, restoration and re-use of the historic buildings, based on the used construction materials and energy efficiency criteria applied to these historic buildings, in order to make the research findings available to academic communities.

2. Methodology

The following analyses and activities were performed:

- The on-site observation and assessment of the historic buildings and their remains,

- The criteria and the design for building protection, based on the previous analyses of the property condition and chosen methods of restoration and re-use,

- The analyses of the construction materials and methods used in historic buildings,

- The petrographic analysis of the stone sample and the analysis of the stone type used as a building material and extracted from the local quarry for the purpose of restoration,

- The restoration of historic buildings according to the conservation requirements: maintaining the authenticity of structural elements (façade walls, roof structure),

- The restoration of the façade walls and roof structure using the authentic material extracted from the local quarry, without changing the original architectural value of the building,

- The use of stone as an authentic roofing material and the maintenance of authenticity of the historic building is ensured by following modern requirements and standards of construction physics—in this case, it was the implementation of double layered ventilating roof,

- The case study as a method of restoration and energy retrofit analyses, performed for the Haybarn complex of the monastery Hilandar on Mount Athos, in order to provide scientifically-based findings,

- Planning and designing the restoration of historic buildings in order to adapt the building for a new use, improve energy efficiency and reduce energy demands,
- All the interventions, performed for the purpose of restoration, adaptive re-use or energy efficiency improvement, reversibility according to the conservation requirements,
- The analysis and assessment of the obtained results of the carried out restoration, in relation to energy retrofit, adaptive re-use and renewable energy sources aimed at providing energy supply and environmental protection.

3. The Research and the Results for the Haybarn Complex

3.1. Specific Characteristics of the Hilandar Monastery Property

Mount Athos is situated on a secluded, inaccessible place, in the eastern finger of the Chalkidiki peninsula, in the northern part of Greece. Even nowadays there is no land route, which means that it is accessible only by ferry, over the surrounding sea. Mount Athos is a unique monastic country, the only one in the whole world. Its autonomy is granted by the Constitution of Greece and Mount Athos Statute. Its autonomous status aims to protect spiritual and religious values, measured by hundreds and thousands of years of being the most important cultural and religious center for the entire Eastern Orthodox Church. Such a specific position of Mount Athos is accepted by the European Union, owing to the special declaration – annex to the Agreement between Greece and the European Union. In 1988, the UNESCO decided to inscribe Mount Athos and its monasteries on the World Heritage Liste.

Strict principles and rules of this monastic community have been unchanged for centuries. Women are not allowed to enter into the territory of Mount Athos, which means that the only residents are men. There are 20 sovereign monasteries on Mount Athos. Their number and hierarchy were never changed.

Hilandar Monastery ranks the fourth in the hierarchy of significance and influence, and its territory is the second largest one. Hilandar Monastery is one of the most important testaments of the Serbian architectural heritage. It was founded in 1198, when the Byzantine emperor decided to hand over the remains of the old and abandoned monastery of Helandariy, built at the end on the 10th century, by the Serbian noblemen Saint Sava and his father, Saint Simeon [19] in order to establish a new monastery accommodation for the Serbian monks. Ever since the Serbian monastic brotherhood has lived in this monastic place. Nowadays, the complex of Hilandar Monastery resembles inside its walls a medieval town.

In the founding Charter of Hilandar Monastery, probably issued in the second half of the 1198 by Saint Simeon and Saint Sava it was stated:

"I left my native land and came to Mount Athos where I found a former monastery, Hilandar, the Consecration to the Holy Mother of God, where there was no stone left but ruined stone... and in my old age, with the help of my son, the Gran Zupan Stefan, the Bishop honored me with the duty of a founder. And I started searching for ruined parts and remains, and I renewed it" [20].

This valuable Charter testifies about the first use of the building stone in the Hilandar Monastery property. The stone as a building material was widely used in years to come, including the use of a great variety of rocks found in the geological basement of this site.

In spite of a great number of charters on the history of the holy monastery, there are almost no records on the used construction materials, location, exploitation, and processing methods. The only type of stone that was mentioned is marble, which used to be highly appreciated for its great technical performances.

Later on, the researchers who visited the Hilandar monastery, wrote that the buildings were made of square stones and unbaked bricks. The type of rocks used for square stone is determined by used building material. Since the process of creating stone plastics is difficult, the parts of the demolished building were used for the restoration of the building, in the same way as it had been written in the founding Charter of Hilandar Monastery.

Such organizational effectiveness of the construction project was not only a reflection of painstaking works of builders and artisans living in that time, but it is also a rational review on the use of God-given resources, nowadays known as sustainable use of natural resources.

3.1.1. Geological Framework

Mount Athos represents the farthest southern part of the Serbo-Macedonian mass. This mass is composed of two crystalline complexes—the lower, metamorphosed in the amphibolite group of facies (in the western part), and the upper—metamorphosed in the group of green schist facies (in the eastern part) [21]. The relation with the Vardar zone or the internal Vardar sub-zone is well defined in Macedonia.

The Serbo-Macedonian mass represents the basic mass that formed the central parts of the Internal Hellenides, which were formed by the accumulation of various terranes during the Mesozoic [22]. The Athos Peninsula is at the border of two large tectonic units, the Vertikos terrane in the west and the Kerdilion unit in the east. The contact zone of these two units is located not far from the eastern border of the property of the Hilandar Monastery and forms a narrow marble zone in mélange. The last geological fabric was formed by the embedment of granite during the early Tertiary. The rocks that are present in the Athos area can be divided into three groups: gneiss group, ophiolitic mélange group, and granite group [22]. The rocks from the gneiss group are located south from the property of the Hilandar Monastery. The granite group is predominantly present in the Monastery's area and belongs to the granites of Uranopolis which have embedded into the ophiolitic mélange and gneiss zone. The ophiolitic mélange group is widespread in the north-western and southern parts of the Athos Peninsula and is represented by amphibolites and ultramafites, marbles, eclogites, peridotites, and gneisses. These rocks are covered in places by Neogene sediments.

The geological map with the ophiolitic zone within the property of the Hilandar Monastery is shown in Figure 1.

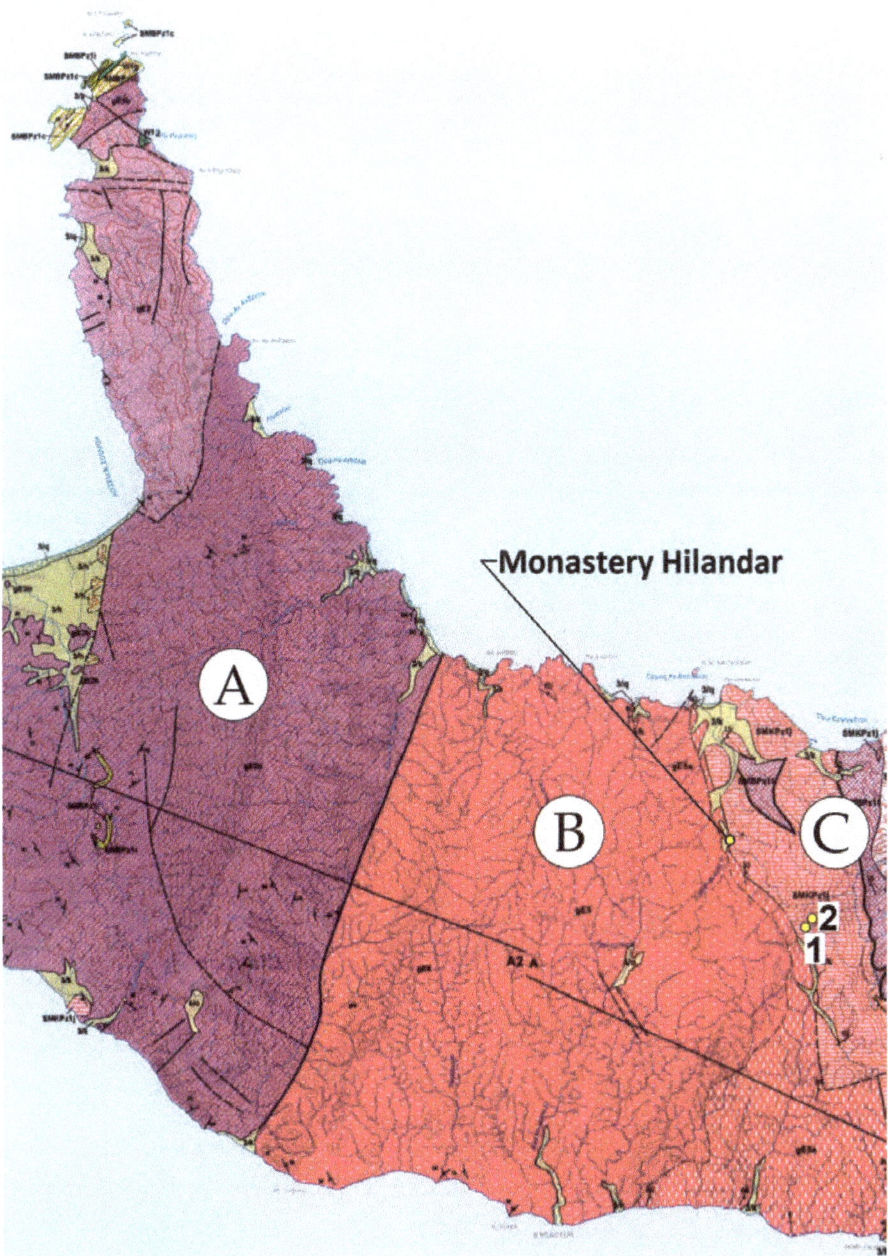

Figure 1. The geological map of the property of the Hilandar Monastery [23]: (A): Rocks of granite group, (B): Rocks of gneiss group, (C): Rocks of ophiolitic mélange group; 1. 2—places where sample 1 and sample 2 were taken.

3.1.2. History Background of the Haybarn Complex

The Haybarn Complex is located near the access road to Hilandar Monastery. Originally it was used to shelter monastery animals and their food (hay). The complex consists of the abandoned and dilapidated buildings, erected in the first part of the 19th century, outside the monastery walls but still close to them. It has been decades since the Haybarn Complex was used, therefore, its buildings were partially demolished.

Before the 2005 and 2006 restoration stages, the Haybarn complex (Figure 1) have included [24]:

- The Stable: a building for keeping mules,
- The Mulekeepers' House: a building for the people taking care of the mules,
- The Haybarn: a huge building for hay storage, hence the name for the entire complex.

The rise and development of modern society significantly increased the number of pilgrims and devotees from all over the world, who wanted to visit this unique monastic country [25].

Prior to the devastating fire of 2004, when a part of the Hilandar complex was torn down, one of the main problems was accommodation capacity—the monastery workers and the ever-increasing number of visitors required more space. Therefore, the restoration of the Haybarn Complex was launched in 2005 in order to provide new accommodation capacities (Figure 2). It successfully ended in 2006. The main and preliminary design for the restoration of the Haybarn Complex were made under the supervision of Prof. M. Kovačević, Prof. N. Šekularac, S. Tripković, B.Sc.Arch., D. Krivokuća, B.Sc.Arch., 2004 and 2005. Prof. N. Šekularac, Member of Expert Council for the Reconstruction of Holy Monastery Hilandar, is the construction manager and structural engineer, one of the authors of this study. The supervision of restoration and conservation works was carried out by S. Barišić, B.Sc.Arch.

Figure 2. The Haybarn Complex: Mulekeepers' House, Stable, Haybarn (from left to right): **(a)** former appearance, **(b)** post-restoration appearance, **(c)** the interior of the Stable after the restoration (provided photos taken by the author).

The purpose of the restoration of the Haybarn Complex was to keep all undamaged parts, remove the demolished ones and to rebuild the damaged walls using the authentic material, using the natural stone as authentic material. During the restoration, the old mud or lime mortar was replaced with a new binding agent, lime cement mortar. All the construction measures were realized in accordance with the conservation requirements; the stone façade walls, roof structure, roof covering and the authentic appearance of the building were not to be changed.

During the restoration of the Haybarn Complex, in addition to the use of the local stone as a building material, the stone taken from the stone deposit in a road, cut below Papa Konaki, near the Hilandar monastery (Figure 3) was also used. The petrographic composition of this stone is the same as the stone used as a building material for the entire complex at the beginning of the 19th century.

Figure 3. The stone deposit in a road-cut below Papa Konaki (provided photos taken by the author).

The mentioned construction measures and the materials used in the restoration, for the purpose of improving the energy efficiency of these buildings (Figure 4), have reversible characteristics, which means that the original condition of the restored buildings can be reestablished without causing any structural damage [26].

In no way should energy efficiency measures change the visual identity of the heritage building and its environment. The concept of sustainability can be achieved through the preservation of buildings, as authentic architectural heritage, and their original function or a new one, along with the improved comfort and energy saving. The restoration of historic buildings and the concept of sustainability are inextricably linked.

Sustainable construction and restoration are basic requirements for heritage building preservation:

- The use of authentic construction materials in restoration and interventions as well as the use of construction materials that are not harmful to the property and its environment,
- Energy efficiency of buildings,
- Waste management during the construction and demolition procedures, and potential re-use of waste materials.

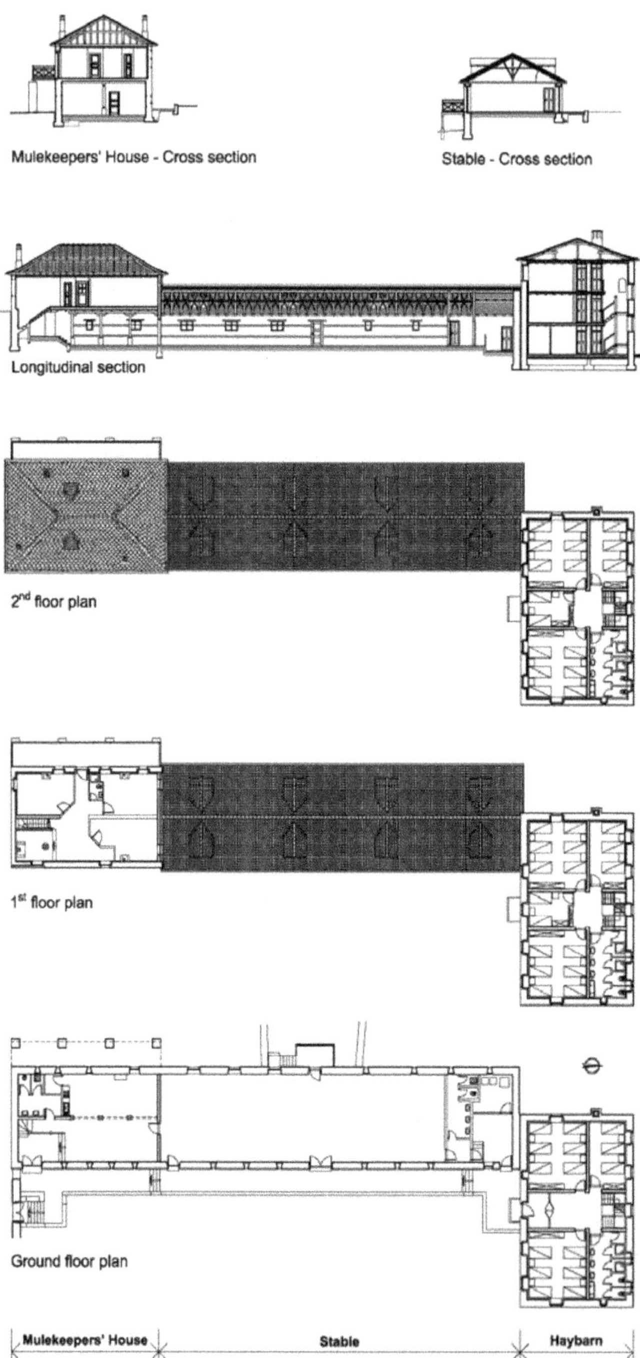

Figure 4. The Haybarn Complex (after the restoration).

3.2. Mineralogical-petrographic Analysis of the Samples of the Stone Used in the Haybarn Complex

Leica DMLSP Polarizing Light Microscope and Leica DFC290 HD digital cameras were used for mineralogical and petrographical analysis. These tests were carried out in the laboratory of the Faculty of Mining and Geology of the University of Belgrade.

The stone outer walls of the Haybarn Complex were built from solid and schistose marble. Sample 1 was taken from the source/ledge near the road in the immediate vicinity of the Monastery (Figure 1). It was established through petrographic analysis that it was formed from densely packed carbonate grains of platelet habitus (Figure 5). In some parts of the sample, it was observed that the rock mass was permeated with post-metamorphic calcite veins. This marble contains a detrital component in the form of very fine grains of quartz and mica, as well.

Figure 5. The photomicrograph of sample 1—massive marble; crossed nichols.

Sample 2 is a schistose marble taken from the source/ledge along the road in the immediate vicinity of the Monastery, 50 m from where sample 1 had been taken (Figure 1). It was established through the petrographic analysis of sample 2 that it was formed from convex carbonate grains which by their elongation define the schistose structure of the rock and fine-grained matrix (Figure 6). This marble contains slightly less detrital components compared to the previously described one.

Figure 6. The photomicrograph of sample 2 rock, schistose marble.

Most of the roofs of the monastery's buildings were covered with traditionally dark-gray plates/slabs. In the petrographic analysis of sample 3 from the remains of stone slabs taken from the collapsed roof of the Haybarn complex during reconstruction, it was concluded that sample 3 represents a schistose marble containing up to 15% volume of mica.

This rock belongs to the schistose marble series, as well. The nature of its schistosity is so different because the carbonate protolith has most likely had the most detrital components, predominantly represented by mica and clay. During metamorphism, there was a growth of mica, which together with the recrystallized carbonate defined the schistose structure of the rock (Figure 7). Therefore, by the point of view of textural characteristics, the rock looks like half schistose marble, and half mica schist.

Figure 7. The photomicrograph structure of sample 3 rock, schistose marble.

During the restoration of the Haybarn Complex, carried out in 2005 and 2006, the old stone covering, made of laminated marble that contained much of mica, was used, and the damaged tiles were replaced with new, stone tiles of the same sizes. The alternate pattern was applied for the installation of an old and new roofing, in order to achieve the visual harmony of the roof plane.

The new stone tiles used for the restoration of the complex are known as Kavala Grey.

The petrographic analysis of sample 4, taken from this new roof slab, has established that the stone type is an epidotic gneiss-micachist (the old geological name leptinolite), which most likely originates from the Serbo-Macedonian mass. The rock has a lepidoblastic and granoblastic texture and schistose to augen structure. It is formed from more than 85–90% quartz, mica, feldspar and epidote, and the rest consists of chlorite, calcite, sphene, apatite, tourmaline, zircon and iron oxides and hydroxides. Figure 8 shows photomicrographs depicting one thicker stripe which is transformed into a lens formed by quartz and plagioclase, surrounded by a matrix of elongated muscovite lamina and quartz lenses. This is the general appearance of the rock, although some millimeter-thick stripes appear, which are only formed from quartz. In Figure 8 with parallel nichols, there are several epidote porphyroblasts that stand out with a high relief.

Figure 8. The photomicrograph structure of sample 4 rock, epidotic: gneiss-micaschist: (a) With crossed nichols, (b) With parallel nichols.

Based on the analysis of scientific literature data and on the mineralogical-petrographic analysis of the four samples from the field, several observations were drawn:

- The area of the Monastery's property is dominated by the granite group and gneiss group of rocks, while the eastern border (closer to the Monastery) is dominated by the rocks of the ophiolitic mélange group,
- Marbles are widespread and are represented by the following varieties: massive marble, schistose marble, pronouncedly schistose marble, and fine-grained marble. The following granite varieties have been identified: coarse-grained granite, granitoid (quartz/monzonite/diorite) and milonitizedgranitoid,
- The performed baseline studies on the Monastery's property have shown that the construction stones are very common and that it is most likely used as construction material, but the quarries are no longer discernible in the relief; no written information has been identified,
- As for the restoration of the Haybarn Complex, the installation of a partially new roof covering instead of old, deteriorated and demolished stone tiles, is a partial use of another type of stone that does not belong to the Mount Athos area, which is a novelty of the multi-century practice. The novelty does not affect the visual quality of these buildings.

3.2.1. The Use of Stone as a Roofing Material

The roofs of the compound building of the Haybarn Complex were constructed by laying smaller stone slabs, densely packed, over the vaults or slope roofs. The main binding material was mud mortar, lime mortar or lime mortar mixed with Santorini earth (natural porcelain). The slabs are laid almost horizontally, to a slight fall toward the outer edge in order to allow water drainage. The upper slabs are laid over the joints of the lower layered slabs. This slight slope prevented water intrusion. The stone used as a roofing material had to be compact, dense and easy to be cut to shape, in terms of the slabs 2 to 3 cm thick. This stone type is laminated marble called slate. It is usually a grey stone, which is easy to be cut into slabs. Slate has been used a lot as a roofing material since it is generally known as a long-lasting material, easily cut into slabs. This explains its widespread use for coating flat and slope surface: vaults, arches and cupolas.

The stone slabs were laid over the chestnut sheathing, installed over the wooden roof structure which resisted heavy load due to the weight of the used stone. This roof structure had to be strong enough to resist such a heavy roof cover that weighs from 300 to 500 kg/m^2 [27].

Since the monastery is situated on the peninsula of Athos in the Aegean Sea, the strong winds are blowing from the sea and the humidity they bring along penetrated deep through the walls and vaults of these buildings. In most cases, the stone slabs were not resistant to the penetration of rain and humidity in the winter months. Therefore, it was a call for additional reconstruction measures.

3.2.2. Modern Principles of Double-skin Ventilated Roofs and Stone Slate Roofing

It is possible to keep the slate roofing, as an authentic roofing structure, and visual identity of the heritage buildings, if the following modern principles of double-skin ventilated roofs are applied [6]: ventilation of roof space and ventilation of stone-coated roof (above the waterproof layer), i.e., the condensation drying [28] that occurs on the lower sides of the stone slabs at dawn, as well as drying out the moisture that is the result of water penetration between the stone slabs.

The roof restoration was the process of applying the layers of roof coating, according to the building engineering physics, as presented in Figure 9.

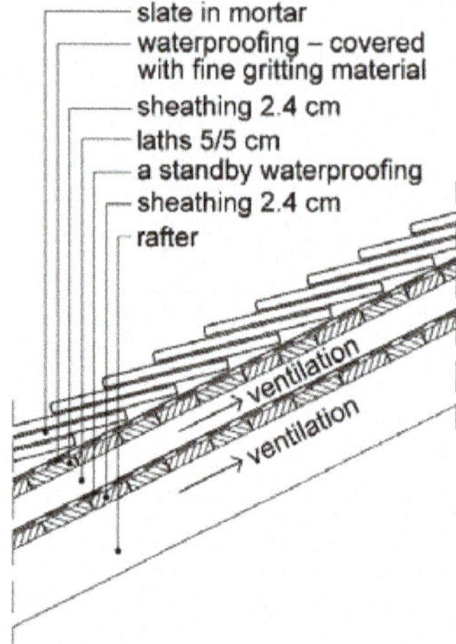

Figure 9. Recommended restoration of the slate roof covering—the applying of the double ventilation modern principles.

The use of stone for the building roofs starts at the eaves, thus enabling the penetration of air and interrupted airflow between the wooden slats, parallel to the eaves. The installation of ridge vents is required as well (Figure 10).

Figure 10. Venting during the restoration of the slate roofing: **(a)** roof appearance, **(b)** roof air vent (the photo taken by the author).

The implementation of modern principles of double-skin ventilated roofs, along with the slate roof covering, ensures a successful restoration of priceless medieval heritage buildings, contributing to the overall world cultural heritage. The use of stone slabs as a roofing material and a double-ventilated roof was first carried out in the area in May 2005, in the restoration of the buildings of the Haybarn Complex [6,29].

3.3. The Research Findings

All three devastated auxiliary buildings of the Haybarn Complex: Mulekeepers' House, Stable, and Haybarn, are massive masonry buildings, with load-bearing masonry walls made of stone, timber roofs and stone roof coverings. All the external walls were made of rubble and roughly dressed stone made of laminated marble.

The case studies regarding energy efficiency improvements of these historic buildings as well as energy retrofit via Building Performance Simulation (BPS) method, were published as journal articles [4,30].

3.3.1. Restoration of the Stable

The Stable is a single-story building. At a time when the building was first erected, it was connected to the ground floor of the Mulekeepers' House and animals were kept there [26]. The façade peripheral walls were made of rubble and roughly dressed stone (case 1). The roof above the peripheral walls was made by timber roof truss: joists, columns, bracing and rafters. Ridge tiles were used for roofing. Dormers were set on the roof. Considering the original function of these buildings, the window openings were scarce, since there was no need for natural light. The original windows were single glazed wooden windows. The window openings were narrow, 50–60cm inside width and 20cm external width, in order to prevent daylight penetration and protect donkeys from vision loss, i.e., the so-called donkey blindness.

The original condition of the building (Figure 2a) and its function could not have been classified into energy classes according to the Regulations on energy efficiency of buildings. If the Stable had been used for accommodating people, along with the applying hygiene measures only, it would not have accomplished the requirements of the regulations (Table 1, case 1).

During the 2005 and 2006 restoration (case 2), the visual quality of the façade walls remained the same and visible stone coating was used for both external and internal walls (Figure 2). The floor restoration was performed on the ground and thermal insulation was added. The restoration of the Stable is significant because the exceptional timber framing of this building remained as it had been before. It was the interior part of the sleeping area for visitors (Figure 2c). New thermal insulation layers were placed over the timber frame structure of the roof. The authentic roofing was made of ridge tiles. The fenestration elements on the façade walls were kept unchanged and the roof dormers were built where they used to be, thus preserving the authentic appearance of the building. The windows were replaced with new 9cm double-glazed wooden frames, and double-glazed, Low-E Glass (4+12+4mm) filled with krypton. Krypton is an inert gas that is used to fill the space between glass panels in order to increase thermal performance and reduce heat loss. The front door was replaced with a new, solid wood door. The structural elements of the building preserved the authentic value after the restoration (Table 1, case 2). The only change was the building's new function. The visual quality of both interior and exterior elements was preserved according to the buildings conservation best practices Nowadays, the restored Stable contains a spacious bedroom for visitors, sanitary facilities and a laundry. The underfloor heating was also installed.

A comparative review for the two analyzed cases of the Stable shows the effects of the restoration and energy efficiency improvement on heat energy reduction on the annual basis (Table 1).

Table 1. Comparative review of upgrading the sheathing elements in terms of thermal protection and energy reducing on the annual basis, for the two analyzed cases of The Stable.

	Case 1		Case 2	
Façade wall on the ground floor $U_{max} = 0.40$ W/m^{2}K	1. stone wall	60.0 cm	1. stone wall	60.0 cm
U (W/m^{2}K)	1.912		1.912	
Roof $U_{max} = 0.20$ W/m^{2}K	1. ridge tiles 2. wood decking 3. rafter—timber construction	2.2 cm	1. ridge tiles 2. air layer 3. waterproofing 4. wood decking 5. air layer 6. thermal insulation polystyrene 7. wood ceiling 8. rafter—timber construction	2.2 cm 5.0 cm 15.0 cm 2.2 cm
U (W/m^{2}K)	1.930		0.192	
Ground floor $U_{max} = 0.40$ W/m^{2}K	1. pavin stone 2. sand 3. ground	10.0 cm 5.0 cm	1. brick 2. mortar 3. concrete 4. thermal insulation polystyrene 5. waterproofing 6. concrete 7. gravel 8. ground	4.0 cm 4.0 cm 4.0 cm 5.0 cm 12.0 cm 6.0 cm
U (W/m^{2}K)	1.632		0.435	

Table 1. *Cont.*

	Case 1	Case 2
Specific heating energy demand per year (kWh/m²/year)	541.84	442.50
Energy class	G	G
Specific heating energy demand per year / Energy savings [(kWh/m²/year) (%)]	541.84 kWh/m² (100.0%)	442.50 kWh/m² (81.70%) / Energy savings 99.34 kWh/m² (18.30%)

3.3.2. The Restoration of the Mulekeeper's House

This building consists of a ground floor and first floor. The ground floor, where the animals were kept, used to be connected to the Stable. The first floor accommodated the mulekeepers. In the winter months, the body temperature of the animals would keep the room and the first floor heated. This was one of the efficient uses of all energy resources a few centuries ago.

All exterior walls were made of rubble and roughly dressed stone [26]. The peripheral walls of the ground floor were coated with visible stone (case 1). The first floor interior walls were all plastered. The original wood windows were single glazed. If this building had been used to accommodate people, along with the applied hygiene measures, (Table 2, case 1), it would not have accomplished the requirements of the stated Regulations on energy efficiency of buildings.

The 2005 and 2006 restoration of the Mulekeepers' House (case 2) maintained the authenticity of its façade walls. The stone coating was used for both external and internal ground floor walls (Figure 2b). Since the first floor façade walls were plastered, thermal insulation was added to the existing façade wall from the inside (Table 2, case 2). As for the ground floor, the floor was restored and thermally insulated. The joists between storeys were kept. However, new fillings and authentic coating were applied. Above the first floor, toward the attic area, the thermal insulation and coating were added. The stone coating as an authentic roof covering was carried out in accordance with the modern requirements of ventilated roofs.

During the restoration, the old windows were replaced with new wooden windows, having 9cm thick, double-glazed Low-E Glass (4+12+4mm) filled with krypton gas. The front door was also replaced with a new, solid wood door.

Today, the ground floor of the restored Mulekeepers' House consists of a sitting room, a kitchenette, and a sanitary block. The first floor consists of a smaller sitting room, three bedrooms, and a sanitary block. Underfloor heating was installed on the ground floor and low-temperature radiators were added to the first floor.

The comparative review for the two analyzed cases of the Mulekeepers' House shows the effects of the restoration and energy efficiency improvement on heat energy reduction on the annual basis (Table 2).

Table 2. Comparative review of upgrading the sheathing elements in terms of thermal protection and heating energy reducing on the annual basis, for the two analyzed cases of The Mulekeepers' House.

	Case 1		Case 2	
Façade wall on the ground floor Umax = 0.40 W/m²K				
	1. stone wall	60.0 cm	1. stone wall	60.0 cm
U (W/m²K)	1.912		1.912	
Façade wall on the first floor Umax = 0.40 W/m²K	1. stone wall	40.0 cm	1. stone wall	40.0 cm
			2. thermal insulation polystyrene	5.0 cm
			3. brick	6.5 cm
			4. mortar coating	1.5 cm
			5. heated space	
U (W/m²K)	2.467		0.517	
Attic floor Umax = 0.40 W/m²K	1. attic area		1. attic area	
	2. mud layer	10.0 cm	2. wood decking	2.5 cm
	3. wood decking	2.5 cm	3. air layer	5.0 cm
	4. timber construction		timber construction	
	5. heated space		4. thermal insulation polystyrene	10.0 cm
			5. gypsum boards	2.5 cm
			6. wood ceiling	2.0 cm
			7. heated space	
U (W/m²K)	2.340		0.289	
Ground floor Umax = 0.40 W/m²K	1. paving stone	10.0 cm	1. brick	4.0 cm
	2. sand	5.0 cm	2. mortar	4.0 cm
	3. ground		3. concrete	4.0 cm
			4. thermal insulation polystyrene	5.0 cm
			5. waterproofing	
			6. concrete	12.0 cm
			7. gravel	6.0 cm
			8. ground	

Table 2. *Cont.*

	Case 1	Case 2
U (W/m^{2}K)	1.632	0.435
Specific heating energy demand per year (kWh/m^2/year)	380.64	173.62
Energy class	G	E
Specific heating energy demand per year / Energy savings [(kWh/m^2/year) (%)]		

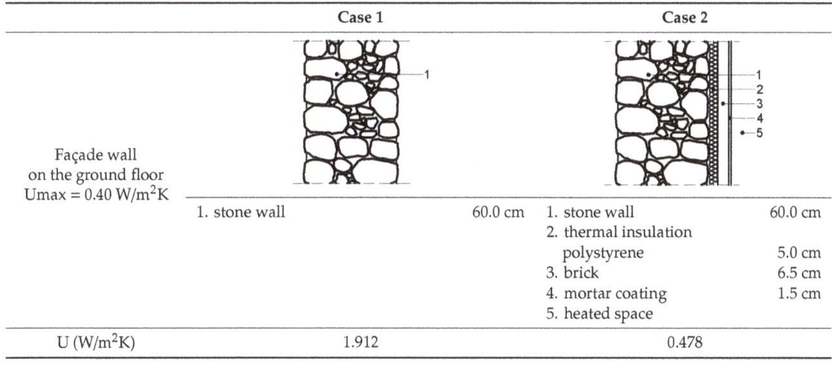

3.3.3. The Restoration of the Haybarn

The Haybarn was a very tall building used for hay storage (case 1). The exceptionally high peripheral walls were made of rubble and roughly dressed stone. The building had the stone coated façade and stone slab roof (Figure 2a). The original appearance and function of this building could not be classified into energy classes according to the regulations on energy efficiency. If the purpose of the building had been to accommodate people, applying the hygiene measures only (Table 3, case 1), it could never have met the requirements of the stated Regulations on energy efficiency of buildings.

During the restoration, the entire space was horizontally divided into three floors by two intermediate floor structures (case 2). All the façade walls were thermally insulated from the inside [26]. The floor of the ground floor was restored and thermally insulated (Table 3, case 2). Thermal insulation was set between the joists, above the top floor, toward the attic area. The wooden framed windows, 9cm thick, with double-glazing and Low–E Glass (4+12+4mm) filled with krypton were installed. The front door was replaced with a new, solid wood door. The roof structure with timber trusses and stone roofing were built according to the modern requirements of ventilated roofs, in the same way as it was done in the Mulekeepers' House. The authentic appearance of the building was completely preserved in terms of building materials, stone coated facades, wooden frames, roof coating as well as the height, structure, and shape of the roof (Figure 2b). Traditional radiator heating was also installed in the entire building. After the restoration, the building was turned into a dormitory for visitors. The restored three-storey building consists of bedrooms and sanitary facilities.

The comparative review of the two analyzed cases of the Haybarn shows the effects of the restoration and energy efficiency improvement on heat energy reduction on the annual basis (Table 3).

Table 3. Comparative review of upgrading the sheathing elements in terms of thermal protection and heating energy reducing on the annual basis, in the analyzed cases of The Haybarn.

	Case 1	Case 2
Façade wall on the ground floor Umax = 0.40 W/m^{2}K	1. stone wall 60.0 cm	1. stone wall 60.0 cm 2. thermal insulation polystyrene 5.0 cm 3. brick 6.5 cm 4. mortar coating 1.5 cm 5. heated space
U (W/m^{2}K)	1.912	0.478

Table 3. *Cont.*

	Case 1	Case 2
	Roof	Attic floor
Roof / Attic area U_{max} = 0.40 W/m²K	1 2 3 4	1 2 3 4 5 6 7
	1. stone slabs in mortar — 12.0 cm 2. wood decking — 2.5 cm 3. rafter—timber construction 4. interior	1. attic area 2. wood decking — 2.5 cm 3. air layer — 5.0 cm timber construction 4. thermal insulation polystyrene — 10.0 cm 5. gypsum boards — 2.5 cm 6. timber construction 7. heated space
U (W/m²K)	2.490	0.301
Ground floor U_{max} = 0.40 W/m²K	1 2 3	1 2 3 4 5 6 7
	1. pavin stone — 10.0 cm 2. sand — 5.0 cm 3. ground	1. decking — 2.2 cm 2. plywood — 2.2 cm 3. thermal insulation polystyrene — 5.0 cm 4. waterproofing 5. concrete — 12.0 cm 6. gravel — 6.0 cm 7. ground
U (W/m²K)	1.632	0.397
Specific heating energy demand per year (kWh/m²/year)	775.70	64.33
Energy class	G	C
Specific heating energy demand per year / Energy savings [(kWh/m²/year) (%)]	775.70 kWh/m² (100.0%)	64.33 kWh/m² (8.30%) Energy savings 711.37 kWh/m² (91.70%)

3.4. Recommendations for Improving Energy Efficiency During the Restoration of Historic Buildings for the Purpose of their Adaptive Re-Use

In order to improve the energy efficiency of heritage buildings, it is recommended to implement all the conservation requirements and the following specific measures:

- Adequate thermal insulation of the building envelope,
- Heat loss reduction,
- Energy efficient systems of cooling, heating, and ventilation, with the use of alternative renewable energy sources.

Based on the conservation requirements, in order to maintain the authentic characteristics of historic buildings, their architectural value and their structure, the standard thermal insulation is not recommended for the external façade walls and neither changing external elements that could diminish the concept of authenticity.

In order to ensure the re-use and the sustainable energy retrofit for the restored historic buildings of the Haybarn Complex, the following energy efficiency measures were recommended and carried out:

- Thermal insulation of façade walls, only at the inside,
- Thermal insulation of floors on the ground floor,
- Thermal insulation of the mezzanine structure toward the unheated attic,
- Thermal insulation of the roof layers above the attic area that is being used, or above the attic area integrated into stores,
- Replacement of façade joinery (windows and front doors),
- New heating and cooling systems, along with the use of renewable energy sources and the applying of the environmental protection requirements,
- Energy efficiency management.

As for the cases where the façade walls were thermally insulated from the inside, it was important to implement additional fire protection measures and prevent the following potential defects: thermal bridge and water vapor diffusion (condensation). All the mentioned construction methods and techniques applied during the restoration aimed to improve the energy performance of the buildings.

Meeting the energy efficiency requirements and carrying out the energy efficiency improvement plans ensure energy saving for this isolated site. Along with the conducted analysis related to the existing conditions on the Mount Athos area and with the decisions on implementing of adequate measures for improving thermal performance, it was necessary to improve energy performance in heating systems as well, according to the environmental protection requirements of the Mount Athos area. Two heat-distribution systems were applied in the complex: radiators and water-based underfloor heating. Wood-fired boilers are also used. The heat distribution system and the use of wood are environmentally friendly choices, in accordance with the requirements for reducing the greenhouse gas emission. The wood is a renewable and sustainable material. Since the Hilandar Monastery is situated in a large forested area, all the required amounts of wood are obtained by their own trees; this fact contributes to a significant cost-reducing, which also enables local access to energy sources in this remoted site, knowing that boats are the only means of transport.

Today, diesel equipment is supplying energy for this area. Energy generating products are delivered by boats. Other studies dealt with the analysis of using solar energy in this area in order to provide energy supply by implementing PV panels in the future [4].

4. Discussion

The main condition that has guided the restoration of the Haybarn Complex was the maintenance of the original architectural character as much as possible, along with the preservation of the authentic integrity of the buildings. At the same time, attention was paid to keep the distinctive architectural elements and decorative features that reflect the historical period and the construction phases, according to the environmental characteristics of the monastery property.

The presented research is a part of the energy retrofit analysis and the restoration of the Haybarn Complex [4,30]. The study provides potential and practical solutions for energy refurbishment and/or energy retrofitting of historic buildings in the case studies.

The analysis of the presented cases of the Haybarn Complex: the building originally built in the first half of the 19th century and the building restored during 2005 and 2006, including the energy efficiency improvement measures, leads to the conclusion that maximum thermal insulation, added to the building envelope elements in accordance with the conservation requirements in order to preserve the authenticity and stone-coated facades, it is possible to achieve a significant level of energy saving.

Based on a comparative review of the obtained values for the overall heat loss for all the three analyzed buildings: the Mulekeepers' House, the Stable and the Haybarn Complex, in both analyzed cases: (a) the building before the restoration (the condition from the 19th century), and (b) the building restored during 2005 and 2006, for the purpose of heat loss reduction, we found that heat energy consumption was reduced on the annual basis. This analysis and the Energy Efficiency Passport for both analyzed cases of the Haybarn Complex ensured the energy classification of the buildings in accordance with the Regulations on the energy efficiency of buildings (Tables 1–3).

The analysis and evaluation of these historic buildings in terms of the performed restoration of the Haybarn Complex, along with applying the recommended energy efficiency improvement measures in accordance with the conservation requirements, the authors presented the key findings that have never been evaluated in this way in the scientific literature. The authors of this study drew valid conclusions on the basis of real values and the results regarding energy saving, after having obtained the thermal insulation measures, according to the conservation requirements and the principles of environmental protection of the site.

The research findings were provided by on-site observation, along with the methods of valorization of the Haybarn Complex and energy saving analysis for each case. The research has also confirmed that it is possible to obtain energy saving and improving of energy efficiency during the restoration of historic buildings if adequate construction technology is applied according to the conservation requirements for the preserving of the authenticity. All these measures are implemented in order to restore and revitalize the buildings as well as to maintain the authentic values of the heritage site.

Although the restoration from 2005 and 2006 improved the energy performance of the Stable, the building remained in the energy class "G" (Table 1). This study confirms that a large area of peripheral walls is part of the external building envelope that has a big impact on heat loss; this explains why the applied thermal insulation measures failed to improve the energy efficiency of the building.

The Mulekeepers' House is the only building that could have been classified into energy class "G" at the very beginning, considering the original function of the 19th century building (Table 2). After the restoration had carried out in 2005 and 2006, along with the applied measures for improving energy efficiency, the evaluated energy performance of this building was increased by two classes, from energy class "G" to energy class "E". This improvement is the result of the thermal insulation added to the peripheral surface, from the inside of the façade walls on the first floor as well as above the top floor toward the attic. Meanwhile, the façade walls on the ground floor maintained its authentic appearance—the stone coated the internal and external sides of the façade walls.

The 2005 and 2006 restoration improved the energy efficiency of the Haybarn building. The applied construction measures turned its energy class "G" into energy class "C" and the restored building was finally energy efficient (Table 3). The energy efficiency was achieved by upgrading all the peripheral areas of the building in terms of thermal insulation on the interior façade walls as well as above the top floor toward the attic.

The hypotheses were tested by on-site observation and valorization of each building of the Haybarn Complex, in terms of energy saving and energy efficiency, in accordance with the conservation requirements. The obtained results support the hypothesis regarding the use of stone, extracted from the local quarry, as an authentic building material used for façade restoration, and the use of slate for roofing. This approach is the only appropriate choice of building materials for the restoration of massive walls and the use of stone as a roofing material on this isolated heritage site. The hypothesis concerning the implementation of adequate construction measures, in accordance with the conservation requirements and authenticity maintenance, was confirmed. The hypothesis about the potential restoration of the devastated buildings and re-use of the building, for the purpose of revitalization and authenticity maintenance, was also tested and confirmed through on-site observations. The research findings confirmed the hypothesis that the restoration of the devastated building could ensure energy saving and contribute to the energy performance of the building.

The research findings confirmed the assertion made by Bionaz [12], since the energy performance of ancient buildings turned out to be underestimated, it led to overestimated interventions aimed to achieve the required standards in accordance with the regulations.

The analysis of the Haybarn Complex and its re-use confirmed the assertion made by Plevoets and Cleempoel [7], historic buildings and their adaptive re-use prove the importance of the intervention for the conservation of cultural heritage.

Adaptive re-use of the Haybarn Complex confirmed the assertion made by Mısırlısoy and Günçe [1], that most appropriate adaptive re-use strategy has a vital importance for heritage building conservation and maintenance of cultural, economic and social values. The authors of this study concluded the same as Mısırlısoy and Günçe [5] that the materials used for restoration should be in harmony with the original materials and in accordance with the structural integrity of the building and space demands of new function activities.

The assertion made by Murgul was also confirmed [11]. It is reasonable to protect the visual quality of the historic environment along with the unique and environmentally sustainable historic construction, tested by century-long working experiences.

This study once again confirms the significance of identifying and preventing all negative effects on heritage buildings, knowing that performed restoration and energy efficiency of historic buildings can be successfully achieved only without any negative impact on historic character and integrity of these buildings, which confirms the research conducted by Hensley and Aguilar [9]. After the restoration of these buildings and their adaptive re-use, the process of constant observation continued in order to detect any failure or imperfection in terms of condensation, mold or inadequate thermal comfort. It is of vital importance to pay close attention to the mentioned occurrences in order to prevent any irreparable damage to historic buildings and their architecture. This study led to the same conclusions as to the ones presented by Hensley and Aguilar [8] regarding the need for regular maintenance of historic buildings, monitoring and observation of the restored buildings during the carrying out of their new function, and the implementation of long-term protection measures aimed to conserve their historic character.

This study led to the same conclusions as the ones made by Bionaz [12] regarding the restoration of historic buildings having massive stone walls: the interventions in terms of window replacement and shade installation will fail to meet the modern living requirements in terms of thermal comfort and energy efficiency, unless all the peripheral stone walls are thermally insulated from the inside.

The assertion of Martin & Carlos was also confirmed [10] the heritage building can be preserved if retrofitting is carried out in the interior of that building, along with keeping the original aspects of exterior walls.

Once again this research confirms the conclusion made by Todorovic et al. [3] that the dominant characteristic of heritage buildings is bioclimatic design, carefully planned architectural structure and appropriate choice of construction materials, whose endurance and durability performances are tested and measured by the hundreds of years, reflecting their sustainability.

The study once again points out the importance of appropriate analysis of all the aspects that can negatively affect the heritage building during the restoration, keep in mind that energy efficiency of a historic building can be optimal only without any negative effects on architectural character and integrity, which confirms the research findings formulated by Hensley and Aguilar [8].

What makes this study innovative in comparison to other authors is the unique model of the restoration of the auxiliary Haybarn facilities, aimed to carry out retrofitting and adaptive re-use in terms of accommodations for monastery visitors. This study analyzed the principles and measures of energy efficiency for the purpose of energy refurbishment of the heritage building, since it was one of the first restorations of this type that was carried out on Mount Athos, thus proving the professional value and scientific argumentation in relation to practical use of it and to potential applying on other similar heritage sites. All the analyzed cases of the Haybarn Complex were designed and realized following the energy efficiency principles. The research was tested directly through the practical implementation of the secluded and ecological area of Mount Athos, which is regarded as its greatest achievement [31]. The objective of further research in this field is to analyze the possibilities of energy saving in historic buildings, using extensive reconstruction measures included in a large-scale construction project.

5. Conclusions

The need for additional accommodation for a growing number of Hilandar Monastery visitors led to the restoration and revitalization of the damaged buildings of the Haybarn, turning them into lodgings. The creation of the Haybarn Complex linked cultural heritage to modern social changes.

The restoration and adaptive re-use of heritage buildings, along with the implementation of energy efficiency measures, requires an individual approach to each historic building. The main contribution of this study is data analysis and interpretation of the research findings in order to determine retrofitting potentials by implementing energy efficiency measures in the selected heritage buildings, in accordance with the conservation requirements, the preservation of the authentic character and integrity of the Haybarn Complex and the use of stone as an authentic material extracted from the local quarry, thus maintaining the authenticity of the entire heritage site. The research findings could be used for choosing adequate sites to supply the construction stone for the purpose of restorations.

One of the main postulates of heritage conservation is the use of natural resources as authentic materials for restoration and reconstruction. In order to achieve energy saving, according to the conservation requirements, the following construction measures are to be considered and carefully planned: (a) the building insulation and (b) replacement and installation of adequate windows and doors. The mentioned measures and the use of renewable energy sources in order to enable annual energy supply can significantly improve the energy performance of historic buildings.

The restoration of heritage buildings, for the purpose of their adaptive re-use and energy efficiency improvement, should include the following actions:

- Thorough assessment of a building conditions, along with determining the level and type of required restoration, based on the building condition,
- Selection of construction measures to be applied during energy retrofitting and adaptive re-use, in accordance with the conservation requirements,
- Selection of adequate authentic material for walls and roofing (in this case – it is stone), following the conservation requirements and local resources availability—the quarries from where the stone used to be extracted centuries ago,
- Selection of energy refurbishment and/or energy retrofitting interventions for the purpose of re-use, in accordance with the conservation requirements,
- Selection of adequate methods for using renewable energy sources in order to enable annual energy supply, for the purpose of achieving energy saving and ensuring environmental protection of this remote area of Mount Athos, as included in the concept of energy management.

It can be concluded that restoration, reconstruction, and revitalization of heritage buildings, along with the implementation of energy efficiency measures, are regarded as reversible and requires an individual approach to each historic building. In order to achieve energy saving and to reduce heat loss, it is necessary to add thermal insulation on the building envelope, following the requirements of the conservations, since it is the only way to improve the energy performance of historic buildings. All the measures of energy refurbishment and/or energy retrofitting were implemented in a case study at Haybarn Complex in accordance with technical and ethical conservation requirements. The main contribution of this paper is the analysis of the energy refurbishment results, in terms of adaptive re-use, achieved by implementing the energy efficiency measures in the restoration of heritage buildings, according to the concept of maintaining the authenticity for the visual characteristics.

Author Contributions: All authors contributed equally to this paper. Conceptualization and methodology, N.Š., N.D.R, and J.I.-Š.; validation, investigation, N.Š., N.D.R, V.C., D.M., S.B. and J.I.-Š.; writing-original draft preparation: N.Š., N.D.R, D.M, V.C., S.B, and J.I.-Š.; writing-review & editing N.Š., N.D.R, and J.I.-Š. All authors read and approved the final manuscript.

Funding: This paper was written as a part of the projects "Spatial, Environmental, Energy and Social Aspects of Development of Settlements and Climate Change", under the Grant number (TR 36035) financed by the Ministry of Education and Science of the Republic of Serbia for the period 2011–2019.

Acknowledgments: The authors would like to thank Brkić Milan, Geological Engineer, Geoprofesional d.o.o., Belgrade, Serbia, for technical support.

Conflicts of Interest: The authors declare no conflict of interest.

References

1. Mısırlısoy, D.; Günçe, K. Adaptive reuse strategies for heritage buildings: A holistic approach. *Sustain. Cities Soc.* **2016**, *26*, 91–98.
2. Stratton, M. *Industrial Buildings: Conservation and Regeneration*; Taylor & Francis: London, UK, 2003.
3. Todorović, M.S.; Ećim-Đurić, O.; Nikolić, S.; Ristić, S.; Polić-Radovanović, S. Historic building's holistic and sustainable deep energy refurbishment via BPS, energy efficiency and renewable energy—A case study. *Energy Build.* **2015**, *95*, 130–137.
4. Šekularac, N.D.; Šumarac, D.M.; Čikić Tovarović, J.L.; Čokić, M.M.; Ivanović-Šekularac, J.A. Re-use of historic buildings and energy refurbishment analysis via building performance simulation: A case study. *Therm. Sci.* **2018**, *22*, 2335–2354.
5. Günçe, K.; Mısırlısoy, D. Assessment of Adaptive Reuse Practices through User Experiences: Traditional Houses in the Walled City of Nicosia. *Sustainability* **2019**, *11*, 540.
6. Sekularac, N.; Sekularac, J.I.; Tovarovic, J.C. Application of Stone as a Roofing in the Reconstruction and Construction. *J.Civ. Eng. Archit.* **2012**, *6*, 919.
7. Plevoets, B.; Van Cleempoel, K. Adaptive Reuse as a Strategy towards Conservation of Cultural Heritage: A Survey of 19th and 20th Century Theories. 2012. Available online: https://pdfs.semanticscholar.org/ee88/eefed42e089e3cea99d8c6fa41d523c74a3f.pdf (accessed on 29 May 2012).
8. Hensley, J.E.; Aguilar, A. *Improving Energy Efficiency in Historic Buildings*; Government Printing Office: Washington, DC, USA, 2012; Volume 3.
9. ICOMOS, N. The Nara Document on Authenticity. In Proceedings of the ICOMOS, Nara, Japan, 1–6 November 1994.
10. Martins, A.M.; Carlos, J.S. The retrofitting of the Bernardas' Convent in Lisbon. *Energy Build.* **2014**, *68*, 396–402.
11. Murgul, V. Features of energy efficient upgrade of historic buildings: Illustrated with the example of Saint-Petersburg. *J. Appl. Eng. Sci.* **2014**, *12*, 1–10.
12. Bionaz, C. Preservation and Energy Behavior in Aosta Valley's Traditional Buildings. In Proceedings of the VerSus 2014, International Conference on Vernacular Heritage, Sustainability and Earthen Architecture, Valencia, Spain, 11–13 September 2014.
13. Mazzarella, L. Energy retrofit of historic and existing buildings. The legislative and regulatory point of view. *Energy Build.* **2015**, *95*, 23–31.
14. Srbija, R. *Ministarstvo životne sredine, rudarstva i prostornog planiranja (Minister of Environment, Mining and Spatial Planning), Pravilnik o energetskoj efikasnosti zgrada (Regulations on Energy Efficiency of Buildings - Rulebook on energy efficiency of buildings)*; Official Gazette of the Republic of Serbia: Belgrade, Serbia, 2011; pp. 61–2011.
15. Srbija, R. *Ministarstvo životne sredine, rudarstva i prostornog planiranja (Minister of Environment, Mining and Spatial Planning), Pravilnik o uslovima, sadržini i načinu izdavanja sertifikata o energetskim svojstvima zgrada (Regulations on Conditions, Contents, and Methods of Certificate Issuance regarding energy efficiency of buildings)*; Official Gazette of the Republic of Serbia: Belgrade, Serbia, 2012.
16. Parliament, E. Directive 2002/91/EC of the European Parliament and of the Council of 16 December 2002 on the energy performance of buildings. *Off. J. Eur. Communities* **2003**, *1*, 65–71.
17. Recast, E. Directive 2010/31/EU of the European Parliament and of the Council of 19 May 2010 on the energy performance of buildings (recast). *Off. Eur. Union* **2010**, *18*, 2010.
18. Ivanović-Šekularac, J.; Čikić-Tovarović, J.; Šekularac, N. Application of wood as an element of façade cladding in construction and reconstruction of architectural objects to improve their energy efficiency. *Energy Build.* **2016**, *115*, 85–93.
19. Bogdanović, D. *Hilandar*; Republički zavod za zaštitu spomenika kulture Srbije: Belgrade, Serbia, 1978; p. 36.

20. Trifunović, Đ.B.; Bjelogrlić, V.; Brajović, I. Hilandarska osnivačka povelja Svetoga Simeona i Svetoga Save. In *Osam Vekova Studenice*; Episkop žički Stefan, A.J., Kašić, D., Eds.; Sveti arhijerejski sinod Srpske pravoslavne crkve: Belgrade, Serbia, 1986; pp. 194–212.

21. Dimitrijević, M.D. Srpsko-makedonska masa. In *Geologija Jugoslavije*; Geoinstitut & Barex: Belgrade, Serbia, 1995; pp. 115–126.

22. Himmerkus, F.; Zachariadis, P.; Reischmann, T.; Kostopoulos, D. The basement of the Mount Athos peninsula, northern Greece: insights from geochemistry and zircon ages. *Int. J. Earth Sci.* **2012**, *101*, 1467–1485.

23. Kochel, F.; Mollat, H.; Antoniades, P. *Geological Map of Greece, 1:50.000, Ierissos Sheet*; Institute of Geology and Mineral Exploration (IGME): Athens, Greece, 1968.

24. Nenadović, S. *Osam Vekova Hilandara—Građenje i Građevine*; Republički zavod za zaštitu spomenika kulture: Belgrade, Serbia, 1997.

25. Ivanović-Šekularac, J.; Čikić-Tovarović, J.; Šekularac, N. Reconstruction and revitalization of the Complex Senara, within the monastery Hilandar, in order to adapt to modern trends and social changes. In *Keeping up with Technologies to Make Healthy Places*; Faculty of Architecture: Ljubljana, Slovenia, 2015.

26. Ivanović-Šekularac, J.; Čikić-Tovarović, J.; Šekularac, N. Restoration and Conversion to Re-use of Historic Buildings Incorporating Increased Energy Efficiency–A Case Study–The haybarn complex, Hilandar Monastery, Mount Athos. *Therm. Sci.* **2016**, *20*, 131.

27. Šekularac, N.; Ivanović-Šekularac, J.; Čikić-Tovarović, J. Application of stone as a roofing in the reconstruction and construction. In Proceedings of the 2nd WTA-International Ph. D. Symposium, Brno University of Technology, Brno, Czech Republic, 6 October 2011.

28. Fasio, P.; Ge, H.; Rao, J. Urban Building Physics: A CFD study of the Venturi-effect in non-parallel passages between buildings. In *Research in Building Physic and Building Engineering*; Taylor & Francis: Montreal, QC, Canada, 2006.

29. Ivanović-Šekularac, J.; Šekularac, N.; Čikić-Tovarović, J. Introduction of Modern Principles of Design in Covering of Roofs with Stone. In *Harmony of Nature and Spirituality in Stone*; Kamen Srbije: Kragujevac, Serbia, 2012.

30. Ivanović-Šekularac, J.; Čikić-Tovarović, J.; Čokić, M.; Šekularac, N. Restoration of historic buildings and verification of sustainable energy refurbishment. *KGH—Sci. Prof. J. Air Cond. Heat. Refrig.* **2016**, *45*, 49–56.

31. Lidelöw, S.; Örn, T.; Luciani, A.; Rizzo, A. Energy-efficiency measures for heritage buildings: A literature review. *Sustain. Cities Soc.* **2018**, *45*, 231–242.

 *sustainability*

Article

Ultrafast Laser Surface Texturing: A Sustainable Tool to Modify Wettability Properties of Marble

Ana J. López [1],* , Alberto Ramil [1] , José S. Pozo-Antonio [2] , Teresa Rivas [2] and Dolores Pereira [3]

[1] Laboratorio de Aplicacións Industriais do Láser, Centro de Investigacións Tecnolóxicas, Universidade da Coruña, 15471 Ferrol, Spain

[2] Departamento de Enxeñaría de Recursos Naturais e Medio Ambiente, Escola de Enxeñaría de Minas e Enerxía, Universidade de Vigo, 36310 Vigo, Spain

[3] Department of Geology, Universidade de Salamanca, 37008 Salamanca, Spain

* Correspondence: ana.xesus.lopez@udc.es; Tel.: +34-881-01-3250

Received: 20 June 2019; Accepted: 25 July 2019; Published: 28 July 2019

Abstract: Conservation strategies to reduce the degradation of stone caused by the action of water are focusing on increasing the hydrophobicity of the surface by imitating existing solutions in nature (lotus leafs and others). These are mainly based on the existence of hierarchical roughness with micro- and nanoscale structures. In the case of marble, research has focused on protective coatings that sometimes are dangerous for the health and the environment, and with undesirable effects such as color changes or reduction of water vapor permeability of the stone. Laser texturing, however, is an environmentally friendly technique, because no chemicals or toxic waste are added and, moreover, it can process nearly all types of materials. It has been used to change the surface texture of metals and other materials on a micro or even nanometric scale, to meet a specific functional requirement, such as hydrophobicity. The objective of this work was to analyze the feasibility of this technique to provide hydrophobic properties to a marble surface without appreciable changes in its appearance. Therefore, an analysis of the irradiation parameters with ultra-short-pulse laser was performed. Preliminary results demonstrate the ability of this technique to provide hydrophobic character the marble (contact angles well above 90°). Besides, the analysis of the treated surfaces in terms of roughness, color and gloss indicates that changes in the appearance of the surface are minimal when properly selecting the process parameters.

Keywords: marble; wettability properties; hydrophobicity; laser surface texturing; multiscale roughness; cultural heritage; ultrafast pulse laser

---

1. Introduction

Marble has been extensively used in historical architecture and monuments owing to its good mineralogical and microstructural properties, durability, and aesthetic quality. However, exposure to the open air for a long time deteriorates the rock surface. In humid climates, biological factors highly contribute to the weathering through physical and chemical interactions that can cause different kinds of alterations, such as the discoloration of materials, the formation of crusts on surfaces, and the loss of material that can lead to structural damage [1]. Particularly calcareous stones are sensitive to the deterioration effects related to water, and the minimization of water contact with calcareous stone surfaces is thus a key aspect in marble conservation [2].

Conservation strategies to reduce the degradation of stone caused by the action of water are being continuously developed. In this sense, new products and procedures have been recently designed to reduce the wetting ability of stone [2–8]. In the case of marble, research has been focused on the development of protective coatings [9–16]. However, some of these coatings include toxic products or processes that are harmful to human health and the environment; in addition, there are other undesirable effects such as color, modifications, or reduction of water vapor permeability above the safe threshold in cultural heritage interventions. In Europe presently, the use of chemicals in these very sensitive environments is scrutinized and regulated by the European Union.

The main index used to evaluate the wettability of a solid surface is the static contact angle, θ, that describes the behavior of a liquid droplet on a solid surface in air, and is defined as the angle between the tangent at the three-phase point and the solid surface. Solid surfaces with $\theta < 90°$ are considered hydrophilic, while surfaces with $\theta > 90°$ are considered hydrophobic [17]. The wetting behavior of a solid surface is driven by the surface free energy at the interfaces between the solid, liquid, and vapor [18]. Expressions for the equilibrium contact angle of a small liquid droplet are given by Young, Wenzel, and Cassie–Baxter for ideal, rough, and composite surfaces, respectively [17,19]. The Wenzel model implies that wettability can be enhanced by increasing the surface roughness:

$$\cos\theta_W = r \cos\theta_0 \tag{1}$$

where θ_W is the actual contact angle on rough surface, θ_0 is the contact angle on ideal smooth surface defined by Young's equation, and the Wenzel roughness factor, r, is defined as the ratio of the actual surface area wetted by the liquid and the projected planar area ($r > 1$).

The Wenzel equation is based on the hypothesis that liquid enters grooves present on the surface, leading to a homogeneous wetting regime with increased contact area. This model explains why the contact angle of hydrophilic surfaces decreases with increasing roughness whereas contact angle of hydrophobic surfaces increases. However, Wenzel's model can hardly explain water contact angles exceeding 150° that correspond to super-hydrophobic surfaces. This behavior requires the additional model that was given by Cassie and Baxter. Here, the surface tension dominates the system and it is energetically favorable for the liquid to form a droplet that rests on the tips of the surface while air fills the space between solid and liquid, forming air pockets. Consequently, only a small fraction of the solid surface is in contact with the liquid. The Cassie–Baxter contact angle can be calculated according to

$$\cos\theta_{CB} = -1 + f \left(1 + \cos\theta_0\right) \tag{2}$$

with $f = \phi/\gamma$; being γ the surface tension and ϕ the fraction of the solid surface which is in contact with the liquid. Therefore, for a given surface composition and liquid, i.e., fixing θ, the nature of the texture (which is determined by r and ϕ) determines the wetting behavior.

There are a large number of published works concerning the functionalization of surfaces to attain special characteristics of wettability for various environmental engineering applications such as self-cleaning, oil–water separation, anti-icing, anti-fouling, corrosion resistance, and food packaging, among others; see for example the review of Barati et al., and references therein [20]. The solutions proposed in these works imitate the hydrophobic structures existing in nature such as lotus leafs, rose petals—in plantae kingdom—and those of arthropod phylum such as structured wings, antifogging and anti-reflective eyes, and structures allowing a shift over water surfaces [21–23]. In all these cases, the hierarchical roughness of biological systems, with micro and nanoscale structures, seems to play a key role in these special properties; so different techniques have been used to emulate these multiscale textures. In the last decade, laser texturing has been widely used [24].

Laser texturing consists of making geometric structures by means of laser ablation processes to change at the relief of the surface to meet a specific functional requirement. The main advantages of laser texturing are: (1) the ability to produce structures on surface areas from microscale to macroscale; (2) the capability of processing non-planar surfaces; (3) a single-step process performed under normal ambient conditions; (4) an eco-friendly process because no chemicals or toxic waste are added to the environment and (5) the ability to process nearly all types of materials. There are several reported works on nano/micro-texturing of different materials such as the improvement of cell adhesion in metallic bio-implants or tribological properties of mechanical components, self-cleaning, corrosion protection, and other applications [25,26]. The lasers used were short-pulse lasers [27–32] as well as ultra-short pulses [31,33–36]. However, the application of laser texturing to modify the wettability of stones is scarce or even nonexistent [37,38].

The aim of this work was to analyze the feasibility of laser texturing as a new and environmentally friendly process, as an alternative to modify the nature of the surface of marble to provide hydrophobic properties without appreciable changes to its appearance. For this purpose, samples of a commercial marble were structured by means of an ultra-short-pulse laser under different irradiation parameters. Two kinds of texturing patterns consisting of matrices of holes and groove arrays were tested, and the most adequate conditions to obtain hydrophobic surfaces were analyzed in detail. The hydrophobicity of the processed marble was assessed by contact angle measurements and the surfaces were also characterized in terms of roughness, color, and gloss to know the changes caused by the texturing process. The results indicate the ability of laser texturing to increase the hydrophobicity of marble so that, after properly selecting the process parameters, changes in the appearance of the surface are minimal.

2. Materials and Methods

2.1. Stone Characterization

For this study, a marble commercially known as Crystal White, from S.R. of Vietnam, was used. Its highly purity on calcite, its homogeneous white color, its equigranular texture, the barely existence of accessory minerals and the absence of veins make this marble very suitable for the purpose of this work. Polished slabs of dimensions $10\,cm \times 5\,cm \times 2\,cm$ (length, width and height) were prepared. Geochemical, petrographic and mineralogical characterization of the samples was performed. Chemical composition (whole-rock major and trace elements) was obtained by means Inductively Coupled Plasma-Mass Spectrometry (ICP-MS) using an ICP-MS AGILENT 7800. For this analysis, 0.1 g of sample powder is digested with HNO_3 + HF under pressure in high pressure vessels in Milestone Microwave. Mineralogical composition was obtained by X-ray powder diffraction analysis (Bruker D8 Advance) of bulk rock with CuKa radiation and a velocity of 2θ/min. The equipment used is a Siemens D-500 (40 kV and 30 mA) diffractometer using CuKa radiation ($\lambda = 0.154\,37\,nm$) Ni-filter, equipped with the Diffract ATV3 software package. A complete petrographic examination was done following the standard ASTM [39], to describe the mineralogy and textures. A Leica DM2500P microscope under transmitted light was used for this purpose.

2.2. Laser Processing

The laser used was the Spirit system from Spectra Physics with emission wavelength 1040 nm and pulse width < 400 fs. The intensity profile at the laser output was near-Gaussian ($M^2 < 1.2$) and the beam diameter at the exit of the laser head was 1.5 mm. The laser beam presents horizontal polarization (> 100:1). Pulse rate can be selected from single shot to 1 MHz, with maximum pulse energy of 40 µJ at 100 kHz. The maximum mean power output is > 4 W. A two mirror galvanometric scanner (Raylase Superscan

III-15) was used and scanned the laser beam in X–Y directions. The beam was focused by means of a F-theta objective lens, 160 mm focal length, up to a diameter of 30 µm. At the working plane, the beam polarization is parallel to Y direction.

Laser surface treatments were performed in air at atmospheric pressure and two kind of texturing patterns were essayed: matrix of holes and arrays of parallel groves. Hole patterns were obtained through successive laser shoots delivered at the same point of the surface; grooves were obtained by overlapping holes as the laser beam is scanned along X or Y directions.

2.3. Surface Characterization

After laser processing, the samples were analyzed by scanning electron microscopy SEM using a Philips XL30 working in Secondary electrons (SE) mode to obtain a qualitative characterization of the surface. For each sample, 2 cm × 2 cm fragments were carbon coated and visualized with SEM. The optimum conditions of observation were obtained at an accelerating potential of 15–20 kV, a working distance of 9–11 mm and specimen current of $\approx$ 60 mA.

Then, topographical data were acquired with an optical imaging profiler (Sensofar-PLu 2300) and different objectives were used: a 20× EPI objective, with a field of view of 637 µm × 477 µm and a pixel size of 0.83 µm, and a 100× EPI objective with a field of view of 124 µm × 92 µm and a pixel size of 0.17 µm. Data processing, analysis and visualization were implemented by using Scientific Python. After that, the following roughness parameters were obtained in accordance with standard ISO 25178 [40], related to the analysis of 3D areal surface texture: (1) S_a, mean roughness, is the arithmetic mean of the absolute value of the height from the mean plane of the surface; (2) S_{dr}, developed interfacial area ratio, is a hybrid parameter expressed as the percentage of the definition area's additional surface area contributed by the texture as compared to the planar definition area. This parameter is directly related to the parameter r in Wentzel's equation, owing that $r = 1 + S_{dr}$.

To evaluate the wettability of the textured surfaces, static contact angles were measured with a Phoenix-300 equipped with an automatic dosification system and a 6.5× zoom CCD camera. For each tested surface (included the reference marble), static contact angle of three 4 µL droplet were measured, with a precision of ±0.1°, following sessile drop method EN 828:2013 [41].

Color modifications related to texturing of the marble were evaluated. For this, color characterization before and after texturing was obtained by means of a Minolta CM-700D spectrophotometer equipped with SpectraMagicTM NX software. Color was expressed in the CIELAB and CIELCH spaces [42]. A total of 9 measurements were made for each surface to obtain statically representative results. The measurements were made in Specular Component Included (SCI) mode with a spot diameter of 8 mm, D65 illuminant and an observer angle of 10°. The parameters measured were L^*, the lightness, varying from black (0) to white (100); a^*, which varies from +60 (red) to −60 (green), and b^*, which ranges from +60 (yellow) to −60 (blue). Also, C_{ab}^*, chroma or saturation, and H_{ab}, hue, were measured. Color of the non-textured areas was considered to be the reference to calculate color differences ΔL^*, Δa^*, Δb^*, ΔCab^* and ΔH^* and the global color change, ΔE_{ab}^* ($\Delta E_{ab}^* = \sqrt{(\Delta L^*)^2 + (\Delta a^*)^2 + (\Delta b^*)^2}$).

A glossmeter (Novo Gloss MultiGauge Lite from Neurtek Instruments) was used to obtain the gloss before and after texturing of the marble. Three measurements were performed for each tested surface with a reflection angle of 60°.

3. Results and Discussion

3.1. Surface Characterization

The studied marble has a very similar composition as other marbles (e.g., Carrara marble), [43] very important in cultural heritage and the possible implications of our results in terms of conservation-restoration of the architectural heritage. Crystal White marble has a 51.91% of Ca (expressed as CaO) and a loss on ignition of 42.60%. From ICP-MS data. The largest difference observed with Carrara marble is related to the content of dolomite, as the Crystal White marble is slightly more enriched in this phase than others [43]. This is noticed in the X-ray powder diffraction analyses, where traces of dolomite (magnesium carbonate) have been noticed (Figure 1). In the case of the depth, the most significant variable (very small pValue) was the laser power; regarding width, the most significant one was the intercept of the linear fit, which is directly related to the beam spot. The next significant contribution to the width was again the laser power.

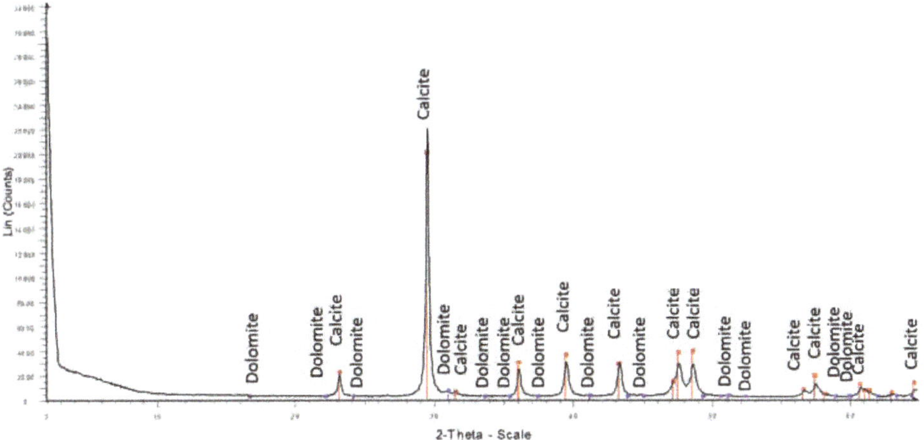

Figure 1. X-ray powder diffraction spectra of the marble selected.

Under petrographic microscopy, the purity of the marble is confirmed as well as the mosaic, granoblastic texture, where individual crystals of calcite meet others with somehow indented boundaries (Figure 2).

3.2. Laser Parameter Analysis

An analysis of the laser processing parameters was performed. The objective was to select the most adequate values to achieve texturing structures that allow the formation of air pockets between the marble surface and the liquid drops, as the Cassie–Baxter model states. The analysis was carried out by means of a multiple linear regression model, which allowed us to determine the sensitivity of the sizes (width and depth) of the structures generated to each of the laser parameters; i.e., pulse frequency, laser power, scanning speed and number of shots (or time of laser shooting).

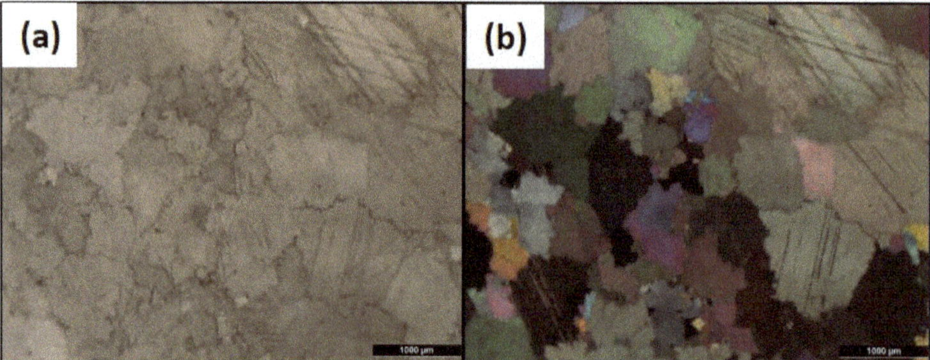

Figure 2. Petrographic micrographs of the Crystal White marble. (**a**) parallel-polarized light, PPL. (**b**) cross-polarized light, CPL.

Owing that the femtosecond laser beam has a Gaussian intensity profile, the profiles of the laser-machined micro-holes will be also like a Gaussian curve. Figure 3 depicts topographic image of a crater and a groove made in the marble surface. Dimensions, depth and width, of these structures were obtained by fitting to a 2D and a 1D Gaussian functions, respectively, the crater and the X-axis projection of the groove. Results of the regression analysis, performed by using the dimensions sizes are summarized in Table 1. Regarding depth, the laser power is the most significant variable for both grooves and holes. In case of width, the most relevant contribution is the intercept of the linear fit whose value is near the beam spot size. The next significant variable is again the laser power.

Table 1. Results of Multiple Linear Regression fit for depth and width. SE is the standard error for the coefficient estimates. pValue is the null hypothesis probability obtained from Student's t distribution. Variables with higher significant result (very small pValue) are in bold.

GROOVES				**HOLES**			
Depth = 1 + Freq + Pow + Speed				Depth = 1 + Freq + Pow + Time			
	Estimate	SE	pValue		Estimate	SE	pValue
(Intercept)	8.2	2.8	1.09×10^{-2}	(Intercept)	8.5	5.7	1.49×10^{-1}
Freq	-0.096	0.030	5.82×10^{-3}	Freq	-0.210	0.063	2.26×10^{-3}
Pow	0.22	0.02	8.21×10^{-9}	**Pow**	0.381	0.053	3.95×10^{-8}
Speed	0.16	0.07	4.49×10^{-2}	Time	-0.00016	0.00090	0.85881
Width = 1 + Freq + Pow + Speed				Width = 1 + Freq + Pow + Time			
	Estimate	SE	pValue		Estimate	SE	pValue
(Intercept)	24.1	3.0	4.40×10^{-7}	**(Intercept)**	27.1	2.0	2.55×10^{-13}
Freq	-0.073	0.031	3.36×10^{-2}	Freq	-0.003	0.016	8.75×10^{-1}
Pow	0.137	0.021	7.48×10^{-6}	**Pow**	0.081	0.021	5.56×10^{-4}
Speed	0.071	0.075	3.58×10^{-1}	Time	-0.000242	0.000297	4.23×10^{-1}

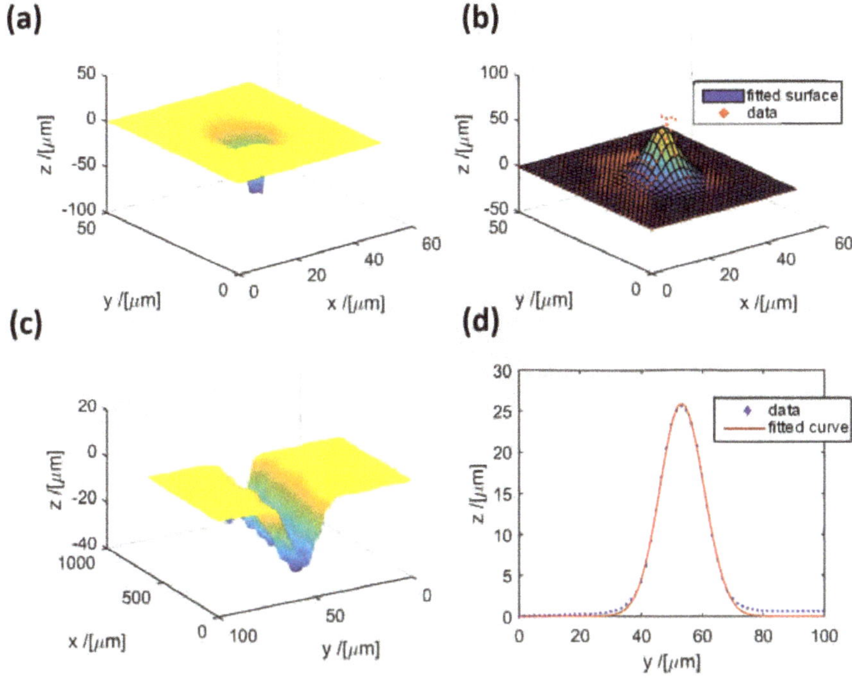

Figure 3. Confocal topographies and fitting results to obtain depth and width dimensions. (**a**) hole topography and (**b**) 2D Gaussian fit. (**c**) groove topography and (**d**) and 1D Gaussian fit.

3.3. Surface Morphology Characterization

The analysis above gave the laser power, P, as the key parameter to control both width and depth of the laser-generated structures. Therefore, in order to investigate the effect of laser power on the surface topography, the marble surface was texturized at a pulse frequency of 50 kHz and laser power ranging from the 100% maximum power, 4 W, to the 40% maximum power, 1.6 W. In the case of textured patterns consisting of a matrix of holes, the time was fixed at 5 ms and, in case of grooves, the scan speed 25 mm s^{-1}. Taking into account the size of the laser spot, $\approx 40\,\mu$m, to obtain different textures the pitch (distance between lines or holes) was varied to cover a wide range of structures; from completely separate ones to overlapping. Table 2 shows the texturing parameters used for each configuration (matrix of grooves and matrix of holes).

Table 2. Texturing parameters.

Grooves	Holes
Frequency: 50 kHz	Frequency: 50 kHz
Speed: 25 mm s^{-1}	Time: 5 ms (250 shots)
Power: 100% (4 W), 80%, 60%, 40%	Power: 100% (4 W), 80%, 60%, 40%
Pitch: 20, 30, 40, 50, 60 µm	Pitch: 20, 30, 40, 50, 60 µm

Figure 4 shows SEM micrographs of the laser textured marble at $P = 4\,\text{W}$, pitch = 60 µm and $P = 4\,\text{W}$, pitch = 20 µm; performed with grooves and holes. As can be observed, two levels of texture can be differentiated: the form induced by the pattern of grooves or holes (Figure 4a,c,e,g) and, superimposed to this, the roughness that can be seen at higher magnification (Figure 4b,d,f,h). This hierarchical structure, is considered to be a key factor for improving surface hydrophobicity, and has been observed by others researchers in both nanosecond and femtosecond laser texturing processes [27,30,31,44,45].

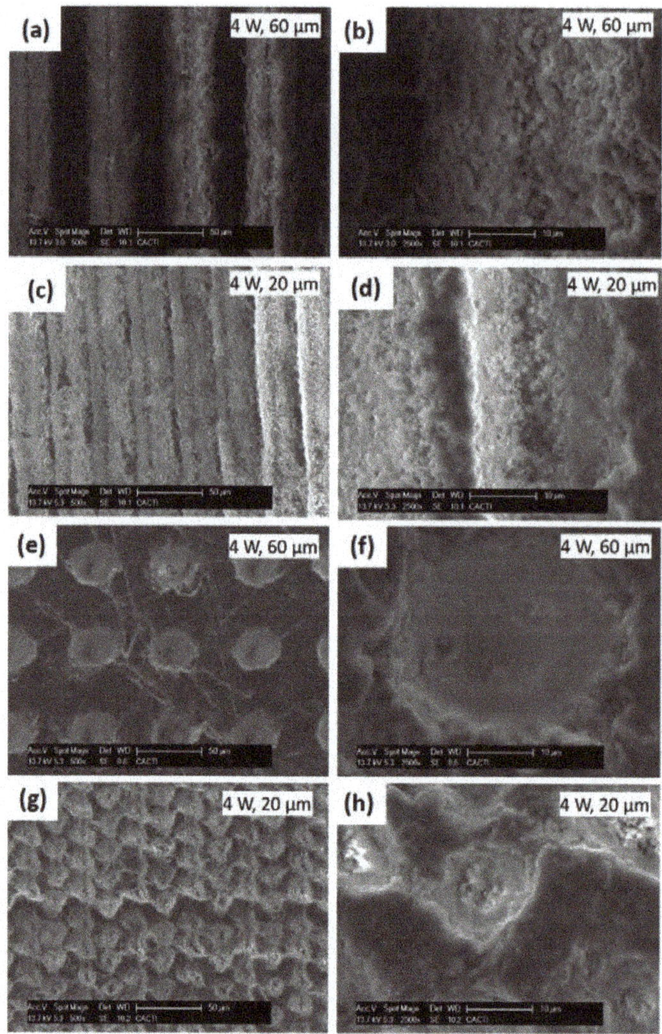

Figure 4. SEM micrographs of some of the laser texturized surfaces treated under different conditions of power and pitch. (**a–d**): treated with grooves. (**e–h**): treated with holes.

The reconstruction of the surface topography obtained with confocal microscope is shown in Figure 5a,b and transverse profiles of the structures induced, with different pitch, are shown in Figure 5c,d. It can be observed how the texture pattern diffuses with the overlapping, i.e.; decreasing pitch, especially in case of the groove patterns (Figure 5c).

From topographic data, the dimensions of the laser induced structures could be obtained. Figure 6 depicts the values of depth and width which correspond to different values of the processing power. It can be seen that separation is in the order of the laser spot (40 µm), that is the overlap becomes appreciable, the depth of the grooves decreases with the pitch. In the case of the hole patterns, similar trend is observed, although the decrease in depth is not as clear as in the case of the grooves, because the overlap in these structures is lower. Besides, the values show higher dispersion, due to the difficulties to obtain high quality data at the bottom of the holes. Regarding width, values remain practically constant but, again, showing large dispersion, especially in the hole pattern textured with the lowest power, 1.6 W.

Results of the surface roughness parameters S_a and S_{dr} are shown in Figure 7. In the case of the grooves, mean roughness, S_a, seems to increase as the pitch increases, but in the case of holes the opposite occurs; S_a increases as the pitch decreases. This behavior of S_a can be attributed to the increment of the texturing density, while maintaining low overlapping in the generated structures.

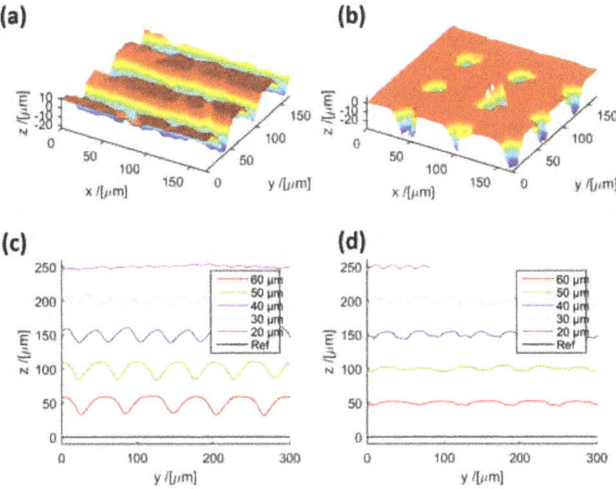

Figure 5. Confocal topographies and profiles of laser texturized surfaces treated under with pitch of 60 µm. (**a**,**c**): treated with grooves. (**b**,**d**): treated with holes.

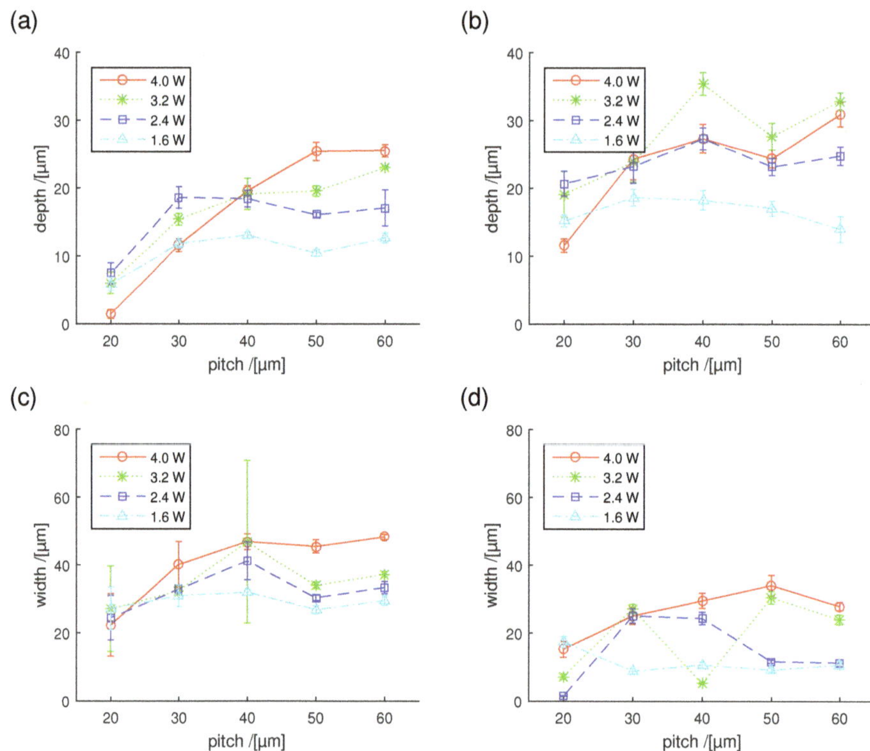

Figure 6. Depth and width versus pitch. (**a**,**c**): treated with grooves. (**b**,**d**): treated with holes.

Regarding S_{dr}, developed interfacial area ratio, this is a more significant parameter from the point of view of the surface wettability. The values obtained, especially in cases of high overlapping, are above the expected ones if only the contribution of the textured density and the overlap, both microscale effects, were considered. These high values of S_{dr} could be attributed to the contribution of the sub-micro scale roughness, which was observed in SEM images of Figure 4. This contribution would increase considerably the textured area with respect to the projected area, thus increasing the S_{dr} values.

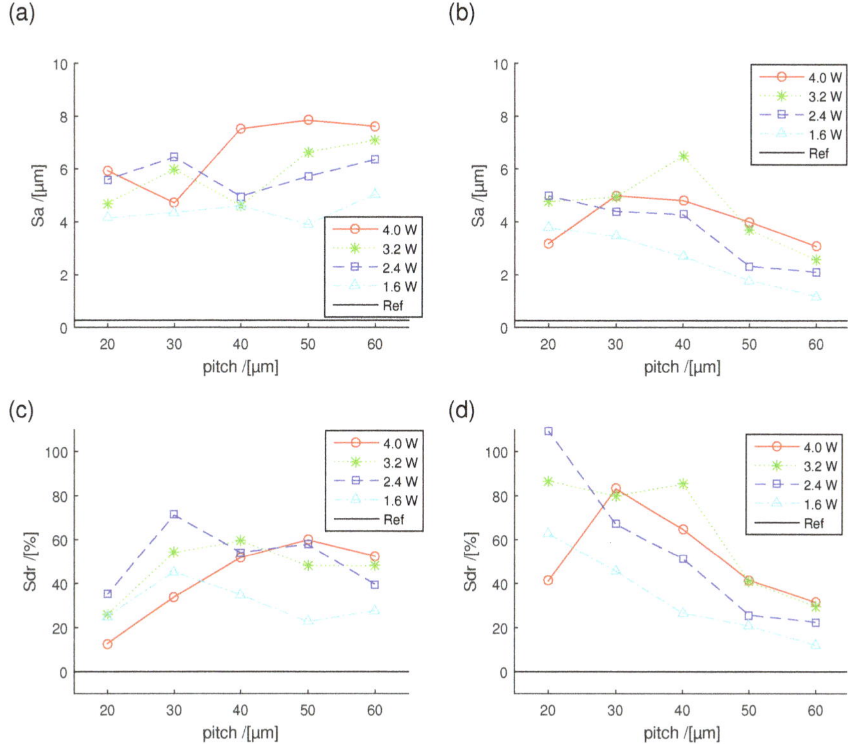

Figure 7. Roughness surface parameters S_a and S_{dr} versus pitch. (**a**,**c**): treated with grooves. (**b**,**d**): treated with holes.

3.4. Wettability

The effect of laser power and pitch on marble wettability was analyzed through the static contact angle, θ. Figure 8 shows the contact angle of the textured surfaces as a function of the pitch, for each value if the laser power used. Gray band corresponds to the static contact angle of the reference marble whose value $\theta_0 = 66.3 \pm 5.8°$ confirms the hydrophilic behavior of this rock. In general terms, after the laser processing, it is observed a significant increase of the contact angles, in most cases well above 90°; that is, the textured marble acquires a clearly hydrophobic character.

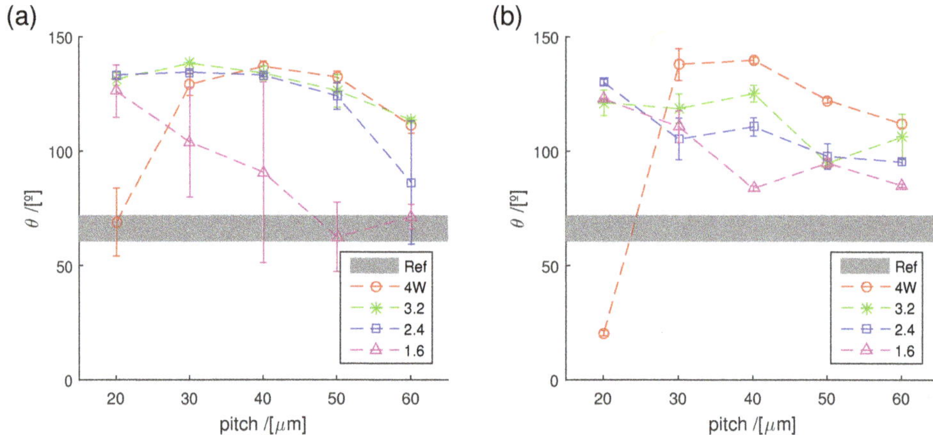

Figure 8. Contact angle of textured surfaces. (**a**): grooves. (**b**): holes.

Focusing on the treatment with grooves, for all the values of the laser power applied (except 1.6 W), the highest contact angles correspond to pitch values in the range 30 µm–50 µm, which correspond to a separation between lines or holes in the order of the laser spot diameter (40 µm). Regarding the case of the lower power used, 1.6 W, the contact angle decreases as the pitch increases; but, it should be taking into account the large dispersion of the data. From Table 3, it can be seen that the power at which the highest contact angle is obtained, corresponds to 3.2 W (30 µm), being $134 \pm 5.5°$ the average of the maximum contact angle values obtained with the four laser power used.

Table 3. Power and pitch values, for each texturing pattern (grooves and holes), which give rise to the maximum value of contact angle θ; in addition, the value of ΔE_{ab}^* obtained under each condition is indicated.

	Grooves			Holes		
Power	**Pitch**	θ	ΔE_{ab}^*	**Pitch**	θ	ΔE_{ab}^*
4.00	40	137.14	3.50	40	139.79	5.05
3.20	30	138.38	4.31	40	125.18	6.17
2.40	30	134.49	7.91	20	130.18	3.63
2.40	20	133.36	4.53			
1.60	20	126.28	4.16	20	122.90	7.62

With regards of the treatment by holes, in case of $P = 4\,\mathrm{W}$ and $P = 3.2\,\mathrm{W}$ the behavior is similar to the case of grooves, and the highest values of contact angle were obtained in the same pitch range 30 µm–40 µm. However, as can be seen, at $P = 2.4\,\mathrm{W}$ and $P = 1.6\,\mathrm{W}$, the contact angle decreases as the pitch increases so that the highest values were obtained at the lowest pitch. As a conclusion, the power that obtains the highest contact angle was 4 W (40 µm) and the average value of the maximum contact angles obtained for all the powers tested is $129.0 \pm 7.4°$, thus a larger dispersion than in the pattern of grooves was obtained.

3.5. Color and Gloss Changes

Regarding color variations caused by laser texturing, the lightness, L^*, was the most affected parameter, showing increases, in all the textured surfaces, in the range 0.2–7.8 CIELAB units. The color coordinates a^* and b^* did not show notable changes: from -0.16 to 3 in Δa and from 0.6 to 1.7 in Δb. Therefore, nor C_{ab}^* and H^* experimented noticeable changes. Then, global color change (ΔE_{ab}^*) was directly related to L^* variations. In Figure 9, the global color change, ΔE_{ab}^* as a function of the textured pitch, at different values of the laser power are depicted. Moreover, in Table 3, ΔE_{ab}^* values that correspond to the textured patterns which gave the highest contact angles are indicated.

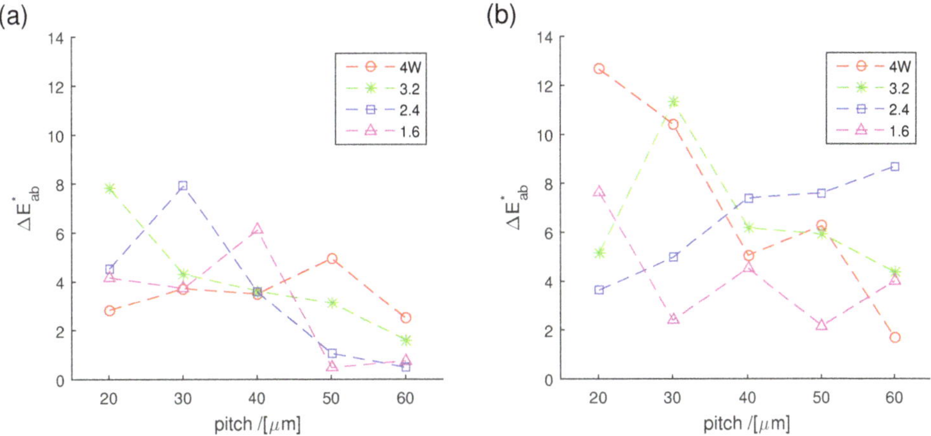

Figure 9. Color difference ΔE_{ab}^* of textured surfaces. (**a**): grooves. (**b**): holes.

Besides, Figure 9 show that neither for groove-type texturing nor for holes-type texturing a clear relationship with pitch can be established; although for groove-type texturing, the lowest values of ΔE_{ab}^* were obtained at the highest pitch, 60 µm. This was an expected result, since the higher the pitch, the lower the textured surface with respect to the original one and, consequently, the potential impact on its color. It is also found that for 60 µm pitch, ΔE_{ab}^* increases with the laser power.

From the literature, it is considered that a global color change ΔE_{ab}^* lower than 3.5 units cannot be observed by an unexperienced human eye [46]. However, it is necessary to interpret color results also from a conservation point of view. In this sense, the rate proposed in [47] was considered. This rate establishes three levels of risk of incompatibility: low risk ($\Delta E_{ab}^* < 3$), medium risk ($3 < \Delta E_{ab}^* < 5$) and high risk ($\Delta E_{ab}^* > 5$).

From the values in Table 3 it can be seen that in surfaces textured with a pattern of holes which lead to the maximum value of contact angle, the color change is higher than in case of grooves; being in all the cases but one, $\Delta E_{ab}^* > 5$ units. Conversely, in grooves-type texturing, ΔE_{ab}^* is always below 5 (this change implies a low to medium risk of incompatibility following [47]), except in the case of 2.4 W; in this case, notice that using a somewhat smaller pitch (20 µm) it was obtained a surface with contact angle just 1° lower and a global color change ΔE_{ab}^* below 5.

Figure 10 shows the gloss values after texturing with grooves and holes applying different laser power and pitch. The value of gloss for the reference surface (gray band in the figure) is 33.25 ± 1.4; this value reflects the polishing finish of the sample. The texturing with both grooves and holes reduces the gloss.

In addition, it is observed, as expected, that the higher the distance between holes or grooves the lower the reduction on the gloss. Furthermore, it can be seen that the laser power applied has a remarkable effect on the gloss; the greater the applied power, the lower the gloss of the textured surfaces.

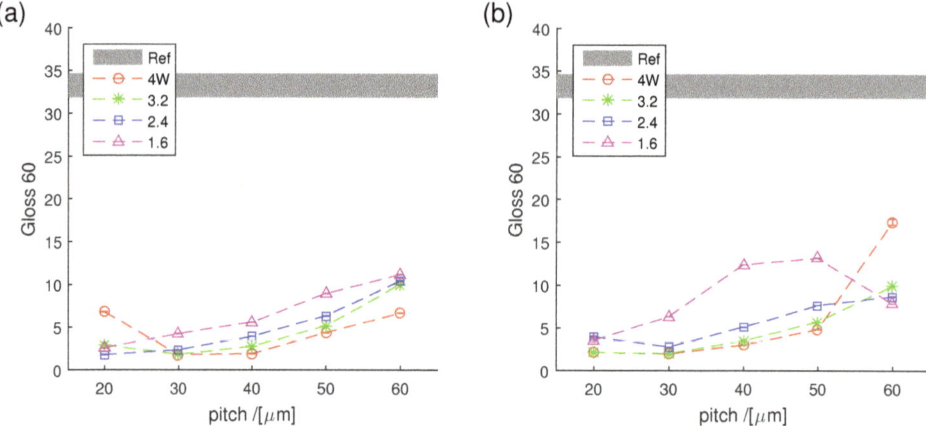

Figure 10. Gloss values of textured surfaces. (**a**): grooves. (**b**): holes.

4. Conclusions

In this work, preliminary results are presented of laser texturing of a marble, a rock widely used in cultural heritage, with hydrophilic character ($\theta = 66.3 \pm 5.8°$), in order to change its wettability characteristics towards an hydrophobic behavior. Surface treatments were performed using a ultra-short-pulse laser and, considering the Cassie–Baxter model of wettability, first an analysis of the irradiation parameters was accomplished to achieve texturing structures that allow the formation of air pockets between the marble surface and the water drops. Therefore, different texturing patterns consisting of matrices of holes and groove arrays were essayed and, from the analysis of irradiation parameters, the laser power has been the most influential variable on the topography of the textured surface, together with the spacing between lines or holes, and pitch, used to fill the surface.

Wettability characteristics of the treated surfaces were measured trough the static contact angle and the results obtained have demonstrated the feasibility of the laser texturing process to modify the marble surface towards a hydrophobic character (contact angle $\theta > 90°$) with, in some cases, values of contact angles close to a super-hydrophobic behavior ($\theta > 150°$). The modification of the wettability depends, however, on the type of structures generated, grooves or holes, the laser power used and the pitch of the texturing patterns. Characterization of areal roughness in terms of the mean roughness, S_a and, especially, the developed interfacial area ratio, S_{dr}, allowed the interpretation of the behavior of the contact angles as a function of the pitch and the power for different texturing patterns. Furthermore, the structured surfaces were evaluated in terms of color parameters and gloss, to detect changes in the appearance of the rock surface that could compromise the aesthetics and, therefore, the applicability of this method in the field of cultural heritage.

The treatment with grooves initially seemed to be the most effective. In this case, values of the contact angle were slightly higher than those obtained with the pattern of holes, and the greatest ones corresponded to pitch in the range 30 µm–50 µm, and they were similar for all the values of the laser power applied ($134 \pm 5.5°$ as the average value). Furthermore, the color changes associated with these texturing conditions were lower than those obtained with the pattern of holes, and they are below 5 CIELAB units. This value would be associated, according to [47], with a low to medium risk of incompatibility. Nevertheless, according to those authors, the value of the change in any particular property (such as color) of a material used in cultural heritage (such a rock, as was the case) must be interpreted jointly with data on changes in other relevant properties and always in the context of the external conditions (climate, pollution, previous treatments . . .) to which this stone will be exposed. Regarding gloss, as was expected, suffered modifications caused by the laser texturing; in this sense, the higher the textured density or the applied power, the lower the surface gloss.

As a conclusion, in this work, we have demonstrated the ability of laser texturing to modify the wettability of marble so that, after properly selecting the process parameters, changes in the appearance of the surface are minimal. One of the next challenges in our research will be the laser texturing of samples with primary cutting surfaces (cut diamond-wire from quarry) or primary sizing (first cut in processing plants of commercial products). This would allow us to incorporate into the study the economic value of laser texturing as an environmentally sustainable alternative stone finishing, with extra properties—hydrophobicity—that would avoid the need to use water-repellent chemicals once the stone is placed on the building.

Author Contributions: The authors confirm contribution to the paper as follows: Conceptualization, A.J.L. and A.R.; methodology, A.J.L., T.R. and D.P.; investigation, A.R., J.S.P.-A., T.R. and D.P.; formal analysis, A.R. and J.S.P.-A.; writing—original draft preparation, A.J.L.; writing—review and editing, A.J.L., A.R., J.S.P.-A., T.R. and D.P.; project administration, A.R.

Funding: This research was funded by the Spanish Ministerio de Economía y Competitividad under grant number BIA2017-85897-R, and the University of Salamanca own financial program.

Conflicts of Interest: The authors declare no conflict of interest.

References

1. Warscheid, T.; Braams, J. Biodeterioration of stone: A review. *Int. Biodeterior. Biodegrad.* **2000**, *46*, 343–368. [CrossRef]
2. Sadat-Shojai, M.; Ershad-Langroudi, A. Polymeric coatings for protection of historic monuments: Opportunities and challenges. *J. Appl. Polym. Sci.* **2019**, *112*, 2535–2551. [CrossRef]
3. Camaiti, M.; Brizi, L.; Bortolotti, V.; Papacchini, A.; Salvini, A.; Fantazzini, P. An Environmental Friendly Fluorinated Oligoamide for Producing Nonwetting Coatings with High Performance on Porous Surfaces. *ACS Appl. Mater. Interfaces* **2017**, *9*, 37279–37288. [CrossRef] [PubMed]
4. Frigione, M.; Lettieri, M.; Frigione, M.; Lettieri, M. Novel Attribute of Organic–Inorganic Hybrid Coatings for Protection and Preservation of Materials (Stone and Wood) Belonging to Cultural Heritage. *Coatings* **2018**, *8*, 319. [CrossRef]
5. Becerra, J.; Mateo, M.; Ortiz, P.; Nicolás, G.; Zaderenko, A.P. Evaluation of the applicability of nano-biocide treatments on limestones used in cultural heritage. *J. Cult. Herit.* **2019**, *38*, 126–135. [CrossRef]
6. Raneri, S.; Barone, G.; Mazzoleni, P.; Alfieri, I.; Bergamonti, L.; De Kock, T.; Cnudde, V.; Lottici, P.P.; Lorenzi, A.; Predieri, G.; et al. Efficiency assessment of hybrid coatings for natural building stones: Advanced and multi-scale laboratory investigation. *Constr. Build. Mater.* **2018**, *180*, 412–424. [CrossRef]
7. Zarzuela, R.; Carbú, M.; Gil, M.A.; Cantoral, J.M.; Mosquera, M.J. CuO/SiO$_2$ nanocomposites: A multifunctional coating for application on building stone. *Mater. Des.* **2017**, *114*, 364–372. [CrossRef]
8. Quagliarini, E.; Graziani, L.; Diso, D.; Licciulli, A.; D'Orazio, M. Is nano-TiO$_2$ alone an effective strategy for the maintenance of stones in Cultural Heritage? *J. Cult. Herit.* **2018**, *30*, 81–91. [CrossRef]
9. Aldoasri, M.; Darwish, S.; Adam, M.; Elmarzugi, N.; Ahmed, S.; Aldoasri, M.A.; Darwish, S.S.; Adam, M.A.; Elmarzugi, N.A.; Ahmed, S.M. Protecting of Marble Stone Facades of Historic Buildings Using Multifunctional TiO$_2$ Nanocoatings. *Sustainability* **2017**, *9*, 2002. [CrossRef]

10. Aldoasri, M.; Darwish, S.; Adam, M.; Elmarzugi, N.; Ahmed, S.; Aldoasri, M.A.; Darwish, S.S.; Adam, M.A.; Elmarzugi, N.A.; Ahmed, S.M. Enhancing the Durability of Calcareous Stone Monuments of Ancient Egypt Using $CaCO_3$ Nanoparticles. *Sustainability* **2017**, *9*, 1392. [CrossRef]

11. Böke, H.; Hale Göktürk, E.; Caner Saltık, E.N. Effect of some surfactants on SO_2–marble reaction. *Mater. Lett.* **2002**, *57*, 935–939. [CrossRef]

12. Matziaris, K.; Panayiotou, C. Tunable wettability on Pendelic marble: Could an inorganic marble surface behave as a "self-cleaning" biological surface? *J. Mater. Sci.* **2014**, *49*, 1931–1946. [CrossRef]

13. Guiamet, P.; Crespo, M.; Lavin, P.; Ponce, B.; Gaylarde, C.; de Saravia, S.G. Biodeterioration of funeral sculptures in La Recoleta Cemetery, Buenos Aires, Argentina: Pre- and post-intervention studies. *Colloids Surfaces B Biointerfaces* **2013**, *101*, 337–342. [CrossRef] [PubMed]

14. Tsakalof, A.; Manoudis, P.; Karapanagiotis, I.; Chryssoulakis, I.; Panayiotou, C. Assessment of synthetic polymeric coatings for the protection and preservation of stone monuments. *J. Cult. Herit.* **2007**, *8*, 69–72. [CrossRef]

15. Cámara, B.; de Buergo, M.Á.; Bethencourt, M.; Fernández-Montblanc, T.; La Russa, M.F.; Ricca, M.; Fort, R. Biodeterioration of marble in an underwater environment. *Sci. Total Environ.* **2017**, *609*, 109–122. [CrossRef] [PubMed]

16. Kronlund, D.; Lindén, M.; Smått, J.H. A sprayable protective coating for marble with water-repellent and anti-graffiti properties. *Prog. Org. Coat.* **2016**, *101*, 359–366. [CrossRef]

17. Bico, J.; Thiele, U.; Quéré, D. Wetting of textured surfaces. *Colloids Surfaces A Physicochem. Eng. Asp.* **2002**, *206*, 41–46. [CrossRef]

18. de Gennes, P.G. Wetting: Statics and dynamics. *Rev. Mod. Phys.* **1985**, *57*, 827–863. [CrossRef]

19. Bico, J.; Marzolin, C.; Quéré, D. Pearl drops. *Europhys. Lett. (EPL)* **1999**, *47*, 220–226. [CrossRef]

20. Barati Darband, G.; Aliofkhazraei, M.; Khorsand, S.; Sokhanvar, S.; Kaboli, A. Science and Engineering of Superhydrophobic Surfaces: Review of Corrosion Resistance, Chemical and Mechanical Stability. *Arab. J. Chem.* **2018**. [CrossRef]

21. Darmanin, T.; Guittard, F. Superhydrophobic and superoleophobic properties in nature. *Mater. Today* **2015**, *18*, 273–285. [CrossRef]

22. Zhang, M.; Feng, S.; Wang, L.; Zheng, Y. Lotus effect in wetting and self-cleaning. *Biotribology* **2016**, *5*, 31–43. [CrossRef]

23. Fadeeva, E.; Chichkov, B.; Fadeeva, E.; Chichkov, B. Biomimetic Liquid-Repellent Surfaces by Ultrafast Laser Processing. *Appl. Sci.* **2018**, *8*, 1424. [CrossRef]

24. Müller, F.; Kunz, C.; Gräf, S.; Müller, F.A.; Kunz, C.; Gräf, S. Bio-Inspired Functional Surfaces Based on Laser-Induced Periodic Surface Structures. *Materials* **2016**, *9*, 476. [CrossRef] [PubMed]

25. Lutey, A.H.A.; Gemini, L.; Romoli, L.; Lazzini, G.; Fuso, F.; Faucon, M.; Kling, R. Towards Laser-Textured Antibacterial Surfaces. *Sci. Rep.* **2018**, *8*, 10112. [CrossRef] [PubMed]

26. Wahab, J.A.; Ghazali, M.J.; Yusoff, W.M.; Sajuri, Z. Enhancing material performance through laser surface texturing: A review. *Trans. Inst. Met. Finish.* **2016**, *94*, 193–198. [CrossRef]

27. Fiorucci, M.P.; López, A.J.; Ramil, A. Comparative study of surface structuring of biometals by UV nanosecond Nd:YVO4 laser. *Int. J. Adv. Manuf. Technol.* **2014**, *75*, 515–521. [CrossRef]

28. Lutey, A.H.A.; Romoli, L. Pulsed laser ablation for enhanced liquid spreading. *Surf. Coat. Technol.* **2019**, *360*, 358–368. [CrossRef]

29. Cai, Y.; Chang, W.; Luo, X.; Sousa, A.M.; Lau, K.H.A.; Qin, Y. Superhydrophobic structures on 316L stainless steel surfaces machined by nanosecond pulsed laser. *Precis. Eng.* **2018**, *52*, 266–275. [CrossRef]

30. Jagdheesh, R.; García-Ballesteros, J.J.; Ocaña, J.L. One-step fabrication of near superhydrophobic aluminum surface by nanosecond laser ablation. *Appl. Surf. Sci.* **2016**, *374*, 2–11. [CrossRef]

31. Huerta-Murillo, D.; García-Girón, A.; Romano, J.M.; Cardoso, J.T.; Cordovilla, F.; Walker, M.; Dimov, S.S.; Ocaña, J.L. Wettability modification of laser-fabricated hierarchical surface structures in Ti-6Al-4V titanium alloy. *Appl. Surf. Sci.* **2019**, *463*, 838–846. [CrossRef]

32. Riveiro, A.; Soto, R.; Del Val, J.; Comesaña, R.; Boutinguiza, M.; Quintero, F.; Lusquiños, F.; Pou, J. Texturing of polypropylene (PP) with nanosecond lasers. *Appl. Surf. Sci.* **2016**, *374*, 379–386. [CrossRef]

33. Arenas, M.; Ahuir-Torres, J.; García, I.; Carvajal, H.; de Damborenea, J. Tribological behaviour of laser textured Ti6Al4V alloy coated with MoS2 and graphene. *Tribol. Int.* **2018**, *128*, 240–247. [CrossRef]

34. Rukosuyev, M.V.; Lee, J.; Cho, S.J.; Lim, G.; Jun, M.B.G. One-step fabrication of superhydrophobic hierarchical structures by femtosecond laser ablation. *Appl. Surf. Sci.* **2014**, *313*, 411–417. [CrossRef]

35. Moradi, S.; Kamal, S.; Englezos, P.; Hatzikiriakos, S.G. Femtosecond laser irradiation of metallic surfaces: Effects of laser parameters on superhydrophobicity. *Nanotechnology* **2013**, *24*, 415302. [CrossRef] [PubMed]

36. Rajab, F.H.; Liauw, C.M.; Benson, P.S.; Li, L.; Whitehead, K.A. Picosecond laser treatment production of hierarchical structured stainless steel to reduce bacterial fouling. *Food Bioprod. Process.* **2018**, *109*, 29–40. [CrossRef]

37. Chantada, A.; Penide, J.; Pou, P.; Riveiro, A.; del Val, J.; Quintero, F.; Soto, R.; Lusquiños, F.; Pou, J. Laser surface texturing of granite. *Procedia Manuf.* **2017**, *13*, 687–693. [CrossRef]

38. Chantada, A.; Penide, J.; Riveiro, A.; del Val, J.; Quintero, F.; Meixus, M.; Soto, R.; Lusquiños, F.; Pou, J. Increasing the hydrophobicity degree of stonework by means of laser surface texturing: An application on Zimbabwe black granites. *Appl. Surf. Sci.* **2017**, *418*, 463–471. [CrossRef]

39. ASTM International. *ASTM C1721—15 Standard Guide for Petrographic Examination of Dimension Stone*; Technical Report; ASTM International: West Conshohocken, PA, USA, 2015. [CrossRef]

40. International Organisation of Standardization. *ISO 25178 Geometric Product Specifications (GPS)—Surface Texture: Areal*; International Organisation of Standardization: Geneva, Switzerland, 2010.

41. AENOR. *UNE-EN 828: 2013—Adhesives - Wettability - Determination by measurement of contact angle and surface free energy of solid surface*; Technical Report; AENOR: Madrid, Spain, 2013.

42. CIE S014-4/E. *2007 Colorimetry—Part 4: CIE 1976 L*a*b* Colour Space*; Technical Report, Commission Internationale de l'Eclairage; CIE Central Bureau: Vienna, Austria, 2007.

43. Herz, N.; Dean, N.E. Stable isotopes and archaeological geology: The Carrara marble, northern Italy. *Appl. Geochem.* **1986**, *1*, 139–151. [CrossRef]

44. Fiorucci, M.P.; López, A.J.; Ramil, A. Multi-scale characterization of topographic modifications on metallic biomaterials induced by nanosecond Nd:YVO 4 laser structuring. *Precis. Eng.* **2018**, *53*, 163–168. [CrossRef]

45. Zheng, B.; Jiang, G.; Wang, W.; Mei, X. Fabrication of superhydrophilic or superhydrophobic self-cleaning metal surfaces using picosecond laser pulses and chemical fluorination. *Radiat. Eff. Defects Solids* **2016**, *171*, 461–473. [CrossRef]

46. Mokrzycki, W.; Tatol, M. Colour difference δE—A survey. *Mach. Graph. Vis.* **2011**, *20*, 383–411.

47. Rodrigues, J.D.; Grossi, A. Indicators and ratings for the compatibility assessment of conservation actions. *J. Cult. Herit.* **2007**, *8*, 32–43. [CrossRef]

 *sustainability*

Article

The Use of Dolostone in Historical Buildings of Coimbra (Central Portugal)

Lidia Catarino [1,2,*] , Roque Figueiredo [2], Fernando Pedro Figueiredo [1,2], Pedro Andrade [1,2] and João Duarte [1]

1   Geosciences Center, University of Coimbra, 3030-074 Coimbra, Portugal
2   Department of Earth Sciences, University of Coimbra, 3030-074 Coimbra, Portugal
*   Correspondence: lidiagil@dct.uc.pt; Tel.: +351-239860530

Received: 20 June 2019; Accepted: 30 July 2019; Published: 1 August 2019

Abstract: In this paper, the importance of the dolostone (a carbonate sedimentary rock where the dominant carbonate mineral is dolomite) in monuments and urban buildings of the city of Coimbra, Portugal, is highlighted. Old quarries are not visible in the nucleus of the city, due to the sequential occupation by houses, and can only be identified by documentation (draws, contract letters and purchase orders). However, on the southern side of Mondego River (Santa Clara) some outcrops can be observed and were exploited until the mid-20th. It is presented a list of the old quarries and monuments made with this rock. The characterization of dolostones from the Coimbra Formation is also presented. It is made the connection between the local geology and the "identity" of Coimbra, putting in evidence the stone as a symbol that characterize the gilded aspect of the buildings. For restoration and rehabilitation works, a small number of blocks could be extracted from the Carvalhais quarry if were eventually necessary.

Keywords: dolostone; quarries; Coimbra Formation; heritage

---

1. Introduction

Ever since the beginning of humanity, stone has always been an essential and emblematic element, constituting one of the supports for the development of society, including its connection to the discovery and control of fire by Homo Erectus, as well as the first utensils used for hunting and gathering food. With the Neolithic Revolution humankind stopped the hunter-gatherer lifestyle adopting sedentary habits, giving rise to great technological advances, one of the most important aspects being the establishment of and in communities. Over time, this new form of organization gave rise to the concept of state, and to the development of large cities, which would become major commercial centers [1]. With this evolutionary explosion, the demand for geological resources became higher, giving rise to the increase of stone exploitation.

Due to the characteristics of different types of rock, and their advantage for construction, stone has always been the most required and used material—from the construction of communication routes to the ostentatious monuments that have been developed over the centuries. However, most of the times, quarries were established in a local exploitation perspective, since transporting stone blocks over hundreds or thousands of kilometers was very difficult to achieve [2]. From this perspective, it is possible to create a link between interconnected local geology and heritage with the identity of a region.

The city of Coimbra, Portugal, is provided with several buildings in the historic center and some of them going back to the Roman occupation made with dolostone (a carbonate sedimentary rock where the dominant carbonate mineral is dolomite) and, sometimes, with dolomitic-limestone rocks [2]. Over the centuries these buildings became monuments and were the object of masonry conservation or modifications. However, from an architectural point of view, most of them preserve some or all the characteristics that marked the periods in which they were built [3].

Stone has always been used in several different ways in construction. One of the biggest uses was stonework to made ashlars when the buildings presented stone facades. However, the carbonated rocks were also calcined to produce lime, used in mortar or renders. In Coimbra, lime was of dolomitic composition, due to the presence of dolomite in the stone used for this purpose. In current residential buildings, stone blocks with smaller dimensions were used in masonry, bonded by mortar and covered with one or two layers of render and generally painted as well.

In fact, stone can be one of the characteristic symbols of a region's singularity, in terms of color and features, but also in terms of the urban landscape.

Coimbra is internationally known as a city where the university presents a great preponderance after the 13th century and, since 2013, integrate the UNESCO World Heritage List with the designation "University of Coimbra—Alta and Sofia".

A tourist will generally observe the monuments from an architectural/emblematic perspective, often forgetting the history behind each block that constitutes the monuments or decorative pieces. This work highlights the use of dolostone and its influence on the building heritage in the municipality of Coimbra. The main objective is to understand the linkage of local geology with the most emblematic monuments that characterize the historical patrimony of the city and its proximity with old quarries. It is also important for the identification of locals with the exploitation potential of blocks to be used in rehabilitation and restoration.

2. Study Area

2.1. Geological Overview

The municipality of Coimbra is situated in the Central region of Portugal, belonging to the Baixo Mondego sub-region, presenting an area of 319.4 km^2. The old city is located in a hill with approximately 25 ha and 99 m higher altitude.

Coimbra municipality is located in the contact between two major lithological groups: The Hesperic Massif with Pre-Cambrian and Paleozoic Formations, and the Meso-Cenozoic western border.

The Hesperic Massif, also denominated as Iberian Massif, is mainly composed by schist and hercinic granites [4]. In Coimbra region, the schists are dominant and impose a significant contrast in the morphology presenting higher altitudes than the sedimentary rocks.

The Meso-Cenozoic western border belongs to the Lusitanian Basin [5], which is located on the Iberian Western Margin and is a non-volcanic passive margin rift [6]. The Basin occupies an area of 20,000 km^2, with a length of 200 km in the NNW-SSE direction and 100 km of extension in the perpendicular direction, corresponding to 2/3 of the continental emerged area [7].

Since the 19th century, the region of Coimbra aroused the curiosity of national and foreign geologists. Over time, the records obtained from the investigations correspond to several designations for the various lithostratigraphic units.

In this work, the main geological object corresponds to the Lower Jurassic dolomitic-limestone succession of Lusitanian Basin outcropping in the old urban area of Coimbra where most of the monuments can be found, commonly named with the informal designation of "Coimbra Group".

In the most recent lithostratigraphic definition of the Lower Jurassic dolomitic-limestone succession from Lusitanian Basin [8], the previous informal Coimbra Group designation [7,9,10] was abandoned and replaced by the joint of two formal units—the essentially dolomitic Coimbra Formation (Fm.), at the base, and the more calcareous and marly-limestone (when not dolomitized) S. Miguel Fm., at the top.

As previously mentioned, the "Coimbra Group" was always subdivided into two diachronic units. However, according to Reference [8] with the normative imposed by the International

Stratigraphic Guide, the transition between both is now placed at a different level when compared to previous publications.

This dolomitic-limestone succession, cropping out in the Coimbra-Penela region, ranging in age from the Early Sinemurian to the Early Pliensbachian ([8]; and references herein), is fit between the essentially pelitic units of the Pereiros Formation (Hetangian) and the marly-limestone units of the Vale das Fontes Formation (Figure 1). It is geographically located between the North of Coimbra and Penela, and occupies an area of 210 km^2 with a combined thickness of ca. 110 m, presenting a W-NW orientation dip, with an irregular relief and tops around 300 m [8].

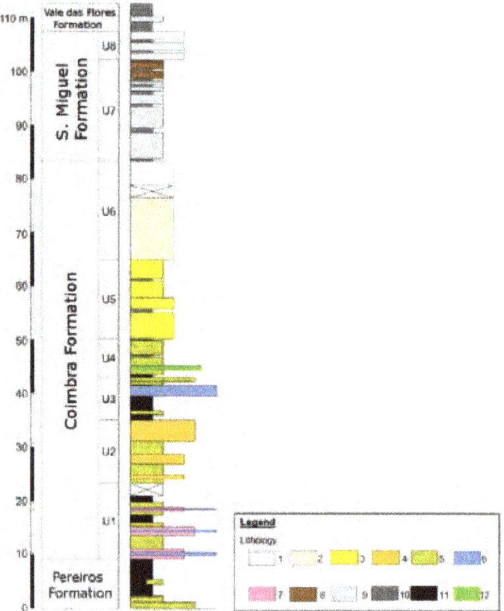

Figure 1. Synthetic lithostratigraphic log for the Coimbra and S. Miguel Formations cropping out in the Coimbra-Penela region: (**1**) Limestone; (**2**) dolomitc-limestone; (**3**) calcareous-dolomite; (**4**) dolostone; (**5**) impure dolostone (argilaceous/ferruginous); (**6**) breccia calco-dolomitic; (**7**) dolomitic sandstone (detritial/biodetritic); (**8**) carbonaceous limestone; (**9**) marly limestone; (**10**) marl; (**11**) pelite/clay; (**12**) quartzarenite (adapted from Reference [8]).

The Coimbra Fm. is dated from the Early to Late Sinemurian, and is on apparent continuity with the Pereiros Fm. [11]. The upper limit of this unit corresponds to the passage between the layers of microcrystalline limestone or dolomitic limestone for the more marly-limestone layers above, defined by a discontinuity surface [8]. From bottom to top, it is possible to identify layers of massive orange dolomites interspersed with gray to black laminate pelitic rocks, levels of marly dolomites and low fossiliferous masses presenting weak bioturbation and impure massive dolomites with a greyish color to the top. Following these are strata of dolomitic limestone and micritic dolostone with an orange-yellowish color, sometimes fossiliferous and bioturbated.

In the Coimbra Fm., it is possible to recognize spathic calcite veins linked to dissolution phenomena with successive (or contemporaneous) recrystallization and reduced content of fossils [8].

The S. Miguel Fm. is dated from the Upper Sinemurian to the base of the Pliensbachian, presenting a thickness of 40 ± 10 m, with more regular strata and apparent coherence when compared with the Coimbra Formation. The upper limit between the S. Miguel Fm. and the Vale das Fontes Fm. represents the main discontinuity surface of the lithostratigraphic succession. From bottom to top, it is composed of gray and whitish dolomitic limestone, sometimes fossiliferous and bioturbated, interspersed by gray or yellowish marls with little thickness (Figure 1). Finally, are identified limestones gray to orange, followed by biodetritic/fossiliferous marls and marly-limestones with a very small thickness [8].

2.2. Historical Urban Development

The city's origin dates back to the Roman occupation, during Emperor Augustus domain (27 BC – AD 14). The civitas Aeminium was organized in an orthogonal configuration, in which the *cardo maximus* (N-S) and the *decumanus maximus* (E-W) intersected in the forum, which is a political, religious and mercantile center of the city [12]. One of the biggest roman remains is a cryptoportic, in the present day integrating the Machado de Castro Museum. After the Roman Empire fell off in the fifth century, Aeminium became part of the Suevian domain and with the fall of these in the sixth century, of the Visigoth domain. With the weakening of the Visigothic kingdom and coinciding with the Muslim campaigns that ran between 711 and 715, the city fell into Muslim rule in 714 and was later conquered by Christians in 1064 [13].

After the formation of the kingdom of Portugal (1139) and until 1260, the city of Coimbra was the capital of the territory. The construction of new churches and rebuilding of the old ones is related to city development. The number of constructions dated from 11–12th century is larger than the precedent and subsequent periods. At the time, the city had a military function but also corresponded to a dynamic commercial town connected with two important trade routes. Therefore, the twelfth century is considered as the golden period of the city's history [13].

In the Middle Ages the existence of a defensive wall was fundamental for the preservation of the people and the city. The Coimbra city walls no longer exist, but they are represented in pictures, and some of the medieval defensive towers are still preserved.

The construction of the Old Cathedral began in the early 1160s, and was completed in the 13th century. It was built in the same place as an existing mosque of the 9th century, due to the privileged location. The main facade presents a defensive structure, organized in a tripartite form, with central body advanced and incorporating military features, such as two towers. In the 16th century the construction of the *Porta Especiosa* by the architect João de Ruão using Ançã Stone [14,15], contributing to the aesthetics of the building. Despite some remodeling, its interior presents the same characteristics of its original configuration [16].

The Santa Cruz Monastery is one of the biggest constructions of the golden period. It was founded in 1131 and completed in 1228. Initially, the church was composed by one nave, reinforced by two lateral naves structured in chapels connected to the central nave and by a tower in the facade. The Santa Cruz Monastery underwent several rebuilding and modifications elaborated throughout the centuries, particularly those of the first half of the 16th century. Between 1522 and 1526, it was included the portal in Ançã Stone, created by Diogo Castilho and produced by Nicolau de Chantereine, and the cloister [17].

On the southern side of the Mondego River, Santa Clara-a-Velha Monastery started being built in 1286. The original building was supervised by the architect Domingos Domingues. The church is constituted by three naves of seven sections, without a transept and with a tripartite chapel of the polygonal apse. Floods caused by the Mondego River over the years led to several problems. Despite some restructuring, life in the monastery became unsustainable which led to its abandonment in 1677, and the nuns were installed in the Santa Clara-a-Nova Monastery, located at a higher altitude on the same side of the river [18].

The Royal Palace, with the original construction as the *Alcáçova* (fortified palace where the governor of the city lived during the period of Muslim rule), was built under the orders of Almançor in the late 10th century [19]. During the 20th century, archaeological excavations were carried out, which identified roman structures and mosaics stressing the importance of the local (Figure 2). After the conquest of the city, the Royal Palace became the residence of Afonso Henriques (the first king of Portugal), and today is the main building of the University of Coimbra, including the S. Miguel chapel and the opulently decorated Baroque Library (1716–23) [20].

Figure 2. Roman mosaics retrieved from the Royal Palace during archaeological excavations. The yellow tesserae correspond to dolostone.

With the permanent settlement of the university in Coimbra in 1537 by order of João III, the urban space registered significant growth. The construction of many colleges to host students that want to become enlightened, was one of the significant developments for the city. One example is the S. Bento College, located near the S. Sebastião Aqueduct, which was rebuilt in the same epoch, but maintained the same layout of the previous one, that dates back to the Roman period [13]. The Botanic Garden was constructed in a nearby area of the S. Bento College for educational purposes. According to its importance, the university was nominated to belong to the UNESCO World Heritage List in 2013 with the designation "University of Coimbra—Alta and Sofia".

The construction of the Santa Clara-a-Nova Monastery began in 1649 and was completed in the 18th century with the conclusion of the cloister, gateway and aqueduct. The plant was designed by Frei João Turriano and supervised by Mateus do Couto. By 1696 occurs the consecration of the church with the works partially completed, but part of the building was already occupied. For the construction of this monument, a simplified plant and straight lines were used in accordance with the Mannerist movement. The rectangular nave is divided into five sections separated by framing Doric pilasters and altarpieces of golden carvings, designed by Mateus do Couto and built by António Gomes and Domingos Nunes in 1692 [21].

In this short description, it is only mentioned few of the principal buildings referred below, due to its relevance in relation with the quarries.

3. Methodology

The methodology used to perform this work is expressed in the flowchart of Figure 3. The identification of monuments was made by bibliographic research and fieldtrip in order to verify the existence of dolostone as stonework in facades or other constructive elements. The quarries identification of the non-visible was made using references, consulting aerial photography and Google Earth Pro application to verify its existence, according to the actual morphology of the surface followed by field survey. Alongside the field research of the abandoned quarries, the distribution of the lithological types was identified, and some characterization assays were carried out—allowing the future to complement the assessment of its potential to be used in rehabilitation and restoration works. The strength, deformability and durability of the selected rocks can be determinate using several tests. Density and porosity can be obtained following EN 1936 (1999) [22] or ISRM (1981, 2007) [23,24]. Sound velocity, Schmidt rebound hardness, slake durability tests can be performed in accordance with

ISRM (1981, 2007) [23,24], uniaxial compressive strength and deformability can be carried out in line with ISRM (1981, 1999) [23,25], point load strength can be obtained following ISRM (1985) [26].

This procedure, in conjunction with the use of other characterization methods, will be important tools to select areas with similar rock types, to those used in the building of monuments, conservation, and restoration.

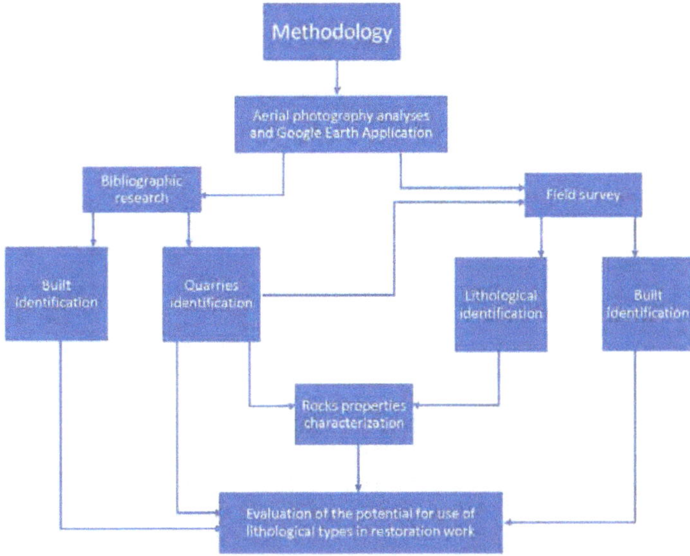

Figure 3. Flowchart of the methodology used.

4. Results

The monuments mentioned above corresponds to some of the more relevant in the city of Coimbra. However, taking in consideration the presence of dolostone in monuments and urban buildings, were identified a total of 27 monuments of particular importance, which were constructed between the seventh and eighteenth centuries, proving the reputation of the dolostone during several centuries in various urban buildings and in other useful or defensive structures on the city and surrounds (Figures 4–6 and Table 1).

4.1. Location of Ancient Quarries

Following the logic that the ancestors take advantage of the local geology, due to logistical and economic aspects, research was carried out in order to identify old quarries in Coimbra. Until the middle of the 20th century, some small quarries were still in progress, but nowadays no dolostone quarry is active in the surrounding area of the city. The red sandstone was also quarried and permitted to obtain blocks used in masonry. One of the quarries was exploited in the city until 1895 when the area was transformed into an arboreal park [27].

Related to this subject and using the local toponymy it can be identified very close to the university (Royal Palace) a place that preserve the word quarry (in Portuguese language: *Pedreira*) in a church (Santo António da Pedreira church), and an alley in the corner of the same building (Figure 7). This quarry was active from 13th century to 1773 [15,28].

Table 1. Principal buildings where dolostone is observed in the city of Coimbra (adapted from Reference [29]).

Name	Construction Date	Conservation Works/Rebuild
Cryptoporticus of Aeminium	1st or 2nd century	
S. Bartolomeu Church	7th century	12th and 13th
Almedina Tower (city wall)	9th century	16th and 20th
Old Cathedral	9th century	several from 11th to 20th centuries
S. Salvador Church	11th century	12th, 16th, 17th and 18th centuries
S. João de Almedina Church	11th century	12th, 17th and 18th centuries
Santa Justa Church (old)	11th century	16th, 18th and 20th centuries
S. Tiago Church	11–12th century	several from 16th to 20th centuries
Mozarabe Arch	12th century	
Almedina Arch and Gate (city wall)	12th century	15th, 16th, 20th and 21st centuries
Santa Cruz Monastery	12th century	15th, 16th, 17th and 19th centuries
S. Miguel Chapel	12th century	several from 16th to 21st centuries
Anto Tower (city wall)	12th century	16th century
Contenda Tower (Sub-ripas Palace) (city wall)	12th century	16th and 20th centuries
Celas Monastery	13th century	several from 16th to 20th centuries
Santa Clara-a-Velha Monastery	13th century	several from 14th to 21st centuries
S. Bento College	16th century	18th and 20th centuries
S. Jerónimo College	16th century	18th, 19th and 20th centuries
S. Sebastião Aqueduct	16th century	20th and 21st centuries
Santo António da Estrela Church *	16th century	18th and 20th centuries
S. Francisco Convent	17th century	19th, 20th and 21st centuries
Santa Clara-a-Nova Monastery	17th century	several from 17th to 21st centuries
Baroque Library	18th century	19th, 20th and 21st centuries
Grilos Palace	18th century	20th century
Santa Clara Aqueduct	18th century	
Catholic Major Seminary of Coimbra	18th century	19th and 20th centuries
Botanic Garden	18th century	19th, 20th and 21st centuries

* Actual parish council building of Sé Nova, Santa Cruz, Almedina and S. Bartolomeu.

Figure 4. Principal monuments and building where dolostone was used. (**1**) Cryptoporticus of Aeminium; (**2**) S. Bartolomeu Church; (**3**) Almedina Tower (city wall); (**4**) Old Cathedral; (**5**) S. Salvador Church; (**6**) Mozarabe Arch; (**7**) Santa Justa Church (old); (**8**) S. Tiago Church; (**9**) Almedina Arch and Gate (city wall); (**10**) S. João de Almedina Church; (**11**) Santa Cruz Monastery; (**12**) S. Miguel Chapel; (**13**) Anto Tower (city wall); (**14**) Contenda Tower (Sub-ripas Palace) (city wall); (**15**) Celas Monastery; (**16**) Santa Clara-a-Velha Monastery; (**17**) S. Bento College; (**18**) S. Sebastião Aqueduct; (**19**) Santo António da Estrela Church; (**20**) S. Francisco Convent; (**21**) Santa Clara-a-Nova Monastery; (**22**) Baroque Library; (**23**) Grilos Palace; (**24**) Catholic Major Seminary of Coimbra; (**25**) Botanic Garden.

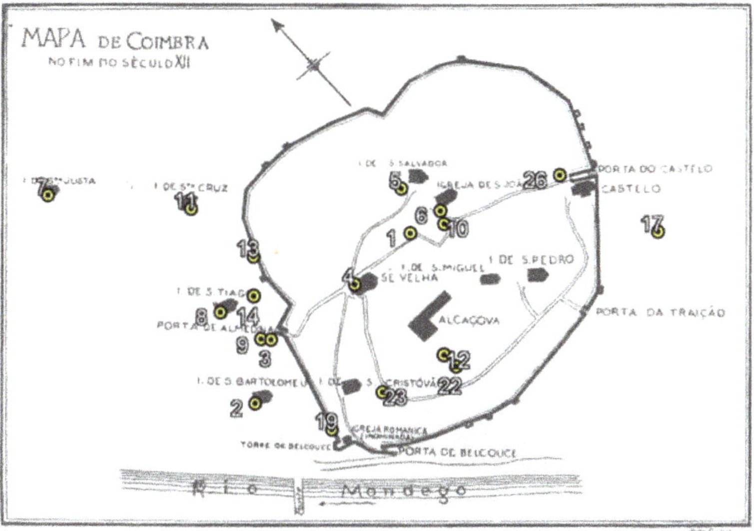

Figure 5. Map of Coimbra at the end of the 12th century. It can be observed most of the important buildings (legend of the same number as Figure 4) and the city wall (adapted from Reference [14]).

Figure 6. Map of the city in the year 1834, with the indication of the existing Colleges and Convents (in brown, legend of the same number as Figure 4) (adapted from Reference [30]).

Figure 7. Illustration of Coimbra made by Pier María Baldi in 1669 (adapted from Reference [31]), where can be observed the area of Santo António da Pedreira quarry close to the city wall.

Alarcão [2,28] based on draws, contract letters and purchase orders mentioned the presence of other quarries located inside the city walls, with the notice that thousands of cubic meters of stone were needed for the construction of several major buildings. Some of the quarries were located very close to the main buildings of the city (Figure 8 and Table 2):

- The stones used for building the Old Cathedral were extracted from the construction site (Figure 9a). Some chapels were carved in the rock massif (12th century) [2,14,15,28];
- The Montarroio quarry was the property of Santa Cruz Monastery since the 12th century [28];
- The S. Sebastião quarry near the castle, in the place where after was constructed the College of Arts and very close to the aqueduct of S. Sebastião (14th century) [28];
- The Penedos Street and the *Largo da Feira* quarries, which disappeared with the University city's remodeling during the second half of 20th century, located in the top of the hill, in front to the New Cathedral, near the Royal Palace and several colleges [2,28] (Figure 9b).
- S. Cristovão quarry, where today, the Sousa Bastos Theater is situated, close to Belcouce Tower inside the city wall [32].

Outside the city walls other references mentioned the Conchada and *Monte Formoso* quarries, probably identified in outcrops still observed at the end of the 20th century [15].

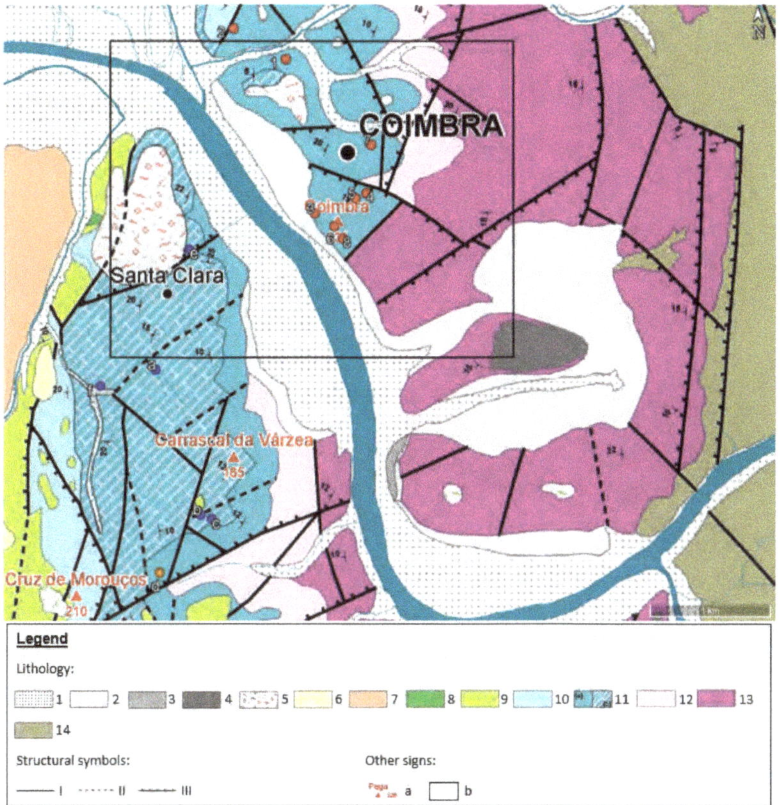

Figure 8. Location of old quarries in the geological map; red dots correspond to non-visible quarries, blue dots correspond to inactive quarries, yellow dot correspond to selected quarry; (1) alluvium; (2) colluvial deposit; (3) alluvial deposit lower terrace; (4) alluvial deposit high terrace; (5) fluvial-torrential deposit (red sands of Ingote); (6) conglomerates and dark pelites; (7) immature conglomeratic sands with silicified horizons and arcosic sandstones; (8) marly sandstone with fossils; (9) conglomerates, arcosarenites and pelites; (10) marly limestones, limestones and marl; (11) dolostones to dolomitc limestones, pelites/clay, limestones to marly limestones and marl (a-Coimbra Formation; b-S. Miguel Formation); (12) gray pelites and dolostones; (13) conglomerate, sandstones and red to whitish pelites; (14) set of ante-mezozoic units; (I) Fault; (II) Probable fault and/or hide; (III) Normal fault; (a) Geodesic vertex; (b) the form corresponds to Figure 4 area (adapted from Reference [9]).

On the left bank of the Mondego River, was used dolostone in both Santa Clara monasteries. In that area are observed six quarries, some of them used for lime production until the mid-20th century (Figure 8). In 1889, a book organized by the Portuguese Industrial Association, about the reality of the mines and quarries exploitation at that time, identified in the urban perimeter of Coimbra the quarry of Bordalo (Santa Clara) and the extracted stone was used to the production of lime, masonry and stonework. It is highlighted the resistance to weathering and the "agreeable" aspect of the stone refereeing that was used in two bridges near the city. The other quarry identified is the one the Montarroio used for masonry and lime production, but with less importance at that time [33].

(a)   (b)

Figure 9. Examples of outcrops: (**a**) Santa Maria chapel, Old Cathedral, where the in situ stone can be observed; (**b**) Outcrop in College of Arts, close to the location of *Largo da Feira* quarry (below these locations) and S. Sebastião quarry.

Table 2. List of old quarries located in Coimbra, in accordance with Figure 8 (based on field survey and [28]).

Quarry	Situation
1 Conchada	Abandoned/non-visible
2 *Monte Formoso*	Abandoned/non-visible
3 S. Cristóvão	Abandoned/non-visible
4 S. Sebastião	Abandoned/non-visible
5 *Largo da Feira*	Abandoned/non-visible
6 Santo António da Pedreira	Abandoned/non-visible
7 Montarroio	Abandoned/non-visible
8 Porta de Belcouce	Abandoned/non-visible
a Alto de Santa Clara	Inactive/abandoned
b Banhos Secos 1	Inactive/abandoned
c Banhos Secos 2	Inactive/abandoned
d Carvalhais	Inactive/abandoned
e Santa Clara-a-Nova	Inactive/abandoned
f Bordalo	Inactive/abandoned

4.2. Dolostone Characteristics

The dolostone quarried in the layers belonging to the Coimbra Fm. is characterized by heterogeneous properties, which provides a differential degradation by its surface [34] and is much more difficult to work than the Ançã Stone, which is often used in the city of Coimbra for decorative pieces, due to the white color and the easy cut. In monuments dating back to the 12th and 13th centuries, dolostone was used as stonework on the exterior walls and also in the interior as vaults and columns. For current buildings, dolostone was adopted in masonry for structural purposes [35].

The chemical composition of the dolostone and calcareous-dolomite of Coimbra Formation obtain from ashlars used in the Old Cathedral presents a content of CaO of 23 to 31%, MgO from 16 to 21%, Fe_2O_3 vary from 0.8 to 4% and residual quantities of other oxides [15]. Analyses of dolostone from outcrops far from the nucleus of the city presents a content in MgO lower than 18%, frequently between 9.5 and 15% [36]. In addition, the high content in Fe_2O_3 gives a yellowish color to the stone [15,36].

With different objectives, several drilling holes were made in the nucleus of the old town close to the location of S. Sebastião and *Largo da Feira* quarries. Physical and mechanical characterization aiming to define strength, deformability and durability (Table 3), was carried out using 85 samples [37]. Dolostone with different degrees of weathering (W1, W2 and W3) was studied and classified according to its results.

Table 3. Results physical and mechanical tests (mean values) of dolostones (adapted from Reference [37]).

	V_p $[m.s^{-1}]$	V_s $[m.s^{-1}]$	Density $[kg.m^{-3}]$	Porosity [%]	R	$Is_{(50)}$ [MPa]	UCS [MPa]	Ed [GPa]	Es [GPa]	Id_2 [%]
W1	3661	2213	2465	13.6	36.7	3.50	67.0	19.74	25.29	99.2
W2	3551	2004	2310	19.0	32.9	1.79	34.6	13.30	16.87	98.3
W3	3214	1657	1954	30.7	25.9	0.96	18.9	8.83	13.48	80.1

Where $Is_{(50)}$ corresponds to point load strength index corrected to a diameter of 50 mm; UCS to uniaxial compressive strength; R to Schmidt rebound hardness; Ed to Dynamic elastic moduli; Es to static elastic moduli; Id_2 to slake-durability index of the second cycle; V_p and V_s to velocity of the longitudinal and transversal waves ultrasonic velocity of propagation, respectively.

Tests performed with samples obtained in existing old quarries in Santa Clara area present values of density and porosity similar to the ones identified as W2 and W3 [38].

Dolostone weathering corresponds to powdering and flacking and presents differential erosion associated with the heterogeneities and the presence of recrystallized calcite filling fractures, more resistant than the dolomite [15,34]. The areas of lower quality present degrees of degradation that are more pronounced.

Seco et al. [39] carried out data in outcrops of the Coimbra Formation from natural gamma-ray spectrometry measurements (total counts, K, Th and U contents). These concentrations can be used for the indirect evaluation of parameters, such as the clay minerals content, grain size variation, porosity and content of total organic carbon, among others. They observed a reduction of the gamma-ray flux from the base to the top. The dolomitic limestone and dolostone (Coimbra Fm.) varying in the base between 140 and 247 cps, in the intermediate part between 61 and 114 cps and at the top between 101 and 106 cps. Gamma-ray flux varies between 107 and 168 cps at the top of the succession (S. Miguel Fm.). In terms of mean values of the radiogenic elements, no significant differences were observed and present values of 0.4% K, 1.3 ppm U and 1.6 ppm Th.

5. Discussion and Conclusions

Alarcão [2,28] highlighted the position of the quarries located in Coimbra in relation to the buildings saying they were inside the city walls and in the highest point in the city close, therefore, to the buildings to be constructed. By observing the geological map and the position of the old quarries inside the wall, it is easy to understand that most of the monuments are located inside the area where the geological units of Coimbra Formation are present (Figures 4, 6 and 8). So the quarries located in the nucleus of the city of Coimbra provided the necessary stone for the monuments and urban buildings located in this area, while the quarries located in Santa Clara provided the needs for the buildings in the other side of the river. Only the quarries located in Santa Clara are still observed (Table 2). Those located in the nucleus of the city, admitted by references in several types of documentation and bibliography, can only be assumed in areas not occupied by houses in illustrations of the time (Figure 7).

Two major architectural/urban phases are identified in the city of Coimbra: The period comprising the 12th and 13th centuries and later between the 16th and 18th centuries. Between the end of the 13th and 15th centuries, there is no evidence of the construction of monuments. This time period corresponds to the consolidation of the city, but also may be related to the period of the Fernandine wars (14th century), as well as to the beginning of the overseas campaigns (15th century).

Dolostone was used in the stonework on the facades and structural pieces in the interior, such as vaults and columns, in the monuments constructed during the 12th and 13th centuries. In this period, it was used with some frequency in decorative pieces, particularly in contrast to white painted walls giving rise to a gilded aspect (Figure 10). Contrast was also created when, centuries later, the Ançã Stone was used due to its very noticeable white colour. In some of these monuments, it is still possible to identify the geological registry of the outcrops in its interior, as is the example of the Old Cathedral (Figure 9a).

Figure 10. Overview of Royal Palace (**a**) and detail of Baroque Library (**b**) and S. Miguel Chapel (**c**).

The dolostone use in the monuments, built between the 16th and 18th centuries, is only visible in the quoins and pilasters. In these buildings, dolostone was also applied in masonry on the outer walls, and lintel around the windows and doors. The gilded aspect also presents a contrast with the surrounding white walls.

The dolostone belonging to the Coimbra Fm. were studied throughout the years from a lithostatigraphic point of view, but only a few publications exist related to the properties as a building material [15,34,37]. This was probably due to the inexistence of active quarries after the middle 20th century when the investigation of that matter presents a great increment. The area of the last quarries of Bordalo and Montarroio, used for blocks and lime production until that time, were also abandoned, due to the building boom. Moreover, the dolomitic lime was replaced by calcite lime and cement.

The velocity of the longitudinal waves ultrasonic velocity of propagation for the W1 and W2 dolostones was considered as medium and low for the W3 dolostones. According to the uniaxial compressive strength and the point load strength index corrected to a diameter of 50 mm values, the rock strength was classified as high for the W1 group, medium for W2 and low for W3. The porosity values of the W1, W2 and W3 dolostones presented a medium, high and very high porosity, respectively. So, it is accepted that the dolostone present mechanical properties that allow over the centuries its utilization in construction, as can be observed.

As presented in Table 2, all the buildings present across the centuries several modifications, reorganizations and changes of use. At the end of 19th century, the Old Cathedral was in conservation/restoration/rehabilitation works for ten years [14]. The Colleges that belong to the University campus, the Baroque Library, S. Miguel Chapel, Botanic Garden and others had also restoration and rehabilitation works [40].

In preservation and rehabilitation of old buildings the use of new stones is often needed to replace the degraded ones, which do not present safe conditions. The exploitation of dolostone around the city of Coimbra has been abandoned for more than twenty years. This conducted to the use of similar stones from other parts of the country when the replacement is need.

An example of this is the S. Francisco Convent. It corresponds to a large compound of different built bodies, with two to three floors each, organized around a cloister. Built in the 17th century, the structure was occupied by the Franciscans until 1843, after which, and until the mid-20th century, several factories were installed, such as a wool factory from 1888 until 1980, with considerable impact on the main structure [41]. The rock used in the facades and pavement of the rehabilitation works (2010–2016) was a dolomitic breccia that came from *Serra de Aires and Candeeiros*, near Moleana, Porto de Mós.

Considering that most of the old buildings of Coimbra are included in the UNESCO World Heritage List, it is important to preserve the heritage and the materiality of the city. Taking into account the existence of several outcrops in the Santa Clara region that corresponds to old quarries, some of them exploited up to the last twenty years for the production of aggregates; it is possible to use, with prudence and vigilance from the authorities, stone from those quarries. The one presented in Figure 11, corresponds to the indicated as *d* (yellow dot) on Figure 8, due to the easy access to trucks and to the occurrence of stone with good quality has great relevance for this use. The other quarries signalized in Figure 6 also present proper stone, but due to the location near houses or monastery (Santa Clara-a-Nova Monastery) and with difficult accessibility, are less recommended for this use.

Figure 11. Example of an old quarry that allows the exploitation of blocks for conservation, restoration and rehabilitation of monuments (40°10′54.4″ N 8°26′37.3″ W).

The "identity" of the city is dependent on its geological location, and the stone corresponds to a symbol that characterizes its gilded aspect. This was highlighted by Vasconcelos in 1930 [14] in relation to the apse of the chapel in Old Cathedral, and can be extended to the chromatic characteristics of the stone buildings and the city.

Author Contributions: All authors contributed equally to this paper and approved the final manuscript. Conceptualization and methodology, F.P.F., J.D., P.A. and L.C.; field work, J.D., F.P.F., R.F. and L.C.; laboratory tests, investigation and validation data, P.A., L.C. and R.F. All authors were involved for the writing and editing of the manuscript.

Funding: This research received no external funding.

Acknowledgments: The financial support of FCT-MEC through national funds and, when applicable, co-financed by FEDER in the ambit of the partnership PT2020, through the research project UID/Multi/00073/2013 of the Geosciences Center is acknowledged.

Conflicts of Interest: The authors declare no conflict of interest.

References

1. Childe, G. *A Evolução Cultural do Homem*; Zahar Editores: Rio de Janeiro, Brazil, 1975.
2. Alarcão, J. Pedreiras antigas de Coimbra. In *A Terra. Conflitos e Ordem. Livro de Homenagem ao Professor Ferreira Soares*; Callapez, P., Rocha, R., Marques, J., Cunha, L., Dinis, P., Eds.; MMGUC: Coimbra, Portugal, 2008; pp. 71–73.
3. Alarcão, J. *A Evolução Urbanística de Coimbra: Das Origens a 1940*; Cadernos de Geografia: Coimbra, Portugal, 1999; pp. 1–10. [CrossRef]
4. Teixeira, C. *Geologia de Portugal, Pré-Câmbrico e Paleozoico*; Fundação Calouste Gulbenkian: Lisboa, Portugal, 1981; Volume I, p. 629.
5. Ribeiro, A.; Silva, J.B.; Cabral, J.; Dias, R.; Fonseca, P.; Kullberg, M.C.; Terrinha, P.; Kullberg, J.C. *Tectonics of the Lusitanian Basin*; Final Report, Proj. MILUPOBAS, Contract n° JOU-CT94-0348, ICTE/GG/GeoFCUL; Universidade de Lisboa: Lisboa, Portugal, 1996; p. 126.
6. Alves, T.M.; Moita, C.; Sandnes, F.; Cunha, T.; Monteiro, J.H.; Pinheiro, L.M. Mesozoic–Cenozoic evolution of North Atlantic continental-slope basins: The Peniche basin, western Iberian margin. *AAPG Bull.* **2006**, *90*, 31–60. [CrossRef]
7. Kullberg, J.C.; Rocha, R.B.; Soares, A.F.; Rey, J.; Terrinha, P.; Azerêdo, A.C.; Callapez, P.; Duarte, L.V.; Kullberg, M.C.; Martins, L.; et al. A Bacia Lusitaniana: Estratigrafia, Paleogeografia e Tectónica. In *Geologia de Portugal*; Dias, R., Araújo, A., Terrrinha, P., Kullberg, J.C., Eds.; Escola Editora: Vila Franca de Xira, Portugal, 2013; Volume II, pp. 195–347.
8. Dimuccio, L.A.; Duarte, L.V.; Cunha, L. Definição litostratigráfica da sucessão calco-dolomítica do Jurássico Inferior da região de Coimbra-Penela (Bacia Lusitânica). *Comun. Geológicas* **2016**, *103*, 77–96.
9. Dimuccio, L.A. A Carsificação nas Colinas Dolomíticas a sul de Coimbra (Portugal Centro-Ocidental). Fácies Deposicionais e Controlos Estratigráficos do (Paleo)Carso no Grupo de Coimbra (Jurássico Inferior). Ph.D. Thesis, FCT University of Coimbra, Coimbra, Portugal, 2014; p. 436.
10. Soares, A.F.; Marques, J.F.; Sequeira, A.D. *Carta Geológica de Portugal, na Escala 1/50 000. Notícia explicativa da Folha 19-D, Coimbra-Lousã*; INETI: Lisboa, Portugal, 2007; pp. 1–71.
11. Soares, A.F.; Kullberg, J.C.; Marques, J.F.; Rocha, R.B.; Callapez, P. Tectono sedimentary model for the evolution of the Silves Group (Triassic, Lusitanian basin, Portugal). *Bull. Soc. Géol. Fr.* **2012**, *183*, 203–216. [CrossRef]
12. Alarcão, J. *O Domínio Romano em Portugal*, 3rd ed.; Publicações Europa-América: Coimbra, Portugal, 1995; p. 244.
13. Mattoso, J. *História de Portugal. Antes de Portugal*; Editorial Estampa: Lisbon, Portugal, 1997; Volume I–III, p. 452.
14. Vasconcelos, A. *A Sé-Velha de Coimbra. Apontamentos para a sua História*; Imprensa da Universidade de Coimbra: Coimbra, Portugal, 1930; Volume I.
15. Quinta-Ferreira, M.; Soares, A.F.; Delgado Rodrigues, J.; Monteiro, B. Carbonate rocks from Sé Velha Cathedral, Coimbra, Portugal. In Proceedings of the 7th International Congress on Deterioration and Conservation of Stone, Coimbra, Portugal, 15–18 June 1992; pp. 947–956.

16. SIPA. Sé Velha de Coimbra. Available online: http://www.monumentos.gov.pt/Site/APP_PagesUser/SIPA.aspx?id=2673 (accessed on 15 May 2019).

17. SIPA. Mosteiro de Santa Cruz. Available online: http://www.monumentos.gov.pt/Site/APP_PagesUser/SIPA.aspx?id=4234 (accessed on 15 May 2019).

18. SIPA. Mosteiro de Santa Clara-a-Velha. Available online: http://www.monumentos.gov.pt/Site/APP_PagesUser/SIPA.aspx?id=2807 (accessed on 15 May 2019).

19. Catarino, H.; Filipe, S. *"Madinat Qulumbriya: Arqueologia Numa Cidade de Fronteira", Al-Ândalus Espaço de Mudança. Balanço de 25 anos de História e Arqueologia Medievais, Seminário de Homenagem a Juan Zozaya Stabel-Hansen (Mértola 2005)*; Campo Arqueológico de Mértola: Mertola, Portugal, 2006; pp. 73–85.

20. Filipe, S. Arqueologia urbana em Coimbra: Um testemunho na Reitoria da Universidade. *Conimbriga Rev. Arqueol.* **2015**, *45*, 337–357. [CrossRef]

21. SIPA. Mosteiro de Santa Clara-a-Nova. Available online: http://www.monumentos.gov.pt/Site/APP_PagesUser/SIPA.aspx?id=2678 (accessed on 15 June 2019).

22. EN 1936. *Natural Stones Test Methods—Determination of Real Density and Apparent Density, and of Total and Open Porosity*; European Committee for Standardization (CEN): Brussels, Belgium, 1999.

23. ISRM. Rock Characterization Testing & Monitoring. In *ISRM Suggested Methods*; Brown, E.T., Ed.; Published for the Commision on Testing Methods; Pergamon Press: Oxford, UK, 1981; p. 211.

24. ISRM. *The Complete ISRM Suggested Methods for Rock Characterization, Testing and Monitoring: 1974–2006*; Ulusay, R., Hudson, J.A., Eds.; ISRM: Ankara, Turkey, 2007; p. 628.

25. ISRM. ISRM suggested method for the complete stress-strain curve for intact rock in uniaxial compression. *Int. J. Rock Mech. Min. Sci.* **1999**, *36*, 279–289.

26. ISRM. ISRM suggested Method for Determining Point Load Strength. *Int. J. Rock Mech. Min. Sci. Geomech.* **1985**, *22*, 51–60. [CrossRef]

27. Rosmaninho, N. *O Poder da arte: O Estado Novo e a Cidade Universitária de Coimbra*; Coimbra University Press: Coimbra, Portugal, 2006; p. 171.

28. Alarcão, J. *Coimbra: A Montagem do Cenário Urbano*; Imprensa da Universidade de Coimbra: Coimbra, Portugal, 2008; p. 308. [CrossRef]

29. SIPA. Available online: http://www.monumentos.gov.pt (accessed on 5 June 2019).

30. Calmeiro, M.R. Apropriação e conversão do Mosteiro de Santa Cruz. Ensejo e pragmatismo na construção da cidade de Coimbra. Monastic architecture and the city. In *Cescontesto, Debates*; Marado, C.A., Ed.; CES: Coimbra, Portugal, 2014; Volume 6, pp. 227–240. Available online: https://www.ces.uc.pt/publicacoes/cescontexto/index.php?id=10019 (accessed on 12 July 2019).

31. Sanchez Rivero, A.; Sanchez Rivero, A.M. *Viaje de Cosme de Médicis por España y Portugal (1668–1669). Laminas*; Junta para Ampliacion de Estudios e Investigaciones Científicos: Madrid, Spain, 1933; Available online: http://purl.pt/12926 (accessed on 10 June 2019).

32. Anjinho, I.M. *Fortificação de Coimbra das Origens à Modernidade*; CEAACP, Universidade de Coimbra: Coimbra, Portugal, 2016; Volume 3, Available online: https://eg.uc.pt/handle/10316/31013 (accessed on 15 June 2019).

33. Monteiro, S.; Barata, J.A. *Catalogo Descriptivo da Secção de Minas*; Imprensa Nacional: Lisboa, Portugal, 1889; p. 373.

34. Delgado-Rodrigues, J. As pedras de Coimbra. Aspectos relativos à sua degradação e conservação. In Proceedings of the Conference International "A imagem dos centros históricos—Bases para a sua salvaguarda", Coimbra, Portugal, 21–23 September 2005; Available online: https://www.academia.edu/34029616/ (accessed on 8 May 2019).

35. Vicente, R.; Silva, J.A.R.M.; Harum, H. Caracterização das Alvenarias dos Edifícios da Baixa de Coimbra. As Suas Anomalias Típicas. In Proceedings of the PATORREB 2006 2.º Encontro Nacional sobre Patologia e Reabilitação de Edifícios, Porto, Portugal, 20–21 March 2006; pp. 509–518. Available online: http://hdl.handle.net/10773/6491 (accessed on 10 May 2019).

36. Manuppella, G.; Moreira, J.C.B.; Romão, M.L. Panorama dos dolomitos e calcários dolomíticos portugueses. *Bol. Minas* **1981**, *17*, 3–15.

37. Faim, R.; Andrade, P.; Figueiredo, F. Physical and mechanical characterization of dolomitic limestone in Coimbra (Central Portugal). In Proceedings of the ISRM Regional Symposium EUROCK 2015 & 64th Geomechanics Colloquium—Future Development of Rock Mechanics, Salzburg, Austria, 7–10 October 2015; Schubert, W., Kluckner, A., Eds.; Austrian Society for Geomechanics: Salzburg, Austria, 2015.

Sustainability **2019**, *11*, 4158

38. Torres, M.I.M. Humidade Ascensional em Paredes de Construções Históricas. Ph.D. Thesis, University of Coimbra, Coimbra, Portugal, 2004; p. 229.
39. Sêco, S.L.R.; Duarte, L.V.; Pereira, A.J.S.C. Gamma-ray spectrometry applications to lowermost Jurassic units in northen sector of the Lusitanian Basin (Portugal): Preliminary data. *Comun. Geol.* **2015**, *102*, 41–44.
40. University of Coimbra. University of Coimbra—Alta and Sofia. 2012. Available online: http://www.uc.pt/ruas/monitoring/protection (accessed on 13 June 2018).
41. SIPA. Convento de S. Francisco. Available online: http://www.monumentos.gov.pt/Site/APP_PagesUser/SIPA.aspx?id=4240 (accessed on 8 May 2019).

 *sustainability*

Article

Minero-Petrographic Characterization of Chianocco Marble Employed for Palazzo Madama Façade in Turin (Northwest Italy)

Francesca Gambino [1,*], Alessandro Borghi [1], Anna d'Atri [1], Luca Martire [1], Martina Cavallo [1], Lorenzo Appolonia [2,3] and Paola Croveri [2]

1 Department of Earth Sciences, University of Turin, Via Valperga Caluso 35, 10125 Torino, Italy
2 Centro Conservazione e Restauro "La Venaria Reale", Via XX Settembre 18, 10078 Venaria Reale (TO), Italy
3 Soprintendenza per i beni e le attività culturali della Regione Autonoma Valle d'Aosta, Piazza Narbonne, n.3-11100 Aosta, Italy
* Correspondence: francesca.gambino@unito.it

Received: 21 June 2019; Accepted: 1 August 2019; Published: 5 August 2019

Abstract: The study of ancient marble plays an important role in the interpretation of historical and archaeological sites and gives interesting information about building materials used in ancient times and their trade routes. The present work focuses on Chianocco marble that represents one of the most important ancient white marbles for cultural heritage exploited in the Piedmont region (Northwest Italy) and employed for the Palazzo Madama façade. A multi-analytical study based on petrographic (optical and scanning electron microscopy), electron microprobe, cathodoluminescence and stable isotope analyses was carried out on these marbles in order to perform an archaeometric study. Chianocco marble was used in Turin during the baroque era by the Savoy architect Filippo Juvarra (1678–1736) in historical buildings, such as the façade of the Palazzo Madama, the plinth of the façade of the town Cathedral and the columns (now plastered) of the portico of Piazza San Carlo. This stone is a dolomitic rock belonging to the Mesozoic cover of the Dora Maira Massif (Pennidic Unit). It shows a vuggy fabric characterized by a vacuolar texture due to tectonic brecciation and subsequent selective dissolution during subaerial exposure. This kind of research is useful to highlight the importance of the use of local stones as building materials and to investigate stone materials for the restoration and maintenance of historical buildings.

Keywords: Chianocco marble; heritage stone; archaeometry; Western Alps; isotopic analysis; SEM-EDS

---

1. Introduction

Ancient buildings, artifacts and findings are mainly made of natural and artificial materials obtained from geological resources. The development of geosciences as applied to cultural heritage highlights how the study of the genesis and characteristics of ornamental stones is primarily a geological matter, and has to be solved by a geologic approach [1]. A proper characterization of these materials requires minero-petrographic studies for defining their provenance, conservation state, and application of good preservation strategies.

In Piedmont, and in particular in Turin, stone has always been largely used for both constructions and decoration, becoming one of the distinctive elements of the local architectural heritage. Statues, city walls, floors, roofs, and other architectural elements, are often made of the many varieties of rocks belonging to the different geological units of the Western Alps [2–4]. Often, the selection of stone materials in architecture is driven by specific values and meanings attributed to the different rocks; moreover, the use of specific lithotypes can be related to aesthetic values, technical progress or even economic circumstances. Because of the ease of cleaning, marble has been widely employed in valuable buildings, from Roman times to the end of the eighteenth century [5].

One of the most prestigious buildings in Turin is certainly the Palazzo Madama (Figure 1a), a historical and architectural complex located in the center of the town. It is an UNESCO World Heritage site and at present is the seat of the city Ancient Art Museum. It is the testimony of two thousand years of history: Originally built by the Romans as a gateway to the town, the building became first a defensive system, and then a symbol of power until the sixteenth century, when it was replaced by the Palazzo Reale as seat of the Duke of Savoy. With King Carlo Alberto, politics also entered Palazzo Madama: In 1848, the king placed the Subalpine Senate in the large hall on the first floor, destined to become one of the places of politics in which Italy's unity was most strongly configured. Considerably embellished under the regency of the two royal ladies also known as "Madame" (hence the name): Maria Cristina of France and Maria Giovanna Battista of Savoy, the old medieval castle was retrained by the work of Filippo Juvarra, who realized (1718–1721) the great façade which dominates the square [6,7]. He chose the Chianocco marble, a yellowish grey marble from the Susa Valley, for coating the façade. The strong deterioration of this marble made necessary, over time, several restorations and replacements by different stone materials recalling the original one, but coming from different sources [8,9]. As a consequence, many archaeometric studies carried out on the façade of Palazzo Madama resulted in contradictory and partially wrong conclusions in the attribution of the stones employed over the centuries [10,11].

Figure 1. Buildings: (**a**) View of the Palazzo Madama façade; (**b**) Study site located in the central area of the façade.

For this reason, and because of many recent conservation issues, the conservation and restoration foundation "La Venaria Reale", in collaboration with the Foundation Torino Musei and under the supervision of the Superintendence of Archeology, Fine Arts and Landscape for the Metropolitan City of Torino promoted several technical and scientific investigations in order to develop a pilot project for the overall conservation and future maintenance of the historical façade (Figure 1b).

The purpose of this paper is to provide a detailed petro-architectonic survey and a minero-petrographic char.

2. Geological Setting

In the central area of the Susa Valley (NW Italy), the metamorphic stratigraphic cover of the Dora Maira Massif crops out and includes marbles. The Dora Maira Massif belongs to the Pennidic Domain of the Western Alps (Figure 2a); it is a continental crust unit. It was involved in the Alpine orogeny (about 50 Ma ago) which resulted in pervasive metamorphism and deformation. The Dora Maira Massif is predominantly made up of Palaeozoic micaschists, gneiss, and rare slices of dolomitic marbles which derived from the Alpine metamorphism of Triassic to Early Jurassic carbonate sediments. The Alpine metamorphism developed under eclogitic conditions in a first event, when peak pressures (P) and temperatures (T) were reached, a retrograde metamorphic event under greenschist facies conditions followed. [12]. Historically, Susa Valley marbles have been distinguished as "Foresto and Chianocco marbles" on the basis of their extraction site [13,14]. In fact, they are two different kind of rocks with different petrographic features resulting from different geological processes. The Foresto marble consists of massive whitish marbles whereas the Chianocco marble shows a vacuolar structure and a yellowish color.

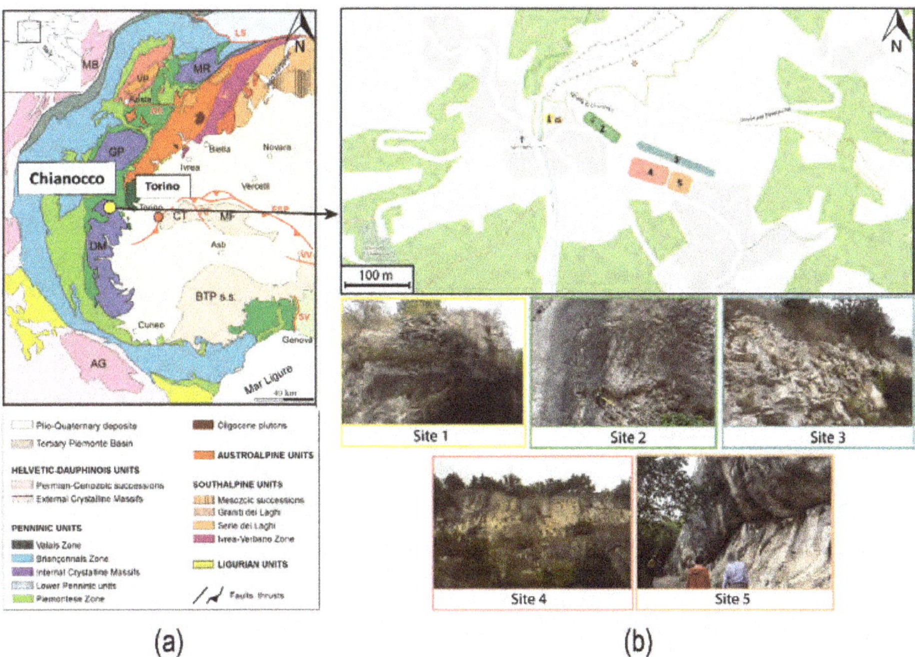

(a)   (b)

Figure 2. The Chianocco municipality: (**a**) Geological setting of the Piedmont region and location of the Chianocco municipality; (**b**) Location of the five quarry sites in the Chianocco municipality.

3. Materials and Methods

The support of the Earth Science Department of Turin to the study of the Palazzo Madama façade consisted in the realization of an architectural-petrographic survey of the façade, in the characterization of its lithotypes, in the diagnosis of the state of preservation, in the study of the degradation causes, and in the definition of a model of the evolution of the marble of the façade from its formation to its employment.

For this kind of study related to buildings, monuments and artefacts constituted in marble (a precious material most used and traded in antiquity) a multi-analytical approach is necessary [15–18].

Starting from the architectonic relief of the façade, a mapping of stone materials in false color, named "petro-architectonic relief" in this paper, was achieved. The fragments detached from the façade were catalogued and, from the data collected, the most representative samples were selected for detailed studies.

In order to understand the properties of the material, the localization of the ancient quarries has been essential. Five significant sites in the Chianocco municipality were individuated and sampling work has been conducted.

Petrographic studies on uncovered thin sections (30μm thick) were carried out by optical microscopy and cathodoluminescence (CL) at the Earth Sciences Department of the University of Turin.

CL observations were performed on polished thin sections using a CITL 8200 mk3 equipment (operating conditions of about 17 kV and 400 μA).

Determination of major elements was performed using a scanning electron microscope (SEM; JEOL JSM-IT300LV) combined with an energy-dispersive X-ray spectrometer (EDX), with a silicon drift detector (SDD) from Oxford Instruments, installed at the Dipartimento di Scienze della Terra of the University of Turin. The following operative conditions were adopted: an accelerating voltage 1 of 5 kV, a counting time of 50 s, a process time 5 μs and working distance of 10 mm. The measurements were performed in high vacuum conditions. A cobalt standard was analyzed for correction and calibration both in energy and in intensity of EDX acquired spectra. Enabling spectra visualization and elements recognition was done by the Microanalysis Suite Oxford INCA Energy 300. A ZAF data reduction program was used for spectra quantification. Astimex Scientific Limited standards were used to obtain full quantitative analyses. All the analyses were formula recalculated using the MINSORT computer program [19]. Representative polished thin sections of the marble were analyzed using the SEM–EDS system, with backscattered electron (BSE) and X-ray signals and it permitted us to define the chemistry of selected minerals. In the BSE images the brightness signal is sensitive to differences among mean atomic number showing different grey levels for different phases (i.e., calcite and dolomite). In fact, the minerals with higher mean atomic numbers (e.g., calcite) are brighter than the ones with lighter forming elements (e.g., dolomite). In addition to carbonates, mica crystals have also been analyzed; a representative selection of mica composition is reported in Table 1. Lastly, mass spectroscopy for the determination of stable isotope ratios was used. The stable isotope analyses (i.e., $\delta^{13}C$ and $\delta^{18}O$) have been performed on calcite and dolomite for the studied marble types. The protocol reported in McCrea (1950) [20] was applied. An amount of 10 mg of powered calcite or dolomite was reacted with 100% orthophosphoric acid under vacuum conditions. The oxygen and carbon isotopic composition produced by CO_2 was analyzed using a Finningan MAT 250 mass spectrometer. The results are expressed as an isotopic ratio in relation to the PDB standard [21], following the convention defined by the International Atomic Energy Agency.

4. The Palazzo Madama Façade

The façade of Palazzo Madama can be considered one of the masterpieces of the architect Filippo Juvarra. Classical and baroque decorative themes coexist; in fact, Juvarra designed a piano nobile with arch-headed windows linked to a mezzanine overhead by a colossal order of pilasters in a composite style. The central three arches are emphasized by the relief offered by the columns attached to the façade. The façade was surmounted by a spectacular balustrade decorated with vases and statues in white marble.

Juvarra's design choice consists in that the façade assumes the function of a transparent grid and through it the interior decorative development can be perceived, in a resulting composition based on the passage of light. Juvarra desired a completely open loggia but weather conditions in Turin had forced him to protect the interior with the screen of large glazed windows [6,7].

The petro-architectonic relief (Figure 3) resulted in the false color representation of the different categories of materials used originally (Chianocco marble, Brossasco marble, Frabosa marble, and Vaje stone), and in the restorations of the façade through time (Carrara marble, Prali marble, Botticino limestone, and Malanaggio stone) (original and restoration stones shown in Table 2).

Table 1. Representative SEM-EDS analysis of white mica from Palazzo Madama façade samples and Chianocco quarries recalculated on the basis of 22 Ox.

| Sample | Façade | | | | | | Quarry | | | | | |
| | | | | | | | Site 3 | | | Site 2–4 | | |
Analysis Number	*Ph 1*	*Ph 2*	*Ph 3*	*Ph 4*	*Ph 5*	*Ph 6*	*Ph 7*	*Ph 8*	*Ph 9*	*Ph 10*	*Ph 11*	*Ph 12*
SiO_2	56.57	56.22	57.4	57.28	56.94	57.21	59.67	54.21	56.57	53.01	59.18	59.25
Al_2O_3	26.83	27.39	25.8	26.05	26.42	26.06	22.64	30.89	26.77	32.52	23.24	23.32
FeO	0	0	0	0	0	0	0	0	0	0	0	0
MgO	5.45	5.35	5.75	5.66	5.61	5.78	7.37	3.79	5.34	3.42	6.99	6.95
CaO	0	0	0	0	0	0	0	0	0	0	0	0
Na_2O	0	0	0	0	0	0	0	0	0	0.72	0	0
K_2O	11.15	11.04	11.05	11.01	11.03	10.95	10.31	11.11	11.32	10.33	10.59	10.48
Total	97.05	96.27	96.86	97	95.77	96.05	95.98	96.47	97.03	95.62	96.07	95.9
Si	7.08	7.03	7.17	7.16	7.12	7.15	7.42	6.79	7.09	6.63	7.37	7.37
Al IV	0.92	0.97	0.83	0.84	0.88	0.85	0.58	1.22	0.92	1.37	0.63	0.63
Al VI	3.04	3.07	2.97	2.99	3.01	2.98	2.74	3.34	3.04	3.42	2.78	2.79
Fe	0	0	0	0	0	0	0	0	0	0	0	0
Mg	1.02	1	1.07	1.05	1.05	1.08	1.37	0.71	1	0.64	1.3	1.29
Ca	0	0	0	0	0	0	0	0	0	0	0	0
Na	0	0	0	0	0	0	0	0	0	0.18	0	0
K	1.78	1.76	1.76	1.76	1.76	1.75	0	1.77	1.81	1.65	1.68	1.66

Figure 3. Petro-architectonic relief in false color representation of the Palazzo Madama façade (architectural drawings courtesy of Foundation Torino Musei-Palazzo Madama).

In particular, Chianocco marble was employed for the entire marble decoration of the façade, including bas relief and ornaments, Brossasco marble for the statues and the vases of the apex, and Vaje stone for the base of the building. Frabosa marble was used for some pillars and the slab in the summit balaustrade. The Carrara marble, light gray in color, was employed for an extensive replacement that involved both parts originally made of Brossasco marble and Chianocco marble elements. Prali marble was used for the first pillar on the left observing the façade and Malanaggio stone replaces numerous elements (pillars, bases and cornices) of the central part of the large summit balaustrade. Botticino limestone was employed for elements of the cornice and of the upper part of the facade and slabs of the balcony between the third and fourth column.

It is worth noting that the Gassino stone, reported by previous authors as a replacement material for the capitals, ledge and balaustrade [10,11], has not been found at the Palazzo Madama.

All kind of stone materials, both original and restoration ones, were exploited in the Piedmont region, except for Carrara marble, that crops out in Tuscany, and Botticino limestone that crops out in Lombardy.

On the façade the following characteristics of the Chianocco marble were observed: a strongly vacuolar structure (Figure 4a), a brecciated fabric with a pervasive vein network (Figure 4b), presence of mortars in the pores (Figure 4c), a reddish alteration of the columns (Figure 4d). Moreover, the local occurrence of a white soft powder on the stone suggests sulphation processes due to acid rains.

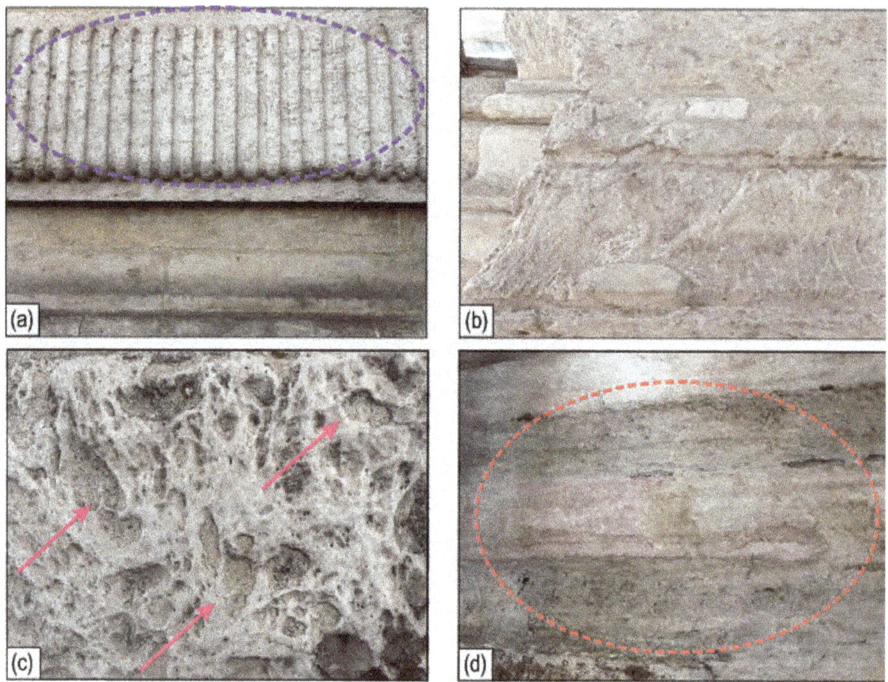

Figure 4. Characteristics of Chianocco marble macroscopically observed on the Palazzo Madama façade: (**a**) Slab of the façade with a strongly vacuolar texture; (**b**) Brecciated fabric with a pervasive vein network observed on a base of column of the façade; (**c**) Presence of mortars in the pores of the stone; (**d**) Detail of the reddish alteration of the column of the façade.

Table 2. The Palazzo Madama façade: (**a**) Original stone materials of architectural elements of the Palazzo Madama façade; (**b**) Replacement stone materials of the architectural elements of the Palazzo Madama façade.

(a)	Original Stones	Use	(b)	Replacement Stones	Replacement of	
					Chianocco Marble	**Brossasco Marble**
	Chianocco marble	Columns, pilasters, ashlars, cornices, portals and summit balustrade		Prali marble	Whole pillar under the first column on the left and slabs of the balcony between first and second column (?)	

Table 2. *Cont.*

(a)	Original Stones	Use	(b)	Replacement Stones	Replacement of	
					Chianocco Marble	Brossasco Marble
	Brossasco marble	Statues and bases on the summit balustrade, balustrade on the windows of the staircase		**Carrara marble**		Several elements of the balustrades on the large windows of the staircase and parts of the vases on the summit balustrade
	Frabosa marble	Staircase and elements of the summit balustrade		**Botticino limestone**	Elements of the cornice and of the upper part of the facade and slabs of the balcony between the third and fourth column	
	Vaie marble	Base of façade		**Malanaggio stone** **Bardiglio marble**	Elements of the great summit balustrade Elements of lower balustrade	

The so called Chianocco marble therefore actually shows a more or less continuous range of fabrics and lithologies from veined marbles to a tectonic carbonate breccia characterized by a high porosity and a vacuolar appearance which is comparable to the cargneules, a historical Alpine term to indicate brecciated carbonate rocks with a vacuolar structure [22].

5. The Chianocco Marble

5.1. Petrography

Petrographic analyses have been conducted on façade samples and on outcrop ones. Specimens from the façade were not sampled directly in situ but they consist of fragments detached from the summit balustrade. Outcrop samples were collected from five sites in the surroundings of Chianocco village (Figure 2b) and four of them (Site 1, 2, 4, 5) resulted as analogous to the façade marble. Conversely, the marble of Site 3 is very massive, fine-grained, white to gray and foliated at macroscopic observation. Similar features were not found in the marble used in the façade of Palazzo Madama.

The marble from Site 3 (Figure 5a) is characterized by a paragenesis consisting of major dolomite (Dol 80–90% in vol) and minor calcite (Cc 10–15% in vol) and some accessory minerals as quartz, white mica, apatite, rutile and opaque minerals. The rock shows a homogeneous grain ranging from homeo- to heteroblastic (average grain size 0.10–0.15 mm) grain. The texture is grano-blastic characterized by a triple point structure; the single crystal shows lobed to irregular edges. It also has a weakly oriented texture defined by some crystals of white mica. A potassium mica, characterized by high Si content (fengitic in composition) and a sodium mica (paragonite) was detected in this marble according to SEM-EDS analyses. This mineral assemblage is indicative of high pressure—low temperature metamorphic conditions. Potassium mica shows a strong zoning characterized by a compositional change in Si content between 6.63 and 7.37 atoms per formula unit (p.f.u.) on the basis of 22 Ox. (Table 1). Sodium mica is characterized by a compositional change in Si content between 5.98 and 6.14 atoms per formula unit (p.f.u.) on the basis of 22 Ox. (Table 3). This implies that the mica has grown under metamorphic conditions of high pressure, typical for the Dora Maira Massif [12]. The zoning of phengite can be ascribed to the effects of partial retrogression of phengite towards muscovite during the second metamorphic event that involved the Dora Maira Massif, which took place in low pressure conditions (Figure 5b–d).

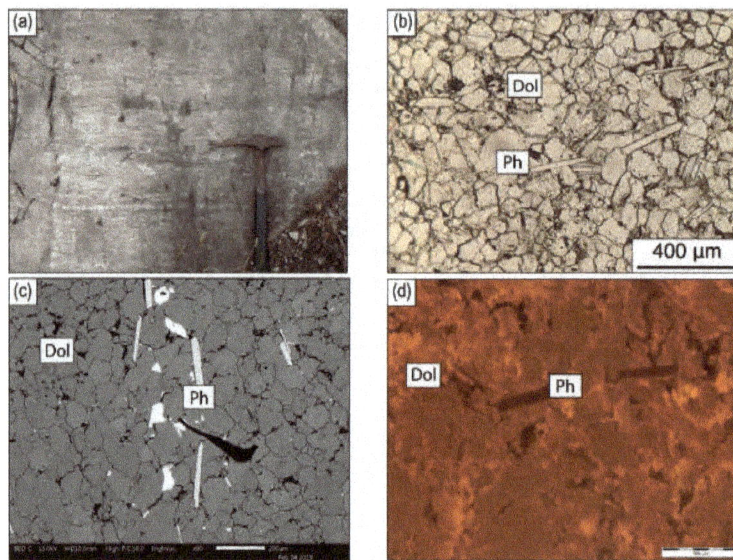

Figure 5. Petrography of the marble of Site 3: (**a**) Macroscopic aspect of the marble at Site 3; (**b**) Photo of optical microscope with only a polarizer, in which dolomite and phengite crystals are indicated; (**c**) SEM backscattered image. Dolomite crystals are dark grey and phengite ones are light-grey; (**d**) Cathodoluminescence image where dolomite crystals appear red and phengite crystals brown.

Table 3. Representative SEM-EDS analysis of paragonite from Chianocco massive marble (Site 3).

Sample	Quarry (Site 3)			
Analysis Number	*Pg 1*	*Pg 2*	*Pg 3*	*Pg 4*
SiO_2	50.57	49.11	50.13	49.64
TiO_2	0.00	0.00	0.00	0.00
Al_2O_3	40.67	42.04	41.07	41.53
FeO	0.00	0.00	0.00	0.00
MgO	0.62	0.00	0.48	0.00
CaO	0.00	0.28	0.00	0.00
Na_2O	7.50	8.17	7.73	7.90
K_2O	0.00	0.00	0.00	0.00
Total	95.53	96.06	96.78	96.38
Si	6.14	5.98	6.09	6.06
Al IV	1.86	2.02	1.91	1.95
Al VI	3.96	4.01	3.97	4.03
Fe	0.00	0.00	0.00	0.00
Mg	0.11	0.00	0.09	0.00
Ca	0.00	0.04	0.00	0.00
Na	1.77	1.93	1.82	1.87
K	0.00	0.00	0.00	0.00

Based on microscopical observations, this marble is strictly comparable to the Foresto marble [23].

Conversely the Chianocco marble (Sites 1, 2, 4, and 5) is characterized by a greater complexity in both structure and composition. Macroscopically it commonly shows a porous and vacuolar texture with irregularly shaped voids up to some centimeters large (Figure 6a,b). Microscopic analyses, and in particular SEM-BSE and cathodoluminescence (CL) imaging clearly show that the rock is dolomitic but calcite may be very abundant (Figure 6c–f). Calcite fills mm-large veins with commonly sharp edges. Crystals are equant, limpid and show an overall dull brown CL colour (Figure 6e).

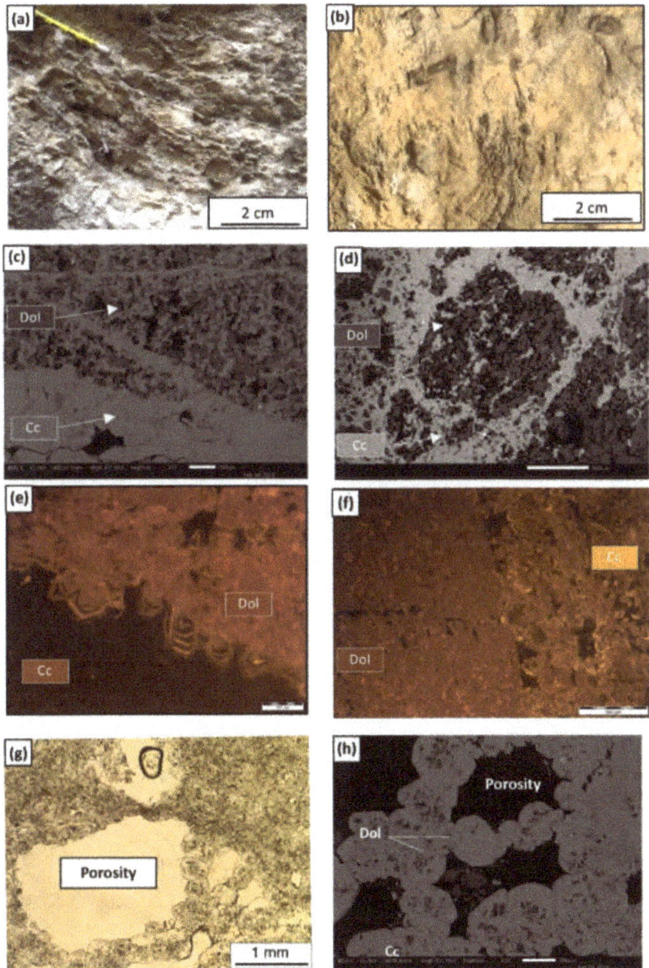

Figure 6. Petrography of Chianocco marble: (**a**, **b**) Macroscopic aspect of Chianocco marble; (**c**) SEM backscattered image where calcite veins are clearly visible; (**d**) SEM backscattered image where cataclasite structure is evident; (**e**) Cathodoluminescence image where in a calcite vein a zoning is recognizable with the very first portion of crystals characterized by thin bands of bright to moderate yellow; (**f**) Cathodoluminescence image where different dolomite, in red, and calcite, in black and yellow, portions are recognizable; (**g**) Optical microscope image with only a polarizer, with evident vacuolar texture; (**h**) SEM backscattered image where the voids are surrounded by calcite septa that originally separated the dolomite clasts.

However, a zoning is recognizable with the very first portion of crystals characterized by thin bands of bright to moderate yellow (Figure 6e,f). This zoning clearly documents a crystal growth in a void in static conditions. Calcite is also present in intimate association with dolomite, clearly distinguishable in CL for the orange colour. Calcite fills spaces among irregularly shaped fragments of dolomite, from few tens of microns to some millimetres large and shows the same zoning observed in veins. This demonstrates that the Chianocco marble is a cataclasite where the original dolomitic marble was fractured and/or comminuted into fragments of heterogeneous grain size; successively the fractures and the open spaces in the cataclasite were cemented by sparry calcite. Moreover, in some portions of the rock, calcite septa that originally separated the dolomite clasts now surround the voids (Figure 6g,h).

Phengite and phlogopite occur in the Chianocco marble and locally are broken and folded (Figure 7).

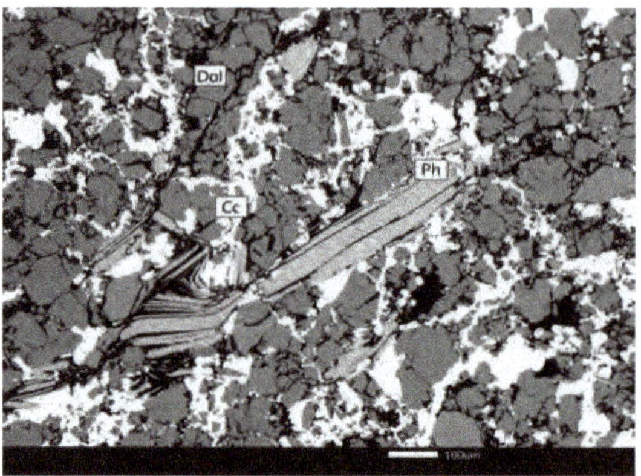

Figure 7. SEM backscattered image of Phengite crystal broken and folded.

The white mica of façade and quarry samples, analyzed by SEM-EDS, plots in the field of phengite and displays a high content of silicon, and index of crystallization at high pressures (Table 1, Figure 8). More in detail, the Si amount, expressed as atoms per formula unit (p.f.u.) based on 22 oxygens, varies between 7.03 and 7.17 for façade samples, and between 6.63 and 7.42 for quarry samples and falls in the field of high pressure. The Mg content is between 1.00 and 1.08 atoms p.f.u. for façade samples and between 0.64 and 1.37 for quarry samples, wheras Fe was always absent, consistently with the carbonate system composition. The composition of phengite of Palazzo Madama samples partially corresponds to the mica of quarry marbles (Figure 8a). Notably the micas of historical quarry samples show a wider range of variation in the Si/Al ratio.

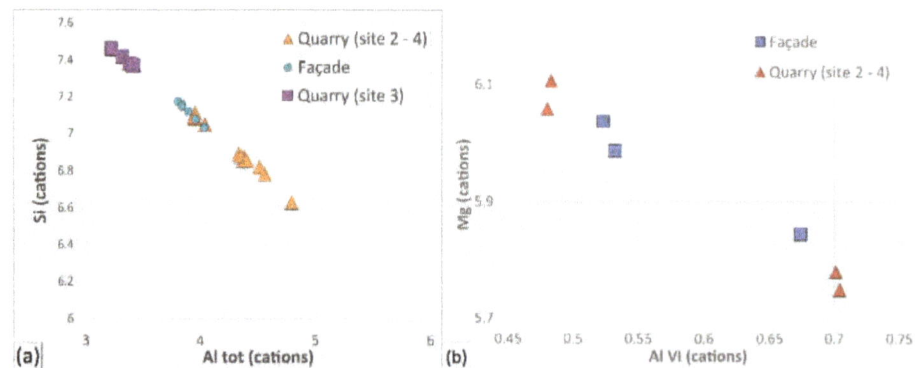

Figure 8. The Palazzo Madama façade and Chianocco quarry: (**a**) Si-Al tot classification diagram for white mica of the Chianocco marble from the Palazzo Madama façade and Chianocco quarry; (**b**) Mg-AlVI classification diagram for phlogopite of the Chianocco marble from the Palazzo Madama façade and Chianocco quarry.

Phlogopite, which is characterized by a much higher Mg:Fe ratio than biotite, was also found. Phlogopite only rarely occurs in marbles and therefore it can be used to characterize the marble variety of the Palazzo Madama façade from the mineralogical point of view. Phlogopite blasts occur rarely in the samples from the Chianocco quarries; Table 4 and Figure 8b show representative SEM-EDS analyses of phlogopite.

Table 4. Representative SEM-EDS analysis of phlogopite from the Palazzo Madama façade samples and Chianocco quarries recalculated on the basis of 22 Ox.

Façade				Quarry (Site 2–4)			
Phl 1	*Phl 2*	*Phl 3*	*Phl 4*	*Phl 5*	*Phl 6*	*Phl 7*	*Phl 8*
46.17	46.51	46.21	47.55	47.27	46.33	46.88	46.46
0.00	0.00	0.00	0.00	0.00	0.00	0.00	0.00
14.87	13.42	13.62	12.67	12.68	14.66	14.41	13.32
0.00	0.00	0.00	0.00	0.00	0.00	0.00	0.00
28.79	29.61	29.30	29.77	29.81	28.34	28.30	30.01
0.00	0.00	0.00	0.00	0.00	0.00	0.00	0.00
0.00	0.00	0.37	0.00	0.00	0.00	0.00	0.00
8.92	9.12	9.21	10.01	8.86	8.63	9.06	8.80
96.84	97.00	96.55	96.70	93.69	95.02	96.68	96.62
6.29	6.36	6.33	6.46	6.44	6.34	6.39	6.34
1.71	1.64	1.67	1.54	1.56	1.66	1.61	1.66
0.67	0.52	0.53	0.49	0.48	0.70	0.70	0.48
0.00	0.00	0.00	0.00	0.00	0.00	0.00	0.00
5.84	6.04	5.99	6.03	6.06	5.78	5.75	6.11
0.00	0.00	0.00	0.00	0.00	0.00	0.00	0.00
0.00	0.00	0.10	0.00	0.00	0.00	0.00	0.00
1.37	1.39	1.41	1.67	1.35	1.31	1.38	1.34

Finally, some samples are characterized by a red zone in calcite veins due to the presence of iron oxides inside calcite crystals (Figure 9a,b). This phenomenon is visible on the macroscopic scale in Site 4 (Figure 9c); the rock on the outcrop is similarly reddish as the column of the façade already mentioned (Figure 4d).

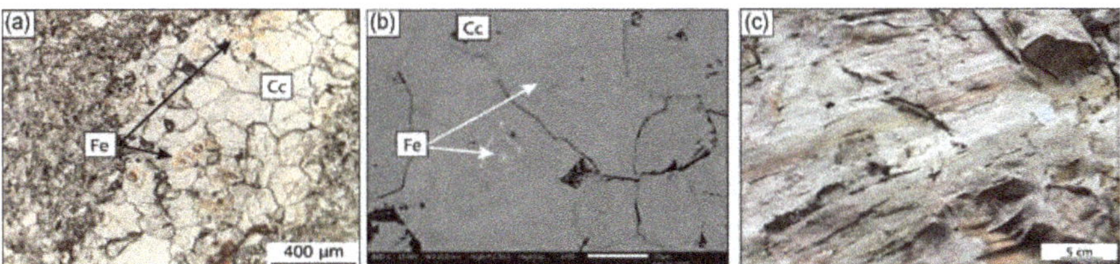

Figure 9. Reddish alteration phenomena: (**a**) Red zones in calcite veins observed by optical microscope with only a polarizer; (**b**) SEM backscattered image where zones of iron oxides in calcite crystals are visible; (**c**) Reddish alteration phenomena visible at the Site 4 quarry.

SEM-EDS analysis also revealed in some areas superficial gypsum with spherical and "rose" morphology (Figure 10a). Also, intergranular gypsum was found (Figure 10b,c).

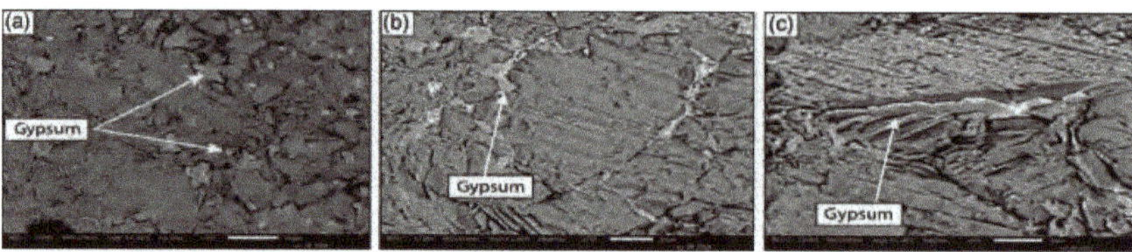

Figure 10. SEM backscattered image with superficial gypsum present in façade samples: (**a**) Spherical and "rose" morphology gypsum; (**b**,**c**) Intergranular gypsum.

5.2. C-O Stable Isotope Analysis

C-O stable isotope analyses have been carried out on two selected samples of Chianocco marble where the characteristic brecciated structure is best represented. One consists of a piece detached from the Palazzo Madama façade and the other comes from Site 4, in the surroundings of Chianocco, village where the rock results are analogous to the façade marble. Values of $\delta^{18}O$ and $\delta^{13}C$ have been determined on calcite and dolomite of both samples. The results, referred to as the PDB standard, are plotted in Table 5 and Figure 11 where data from the literature [18], concerning the massive dolomitic marble, are also reported for comparison. These analyses were not aimed to be a complete archaeometric characterization and provenance of Palazzo Madama marble but only to verify the genetic relationships of calcite and dolomite which, on the basis of petrographic observations, are clearly not in equilibrium. Although the data set is not statistically highly significant, two points are relevant: 1) Isotopic data of dolomite samples coming from the façade and from Chianocco quarries compare well with data referred to the massive dolomitic marble of Chianocco and Foresto quarries [18] with $\delta^{18}O$ values ranging between −7.06 and −6.00 and $\delta^{13}C$ ranging from 0.79 to 1.30; 2) Calcite is significantly more depleted in ^{18}O than dolomite, showing values of −9.2 and −10.8 ‰ PDB i.e., with a shift of up to −4.8 ‰ PDB from dolomite to calcite in the same sample. This establishes marked differences in physico-chemical features of the fluids (temperature, nature and composition of fluids) and thus contrasting geological settings in which dolomite and calcite formed. In particular, the dolomite portion of the rock records the Alpine metamorphic overprint of Triassic sediments whereas the calcite portion is likely due to post-metamorphic evolution of the marble after exhumation and interaction with meteoric waters.

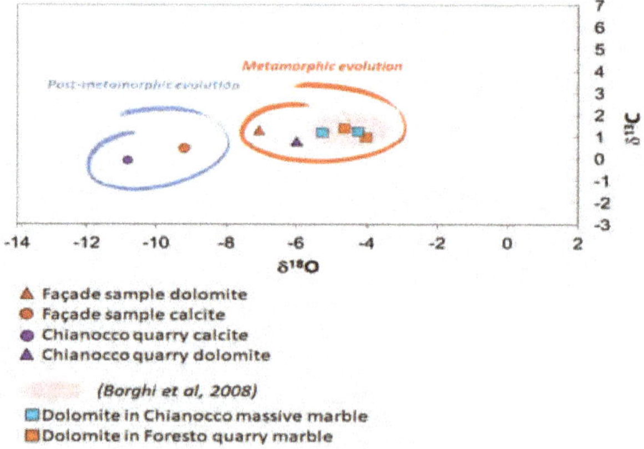

Figure 11. The $\delta^{18}O$ and $\delta^{13}C$ diagram of calcite and dolomite of the investigated Chianocco marble. The isotopic reference of Chianocco and Foresto dolomite according to Borghi et al., 2008 is also reported.

Table 5. Calcite and dolomite stable isotope (C, O) data of the Palazzo Madama façade sample and Chianocco quarry sample.

Sample	Calcite		Dolomite	
	$\delta^{13}C$	$\delta^{18}O$	$\delta^{13}C$	$\delta^{18}O$
Façade sample	0.45	−9.20	1.30	−7.06
Chianocco quarry	−0.10	−10.82	0.79	−6.00

6. Model Evolution

The petrographic study of the quarry and façade samples allowed us to define a model of the evolution of the rock from its formation to its employment. This model is articulated in six steps as shown in Figure 12, starting from deposition of the dolostone, through Alpine metamorphism and brittle deformation and brecciation to superficial partial dissolution, with only the very last step being related to the recent exposure of the stone to atmospheric agents as a facing of the Palazzo Madama. In the following, each step will be commented on in detail.

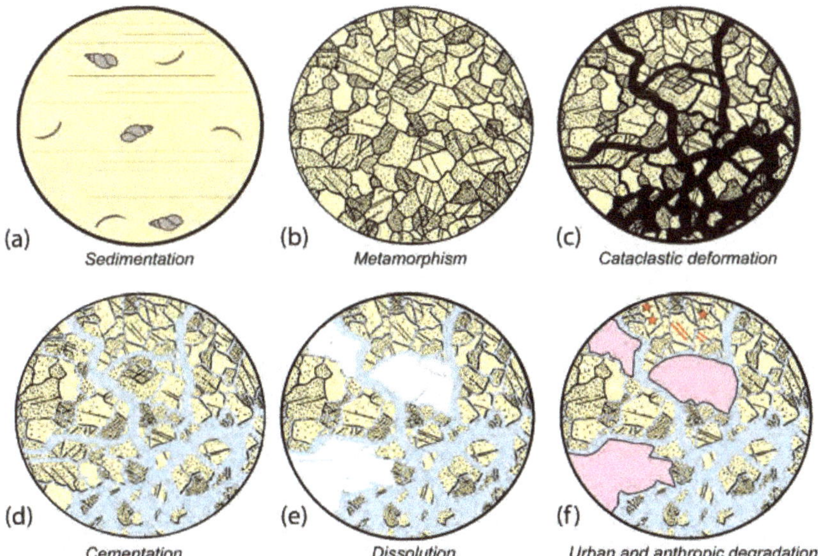

Figure 12. Representation of the model evolution of Chianocco marble from its formation to its employment: (**a**) Original dolostone; (**b**) Dolomitic marble; (**c**) Brittle deformation indicated by black fractures; (**d**) Cementation indicated in light blue color (tectonic carbonate breccia with a complex and pervasive cataclastic fabric); (**e**) Selective dissolution of dolomite marble clasts indicated in white (vacuolar texture); (**f**) Mortars in the pores indicated in pink and sulphation indicated in red stars, circles and lines in the upper part of the round.

Step 1

In the Triassic, a carbonate sediment was deposited in a peritidal environment and was very early dolomitized. No fossil nor sedimentary structures are preserved in the Chianocco marble but it is clearly established in the geological literature that an extensive carbonate platform existed in the Triassic in all the units presently involved in the Alpine chain.

Step 2

During the first part of the Alpine orogenesis (Late Cretaceous-Eocene) oceanic and continental units were involved in subduction processes. The presence of phengite indicates high pressure conditions in Site 3 samples, therefore attesting that it is a marble formed in a metamorphic process in a pressure and temperature range corresponding to eclogitic facies. These characteristics reveal that these samples are comparable to the Foresto Marble, extracted since ancient times a few kilometers from Chianocco, and used in 9 BC for the Arco di Susa [23] and for the façade of the Cathedral of Turin. It is in fact a fine-grained, very compact dolomite marble.

The proximity of Foresto quarries with Chianocco lead to merge Foresto and Chianocco Marbles in a unique lithotype.

Steps 3 and 4

In a later, post-metamorphic, stage which is not possible to date precisely, brittle deformation took place at high crustal levels, probably not far from the surface. This event caused a strong grain reduction of the dolomitic marble and its transformation into a tectonic carbonate breccia with a complex and pervasive cataclastic fabric. The $\delta^{18}O$ values of calcite veins and cement, lighter than marble dolomite, possibly document meteoric waters percolating down and feeding a fracture-related circulation system.

Step 5

A process of selective dissolution of the dolomite marble clasts explains the origin of the vacuolar texture. This process took place when the calcite-cemented breccias were exposed to weathering at or very close to the topographic surface in a very recent past (possibly the Pleistocene) giving rise to a vacuolar structure comparable to that shown by the so called cargneules well known in the Alpine literature [22]. In some portions of the rock, calcite septa that originally separated the dolomite clasts now surround the voids.

Step 6

Regarding the environmental degradation, superficial sulphation of carbonate rocks is typical of degradation due to acid rains, in particular for formation of gypsum crystals with spherical and "rose" morphology. The comparison with stone samples from outcrop shows that gypsum is not present in the Chianocco massive marble, insofar supporting the environmental degradation hypothesis.

Moreover, past restoration interventions carried out with not suitable materials (like cement-based mortars) contributed to accelerated deterioration.

7. Conclusions

A multidisciplinary geological approach was applied to the façade of the Palazzo Madama, one of the most important historical monuments in Turin and a UNESCO World Heritage site, which has been recently affected by environmental degradation. A detailed architectural-petrographic relief and minero-petrographic and isotopic analyses were carried out comparing quarry samples coming from the historic sites of exploitation with selected fragments detached from the façade. The main results may be summarized as follows:

- The kind of ornamental stone used and their precise distribution on the façade were defined distinguishing the original stone materials from the ones used during historical restorations;
- The originally used material, the Chianocco marble, is still the most abundant and the one which shows the greatest degradation;
- The minero-petrographic study of the Chianocco marble and the comparison with the same material cropping out in the historical quarries shows that some features observed on the Palazzo Madama façade such as a vacuolar structure and local reddenings, usually absent in ornamental

marbles, are primary features of the rock itself and are not due to degradation in an urban context. They are conversely related to the very complex history of the rock which started in the Triassic age as deposition of a carbonate sediment, evolved through Alpine metamorphism and deformation, and finished with exposure at the surface where dissolution by meteoric waters generated the vacuolar structure. Only gypsum crystals grown in voids and the application of mortars in natural voids, enhancing the physical degradation of the stone, are due to pollution and human interventions.

This research highlights the importance of geological studies in conservation issues in cultural heritage by defining the characteristics of stone materials, and the reasons for their degradation. In particular, this is true for local heritage stones which can be studied not only on the historical buildings but also in the provenance areas.

Author Contributions: Conceptualization, F.G.; Data curation, F.G., A.B., A.A., M.C. and P.C.; Formal analysis, F.G. and A.B.; Investigation, F.G., A.B., A.A., L.M., M.C. and L.A.; Methodology, F.G., A.B., A.A., L.M. and L.A.; Project administration, F.G.; Supervision, A.B., A.A., L.M., L.A. and P.C.; Validation, P.C.; Writing—original draft, F.G., A.B., A.A. and L.M.; Writing—review & editing, F.G., A.B., A.A., L.M. and P.C.

Funding: The research at the Department of Earth Science has been funded by Compagnia San Paolo and University of Turin in the frame of the GeoDIVE Project.

Acknowledgments: This research was done within the pilot investigation project of the Foundation of Conservation and Restoration Centre of "La Venaria Reale", we would like to thank especially Marta Gomez, Michela Cardinali and Daniela Russo. Thanks to Paola Marini, Claudio De Regibus for the localization of ancient Chianocco quarries. Acknowledgments to Foundation Torino Musei-Palazzo Madama and Superintendence of Archeology, Fine Arts and Landscape for the Metropolitan City of Torino.

Conflicts of Interest: The authors declare that there is not conflict of interest.

References

1. Lazzarini, L. Archaeometric aspects of white and coloured marbles used in antiquity: The state of the art. *Per. Mineral.* **2004**, *73*, 113–125.
2. Sacco, F. Geologia applicata della Città di Turin. *Giornale Geol. Prat.* **1907**, *5*, 121–162.
3. Borghi, A.; d'Atri, A.; Martire, L.; Castelli, D.; Costa, E.; Dino, G.; Favero Longo, S.E.; Ferrando, S.; Gallo, L.M.; Giardino, M.; et al. Fragments of the Western Alpine chain as historic ornamental stones in Turin (Italy): Enhancement of urban geological heritage through geotourism. *Geoheritage* **2014**, *6*, 41–55. [CrossRef]
4. Gambino, F.; Borghi, A.; d'Atri, A.; Gallo, L.M.; Ghiraldi, L.; Giardino, M.; Martire, L.; Palomba, M.; Perotti, L. TOURinSTONES: A free mobile application for promoting geological heritage in the city of Turin (NW Italy). *Geoheritage* **2018**, *11*, 3–17. [CrossRef]
5. Borghi, A.; Berra, V.; d'Atri, A.; Dino, G.A.; Gallo, L.M.; Giacobino, E.; Martire, L.; Massaro, G.; Vaggelli, G.; Bertok, C.; et al. Stone materials employed for monumental buildings in the historical centre of Turin (NW Italy): Architectonical survey and petrographic characterization of via Roma. In *Global Heritage Stone: Towards International*; Pereira, D., Marker, B.R., Kramar, S., Cooper, B.J., Schouenborg, B.E., Eds.; Special Publications; The Geological Society of London: London, UK, 2015; Volume 407, pp. 201–218.
6. Palazzo Madama Official Website. Available online: https://www.palazzomadamaTurin.it/en/node/1080 (accessed on 4 August 2019).
7. Telluccini, A. *Il Palazzo Madama di Turin*; Lattes Editori: Turin, Italy, 1928; 168p.
8. Arnaldi di Balme, C. Consolidamento e manutenzione della facciata di Palazzo Madama (2010–2011). *Palazzo Madama. Studi Notizie. Riv. Annu. Mus. Civ. d'Arte Antica Turin* **2013**, *2*, 144–157.
9. Fratini, F.; Mattone, M.; Rescic, S. Il restauro delle superfici di Palazzo Madama a Turin: Metodi ed esiti. In *Intervenire Sulle Superfici dell'Architettura tra Bilanci e Prospettive*; Ricerche, A., Ed.; Scienza e Beni Culturali: Bressanone, Italy, 2018; pp. 109–119.
10. Berti, C. Il Palazzo di marmo. In *Ricerche sui Materiali Lapidei Della Facciata di Palazzo Madama in Turin*; Associazione Georisorse e Ambiente: Turiiin, Italy, 1998; Volume 95, pp. 237–242.
11. Berti, C.; Gomez Serito, M. *I Marmi Della Facciata di Palazzo Madama a Turin*; Associazione Georisorse e Ambiente: Turiiin, Italy, 1999; Volume 96, pp. 13–22.

12. Gasco, I.; Gattiglio, M.; Borghi, A. New insight on the lithostratigraphic setting and on tectono-metamorphic evolution of the Dora Maira vs Piedmont Zone boundary (middle Susa Valley). *J. Earth Sci.* **2011**, *100*, 1065–1085.

13. Fiora, L.; Audagnotti, S. Chianocco and Foresto Marble: Mineropetrographic characterization and uses. In Proceedings of the International Workshop "Dimension stones of the European Mountains", Torre Pellice, Italy, 10–12 June 2001; pp. 243–250.

14. Fiora, L.; Gambelli, E. Le principali pietre da costruzione e da ornamento in Valle Susa (Piedmont). In *Restauro Archeologico*, 1st ed.; Firenze University Press: Firenze, Italy, 2006; pp. 18–20.

15. Matthews, K.J. The establishment of a data base of neutron activation analyses of white marble. *Archaeometry* **1997**, *39*, 321–332. [CrossRef]

16. Gorgoni, C.; Lazzarini, L.; Pallante, P.; Turi, B. An updated and detailed minero-petrographic and C–O stable isotopic reference database for the main Mediterranean marbles used in antiquity. In *ASMOSIA 5: Interdisciplinary Studies on Ancient Stone, Proceedings of the Fifth International Conference of the Association for the Study of Marble and Other Stones in Antiquity, Museum of Fine Arts, Boston, MA, USA,11-15/06/ 1998*; Herrmann, J.J., Herz, N., Newman, R., Eds.; Archetype: London, UK, 2002; pp. 115–131.

17. Polikreti, K. Detection of ancient marble forgery: Techniques and limitations. *Archaeometry* **2007**, *49*, 603–619. [CrossRef]

18. Borghi, A.; Fiora, L.; Marcon, C.; Vaggelli, G. The Piedmont White Marbles Used in Antiquity: An Archaeometric Distinction Inferred by A Minero-Petrographic And C-O Stable Isotope Study. *Archaeometry* **2009**, *51*, 913–931. [CrossRef]

19. Petrakakis, K.; Dietrich, H. MINSORT: A program for the processing and archivation of microprobe analyses of silicate and oxide minerals. *Neues Jahrb. Mineral. Monatshefte* **1985**, *8*, 379–384.

20. McCrea, J. On the isotopic chemistry of carbonates and a paleotemperature scale. *J. Chem. Phys.* **1950**, *18*, 849–857. [CrossRef]

21. Craig, H. Isotopic standards for carbon and oxygen and correction factor for mass-spectrometric analysis of carbon dioxide. *Geochim. Cosmochim. Acta* **1957**, *12*, 133–149. [CrossRef]

22. Amieux, P.; Jeanbourquin, P. Cathodoluminescence et origine diagénétique tardive des cargneules du massif des Aiguilles Rouges (Valais, Suisse). *Bull. Soc. Géol. France* **1989**, *1*, 123–132. [CrossRef]

23. Agostoni, A.; Barello, F.; Borghi, A.; Compagnoni, R. The white marble of the Arch of Augustus (Susa, North-Western Italy): Mineralogical and petrographic analysis for the definition of its origin. *Archaeometry* **2017**, *59*, 395–416. [CrossRef]

 *sustainability*

Article

The *Bargiolina*, a Striking Historical Stone from Monte Bracco (Piedmont, NW Italy) and a Possible Source of Industrial Minerals

Alessandro Cavallo [1],* and Giovanna Antonella Dino [2]

[1] Department of Earth and Environmental Sciences—DISAT, University of Milano-Bicocca, Piazza della Scienza, 1-4, 20126 Milano, Italy
[2] Department of Earth Sciences, University of Turin, Via Valperga Caluso, 35, 10125 Torino, Italy
*   Correspondence: alessandro.cavallo@unimib.it; Tel.: +39-338-2343834

Received: 22 June 2019; Accepted: 6 August 2019; Published: 8 August 2019

Abstract: The *Bargiolina* quartzite from Monte Bracco (western Alps, northern Italy) represents one of the most important historical ornamental stones of the Piedmont region. Known and used since the prehistoric age as substituting material for chert, it was celebrated by Leonardo da Vinci, and exploited at least since the XIII century, peaking in the XX century. It was extensively used in the construction of basilicas and noble palaces by famous architects of Piedmontese Baroque, for internal and external stone cladding. There are four main commercial and chromatic varieties, and the main technical feature is the regular schistosity, to obtain very thin natural split slabs. The different varieties have a homogeneous mineralogical composition and microstructure: A fine and homeoblastic grain size, and a granular—lepidoblastic texture, with regularly spaced schistose domains. The main rock-forming minerals are quartz, phengite, small amounts of K-feldspar and traces of plagioclase and chlorite. The yield rate of quarries is about 20%, and the poor exploitation planning of the past led to only partly exploited quarry benches, with a very poor residual yield. The large amount of quartz-rich quarry waste and the presence of kaolin-rich gneisses suggests the potential for novel applications in the field of industrial minerals.

Keywords: heritage stone; quartzite; *Bargiolina*; quarries; dimension stone; kaolin; industrial minerals

---

1. Introduction

Bargiolina is the term that defines a group of four main varieties of quartzites extracted from Monte Bracco (1306 m a.s.l.), an isolated mountain close to the village of Barge (western Alps, northern Italy), approximately 40 km from Turin and Cuneo (Figure 1). Known and used since the prehistoric age as substituting material for chert, the *Bargiolina* has been exploited in slabs at least since the XIII century, peaking in the XX century (up to 300 kt/year), used as internal and external facing because of its excellent technical proprieties [1,2]. The first reliable report on its use date back to 1374, to the statutes granted by Amedeo VI of Savoy to the town of Barge, in which it was ordered to maintain the paths that led from the city to the quarries on Monte Bracco in good condition. The use was increased by another Savoyard, Ludovico (1413–1465): The one who bought the Holy Shroud (*Sindone*) and forbade the citizens of Barge the roof covers in wood or straw. Leonardo Da Vinci, on 5 January 1511, celebrated the quartzite and described Monte Bracco as " ... rich in a stratified which is white as the Carrara marble, is speckled and hard as a porphyry ... " [3]. Between the XVII and XVIII centuries the quarries worked at full speed to meet the needs of great architects of the Piedmontese Baroque such as Guarino Guarini (1624–1683), Filippo Juvarra (1678–1736), Francesco Gallo (1672–1750) and Benedetto Alfieri (1699–1767) in the construction of basilicas and noble palaces (Figure 2): From the

Guarini towers of the Racconigi Castle (Cuneo) to the salon of the first subalpine parliament in Palazzo Madama (Turin), and from the Stupinigi Castle (in the district of Nichelino in Turin) to the Reggia di Venaria (Venaria Reale, Turin). Filippo Juvarra used the *Bargiolina* in the flooring of the side wings of the Stupinigi hunting lodge and in the cloister of the Superga Basilica (Turin). In the second half of the XIX century and at the beginning of the XX century, the quartzite was marketed in northern Italy and exported to Russia and South America [1]. Waste materials were milled to be transformed into powder and used as a coating material for the cylinders of cement mills, as a powder for cleaning cannon barrels and even as a component of toothpaste. Nowadays, only two quarries are involved in the occasional exploitation of very small amounts of quartzite. The aim of this paper is to highlight the features of this stone and to characterize its potential in the industrial minerals sector.

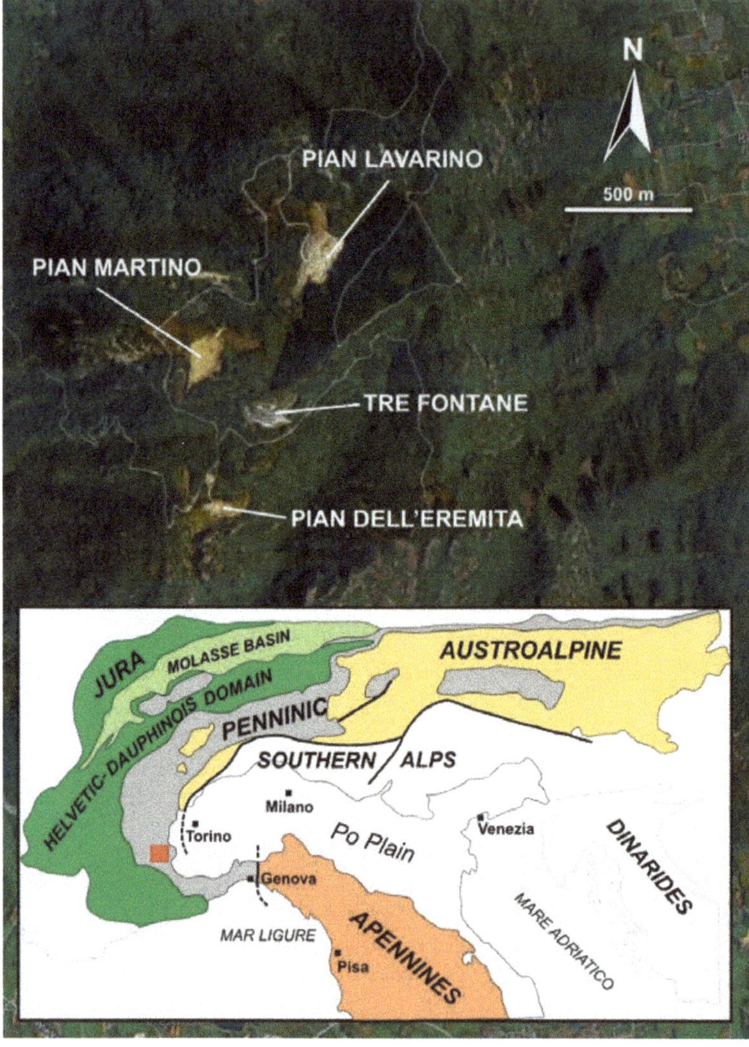

Figure 1. Geological framework of the Alps and aerial view (from Google Earth) of the Monte Bracco main quarry areas.

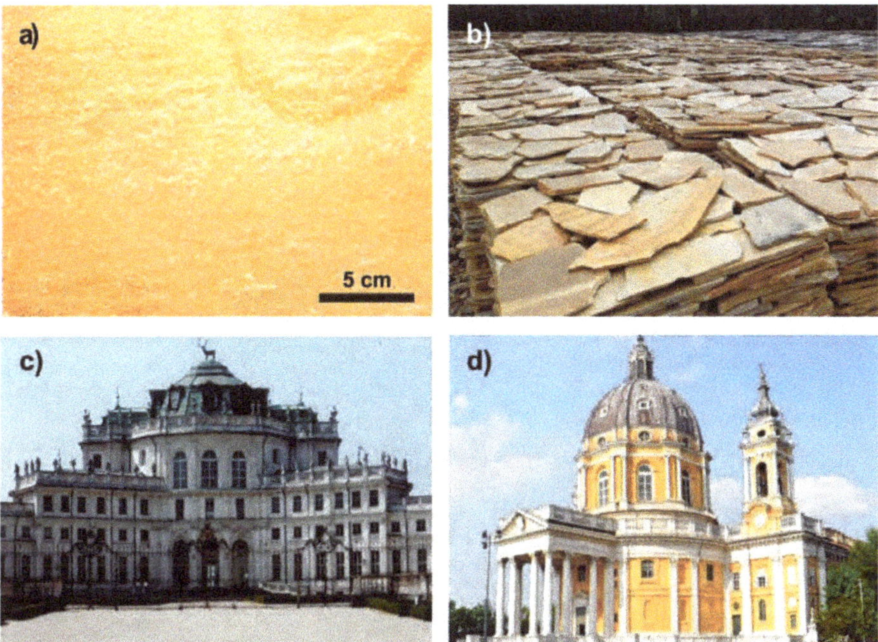

Figure 2. (**a**) "Golden" quartzite slab; (**b**) stacked quartzite slabs on the quarry floor; (**c**) Stupinigi hunting lodge (district of Nichelino in Turin); and (**d**) Superga basilica (Turin).

2. Materials and Methods

A total of 18 representative quartzite samples from the main quarry areas (Barge and Sanfront municipalities), as well as 5 kaolinitic gneisses (the host rock of quartzites) were characterized by polarized light optical microscopy on thin sections, X-ray powder diffraction (XRD) and bulk chemistry (X-ray fluorescence—XRF). The XRD analyses were performed using a PANalytical X'Pert PRO PW3040/60 X-ray diffractometer with Ni-filtered Cu Kα radiation at 40 kV and 40 mA, $\frac{1}{2}°$ divergence and receiving slits, and step scan of 0.02° 2θ, in the 3–80° 2θ range. The limit of detection (LOD) of XRD depends on the mineral phase and is generally comprised between 0.1 wt. % for highly crystalline phases and 2 wt. %. The qualitative XRPD analysis was performed running the X'Pert High-Score® software (Malvern Panalytical Ltd., Almelo, the Netherlands) using the ICSD PDF2 database. A semi-quantitative evaluation of the relative abundance of single minerals was obtained with the internal standard technique (by adding 20 wt. % of NIST SRM 676a α-corundum powder) and the reference intensity ratio (RIR) method [4]. Whole-rock geochemistry (major elements) was assessed by energy-dispersive XRF (Panalytical Epsilon 3-XL spectrometer), whereas physical and technical properties were taken from the literature [5,6].

3. Geological Framework

The Monte Bracco area is located in the Dora-Maira Massif (DMM), a crystalline massif (Paleozoic basement and a thin Mesozoic cover) belonging to the inner part of the Penninic Domain, together with the Monte Rosa and the Gran Paradiso Massifs, extending more than 1000 km^2 in the central sector of the Cottian Alps (western Alps) [7,8]. The southern part of DMM is composed of four main tectonometamorphic units, from the lower to the upper structural levels: The Pinerolo, San Chiaffredo, Brossasco-Isasca and Rocca Solei units [9]. In its northern sector, the DMM comprises two main units that, during Alpine orogeny, were metamorphosed under different pressure–temperature (P–T) peak conditions: The upper is an eclogite-facies polymetamorphic complex (metasediments and upper Ordovician meta-intrusives, covered by thin Mesozoic carbonate metasediments), whereas the lower one consists of a blueschist-facies Permo-Carboniferous complex. The DMM is considered a slab of paleo-European continental crust, involved in Alpine-related E-dipping subduction, W-verging continental collision and deep crust/mantle indentation [10,11], and is now stacked in the axial sector of the Western Alps and tectonically overlain by blueschist-facies and eclogite-facies meta-ophiolite units. Monte Bracco consists of phengite-bearing orthogneisses (*Pietra di Luserna*) and paragneisses, whose foliation is gently NW–WNW dipping (between 0° and 20°), giving rise to an asymmetrical morphology, from the hilly relief on the eastern side, to the sub-vertical walls on the western side. Ortho- and paragneiss-hosted lenses and seams of quartzites (the alpine metamorphic product of the quartz-arenitic Permo-Triassic cover), are cropping out only towards the uppermost part of Monte Bracco (between 1100 and 1300 m a.s.l., an area of approximately 1.5 km^2) and have been quarried as ornamental stones (*Bargiolina*). The thickness of the quartzite benches is between 2 and 10 m, in sub-parallel lenses, from the lower to the upper: Banco Barmalunga (up to 10 m), Banco Combale Rinaudo (7 m), Banco Savoia (9 m), Banco Tre Fontane (2–2.5 m) and Banco Quarzite di Tetto (6–10 m) [12]. The five quartzite lenses are interpreted as the product of isoclinal polyphasic folding and tectonic detachments [13]. The regional foliation is sometimes disturbed by low-angle brittle—ductile tectonic discontinuities and hydrothermal veins (quartz and tourmaline). The host gneisses are locally strongly altered to clay, especially close to the Banco Tre Fontane. Due to the abundance and good quality of kaolin, extensive extraction occurred in the past (up to 50 kt/y till 1997).

4. Exploitation Setting

From the XVI century the *Bargiolina* was quarried in a limited and discontinuous way, but a significant increase in its exploitation occurred from the 1794 thanks to the Trappist monks, who became suppliers of quartzite for many churches built at that time. If at the beginning of the XIX century the quartzite quarries on Monte Bracco were almost 40 in number, at the end of the XX century beginning of the XXI century only two companies from Barge municipality were exploiting the *Bargiolina*: Cave Gontero s.n.c. in Pian Lavarino and A.T.I. Tinarelli S.p.A. in the Pian Martino locality [12]. The quarry La Quarzite di Sanfront in the Sanfront municipality was stopped in the 80s of the XX century, due to the Banco Tre Fontane bench exhaustion. The quarrying method was based on explosives (detonating cord and gunpowder) or mechanical hammer (Figure 3). The yield rate of the quartzite was about 20%, thus 80% of the exploited material represented waste rocks dumped on the slopes of Monte Bracco. Poor exploitation planning in the XX century, which involved the best portions of the rock body, caused the presence of partly exploited quarry benches, characterized by a yield rate (net production) of about 4%–8%. The good material coming from the quarries was then processed to obtain stone for mosaic (nearly 50%), squared slabs (8%), wall blocks (15%) and waste from working activity (about 30%). At the beginning of the XXI century, about 70% of the products were sold in European countries, and the major markets were France and Belgium. The low yield of the quartzite for slab production caused, and potentially can still cause, the production of a large quantity of waste, which must be well managed and potentially recovered (see Section 6). This factor, together with the strong competition by the widespread "golden quartzite" from Brazil (e.g., "Fantasy gold" quartzite, which shows similar aesthetical characteristics compared to *Bargiolina* (https://www.stonecontact.com/fantasy-gold-quartzite/s19476)) caused the progressive reduction of the quarrying activities to nearly zero production, which characterized the last decade.

Figure 3. (**a**) Manual splitting of quartzite slabs at the end of the XIX century; (**b**) Pian Lavarino quarry front; (**c**) Tre Fontane quarry front; and (**d**) blasting at the Pian Lavarino quarry front.

5. Commercial Varieties, Mineralogy, Petrography, Physical and Technical Properties

There are four main commercial and chromatic varieties of *Bargiolina*: "golden" yellow (the most valuable, Figure 2a), pale yellow, olive green, grey and white (*Marmorina*, the most common). The main technical characteristic of the *Bargiolina* is the regular schistosity and the perfect fissility, due to the presence of thin, plane parallel white mica (phengite) layers. Because of this feature, it is possible to obtain very thin natural split slabs (1–2 cm). Under the optical microscope, the quartzite varieties show a quite homogeneous mineralogy and microstructure: A fine and homeoblastic grain size, a granular, lepidoblastic texture (Figure 4) with regularly spaced schistose domains (up to mylonitic); and small amounts of slightly altered K-feldspar porphyroclasts (orthoclase, 1–2 mm) give to the rock a micro-augen texture. The main rock-forming minerals (Table 1) are quartz (65–85 wt. %, interpenetrating textures), white mica (phengite, up to 15 wt. %), K-feldspar (orthoclase, 5–10 wt. %, frequently altered to kaolinite) and traces of plagioclase (albite) and chlorite. Typical accessory minerals (≤1 wt. %) are zircon, rutile, titanite, hematite, tourmaline (schörlite) and limonite. The different colors are linked to the different abundance and degree of oxidation of secondary minerals (goethite and/or limonite), presumably due to weathering and guided by fracture density and distribution. The kaolinitic host gneisses are rich in kaolin (up to 35 wt. %) and quartz, with small amounts of phengite and chlorite; the whole-rock Fe_2O_3 content is generally ≤0.30 wt. %. The chemical characteristics of quartzites show an SiO_2 content ranging between 90–97 wt. %, Fe_2O_3 concentrations between 0.17–0.48 wt. %, 0.3–0.13 wt. % CaO, 0.05–0.35 wt. % MgO, 2.84–8.17 wt. % K_2O and Na_2O around 0.30 wt. %. Loss on ignition (LOI) is generally between 0.3 and 0.7 wt. %. Usually the quartzite slabs are manually split, and the natural roughness of the surfaces is considered a quality for the application as paving, reaching the best slip resistance. The high values in compressive and flexural strength, as well as the low water absorption (Table 2) make these materials optimal for outdoor use, roofing and cladding (used, as above reported, also in important historical buildings).

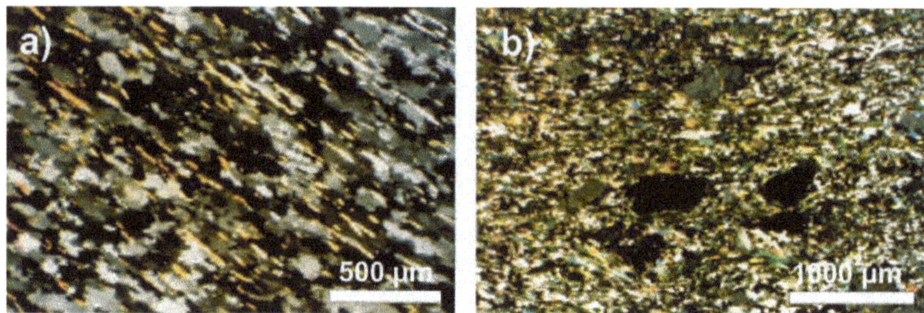

Figure 4. Thin section photomicrographs (cross polarized light) of (**a**) "golden" quartzite and (**b**) the *Marmorina* variety.

Table 1. Main rock-forming minerals (range) of the different quartzite varieties and kaolinitic gneisses. Abundances expressed as wt. % from semi-quantitative XRD analysis.

	"Golden" Quartzite	Pale Yellow	Olive Green	*Marmorina*	Kaolinitic Gneiss
Quartz	90–95	85–90	85–90	70–80	50–70
Phengite	2–6	2–8	5–10	10–15	5–15
K-feldspar	1–2	1–5	2–5	2–10	<LOD
Plagioclase	1–2	1–2	2–4	2–4	<LOD
Chlorite	<LOD	<LOD	traces	traces	3–8
Kaolin	<LOD	<LOD	traces	traces	8–35

Table 2. Physical–technical characteristics of the *Bargiolina* ([5,6]; testing methods are reported in the two mentioned references).

	On Average [5]	*Marmorina* [6]	"Golden" Quartzite [6]
Bulk density (kg/m^3)	2579	2695	2695
Water absorption coefficient (%)	0.21	0.28	0.30
Flexural strength (MPa)	40.5	55.5	42.3
Impact strength (J)	n.a.	7.4	7.4
Impact strength (cm)	102	n.a.	n.a.
Knoop microhardness (MPa)	n.a.	9010	9021
Abrasion resistance (coefficient)	0.67	n.a.	n.a.

6. Discussion and Perspectives: Sustainable Mining of the Mineral Resources of the Monte Bracco Area

The extensive historical documentation, the several applications in prestigious buildings and the excellent technical properties make *Bargiolina* a material that deserves a Heritage Stone designation. The Monte Bracco area presents two other potential raw materials: The first consists of *Bargiolina* extractive waste (introduced in Section 4), and the second is represented by a wide area of kaolinitic gneisses. The quarry waste could become a resource: The quartz-rich composition and the abundance of kaolin in the altered host gneisses suggest interesting applications in the field of industrial minerals.

The morphology of Monte Bracco was perfect to hide the visual impacts connected to waste dumping activity, but not to avoid the problems connected to dump and slope stability. Only at the end of the XX century had the Local Authorities started to be interested in *Bargiolina* programmatic exploitation planning and in the waste rock management [14]. Currently, wide areas are occupied by extractive wastes, not anthropically nor naturally rehabilitated (Figure 5). Those extractive waste piles (irregular slabs, splints and chips) hide several quartzite body portions which could be potentially quarried. Thus, the exploitation of the waste might also be useful for future quartzite bench quarrying. Due to the high SiO$_2$ (quartz) and low Fe$_2$O$_3$ content, the quartzite waste could be exploited as a secondary raw material and mineral dressed for ceramics, refractories, abrasives and glass manufacturing. The project planning for the quartzite waste exploitation is based on the physical–chemical characteristics of the rock and on the quantity of wastes, estimated to be 2,250,000 m^3 (about 4120 kt [15]). An interpolation of the data deduced from literature and information collected and processed during the present research (geological survey of quarry dumps) show that, after complete exploitation of the "virgin" quartzite benches, the rock wastes located in the Monte Bracco area should be evaluated as 4,940,000 m^3 (ca. 11,660 kt; [15]). Physical and mineralogical characteristics of the material, waste volume and commercial product designation heavily influence the choice of handling equipment (power loaders, diggers, etc.) and of the mineral waste dressing processes.

Figure 5. Present situation of quarry front and dumps. (**a**,**b**) Pian Lavarino waste-rock dumping area; (**c**) Pian dell'Eremita quarry dump; and (**d**) Tre Fontane abandoned quarry front.

The exploitation of the kaolinitic gneisses in the Banco Tre Fontane area (Figure 1) should also be considered: Kaolin shows proper geochemical and mineralogical characteristics as an industrial mineral, which, however, should be extracted together with the quartzite (which lays on the kaolinitic gneisses bench). Preliminary data (XRD and XRF) on selected samples show a fair amount of kaolin and a very low Fe_2O_3 content, suggesting a hydrothermal origin (quartz veins, presence of tourmaline) and a good suitability for the ceramic industry. Finally, the chance to exploit the other kaolinitic gneisses on the slopes of Monte Bracco should be further evaluated: If the reuse of quartzite waste and kaolin-rich gneisses are feasible, a more targeted field geological survey, physical, chemical and mineralogical characterization of the potential deposit and total volume estimation (field and GIS investigations) are recommended [16]. In the perspective of sustainable mining, it is important to move towards an integrated exploitation of the Monte Bracco area, with contemporary exploiting of the quartzite bench (second class material—quarrying activity), the quartzite waste and the kaolinitic gneisses (first class materials—mining activity for industrial minerals).

Author Contributions: Conceptualization, A.C. and G.A.D.; methodology, A.C.; software, A.C.; validation, A.C. and G.A.D.; formal analysis, A.C.; investigation, A.C. and G.A.D.; resources, A.C. and G.A.D.; data curation, G.A.D. and C.A.; writing—original draft preparation, A.C. and G.A.D.; writing—review and editing, A.C.; visualization, A.C.; supervision, A.C. and G.A.D.; project administration, G.A.D.; funding acquisition, G.A.D.

Funding: This research received no external funding.

Conflicts of Interest: The authors declare no conflict of interest.

References

1. Di Francesco, G. *La Pietra di Luserna a Barge*; Chiaramonte Ed.: Collegno, TO, Italy, 1999; p. 135.

2. Sandrone, R.; Colombo, A.; Fiora, L.; Fornaro, R.; Lovera, E.; Tunesi, A.; Cavallo, A. Contemporary natural stones from the Italian western Alps (Piedmont and Aosta Valley Regions). *Period. Mineral.* **2004**, *73*, 211–226.

3. Da Vinci, L. *Foglio 1 del Manoscritto G*; Biblioteca dell'Istituto di Francia a Parigi: Paris, France, 1511.

4. Snyder, R.L.; Bish, D.L. Quantitative Analysis. In *Modern Powder Diffraction*; Mineralogical Society of America Reviews in Mineralogy; Bish, D.L., Posts, J.E., Eds.; de Gruyter: Berlin, Germany, 1989; Volume 20, pp. 101–144.

5. AA.VV. Pietre del Piemonte. In *Assessorato All'ambiente Cave e Torbiere Della Regione Piemonte*; Regione Piemonte Ed.: Torino, Italy, 1973; p. 50.

6. AA.VV. *Le Pietre Ornamentali del Piemonte*; Regione Piemonte Ed.: Torino, Italy, 2000; p. 125.

7. Cadoppi, P.; Castelletto, M.; Sacchi, R.; Baggio, P.; Carraro, F.; Giraud, V. *Note Illustrative Della Carta Geologica D'italia Alla Scala 1:50.000—Foglio 154, Susa*; Servizio Geologico D'Italia: Roma, Italy, 2002; p. 127.

8. Dal Piaz, G.V.; Bistacchi, A.; Massironi, M. Geological outline of the Alps. *Episodes* **2003**, *26*, 175–180.

9. Compagnoni, R.; Rolfo, F.; Groppo, C.; Hirajima, T.; Turello, R. Geological map of the ultra–high pressure Brossasco–Isasca unit (Western Alps, Italy). *J. Maps* **2012**, *8*, 465–472. [CrossRef]

10. Balestro, G.; Festa, A.; Tartarotti, P. Tectonic significance of different block-in-matrix structures in exhumed convergent plate margins: Examples from oceanic and continental HP rocks in Inner Western Alps (northwest Italy). *Int. Geol. Rev.* **2015**, *57*, 581–605. [CrossRef]

11. Chopin, C.; Henry, C.; Michard, A. Geology and petrology of the coesite–bearing terrane, Dora Maira massif, Western Alps. *Eur. J. Mineral.* **1991**, *3*, 263–291. [CrossRef]

12. Dino, G.A.; Fornaro, M.; Martinetto, V.; Rodeghiero, F.; Sandrone, R. Le risorse estrattive del Monte Bracco: Valorizzazione mineraria e recupero ambientale. *GEAM* **2001**, *104*, 195–202.

13. Vialon, P. Etude géologique du Massif Cristallin Dora-Maira (Alpes Cottiennes internes-Italie). In *Travaux du Laboratoire de Géologie de la Faculté des Sciences de Grenoble—Mémoires n 4*; Université de Grenoble: Saint-Martin-d'Hères, France, 1966.

14. Bergamasco, L.; Fornaro, M.; Mangano, G.M. *Applicazione di un Nuovo Sistema Informativo Territoriale per la Programmazione Delle Coltivazioni Minerarie ed il Recupero Ambientale*; Il caso del Monte Bracco, PEI Ed.: Parma, Italy, 1998.

15. Dino, G.A.; Gioia, A.; Fornaro, M.; Bonetto, S. Monte Bracco quartzite dumps: Chance of recovery as second raw material for glass and ceramic industries. In Proceedings of the First International Conference on the Geology of Tethys, Cairo, Egypt, 12–14 November 2005; El Sayed, A., Youssef, A., Eds.; Tethys Geological Society (Egypt), Cairo University Press: Cairo, Egypt, 2005.

16. Dino, G.A.; Rossetti, P.; Perotti, P.; Alberto, W.; Sarkka, H.; Coulon, F.; Wagland, S.; Griffiths, Z.; Rodeghiero, F. Landfill mining from extractive waste facilities: The importance of a correct site characterization and evaluation of the potentialities. A case study from Italy. *Resour. Policy* **2018**, *59*, 50–61. [CrossRef]

Article

Suitable Re-Use of Abandoned Quarries for Restoration and Conservation of the Old City of Salamanca—World Heritage Site

Luís Sousa [1,2,*], **José Lourenço [1,2]** and **Dolores Pereira [3]**

1   Department of Geology, University of Trás-os-Montes e Alto Douro, Quinta de Prados, 5000-801 Vila Real, Portugal
2   CGeo Research Centre, University of Coimbra—Pólo II, 3030-790 Coimbra, Portugal
3   Departamento de Geología, University of Salamanca, 37008 Salamanca, Spain
*   Correspondence: lsousa@utad.pt

Received: 28 June 2019; Accepted: 7 August 2019; Published: 12 August 2019

Abstract: Martinamor granite has been used for centuries in the monumental buildings of Salamanca city. In this study, the fracturing pattern of the Martinamor granite outcrops was evaluated in order to assess the possibility of supplying material for the restoration of heritage monuments. Several joint sets with a mean joint spacing lower than one meter compose the fracturing pattern, making the massive exploitation of this granite impossible. Only small blocks for restoration can be obtained; therefore, the outcrops should be protected for such purpose. The area of outcrops and ancient quarries, as well as that of mining activities from the same period, should be preserved as examples of historical extraction techniques and as a remembrance of our geological-materials-based society. Several proposals are presented for the geoconservation of the site.

Keywords: natural stone; fracturing pattern; quarrying; geoheritage; Unesco World Heritage Site

---

1. Introduction

The preservation of historic buildings should seek the use of the same original stone material in order to maintain the homogeneity of the aesthetics and the physical–mechanical properties. In addition, using local resources is important for maintaining the cultural heritage, both from the geological and architectural perspectives, by preserving the quarrying and mining landscape and by using the same natural stones that were originally used to build the monuments and historical buildings. Numerous examples highlight building stones as an important part of the tangible heritage of cities, with prominent monuments and historic buildings that can be associated with traditional stones [1–4]. In some cases, the original stone is classified as a heritage stone resource; therefore, the balance between the preservation of the historic building and quarry sites should be promoted. Archeological quarry sites can supply stone pieces for restoration purposes under specific conditions, provided that previous studies have defined quarriable areas which do not affect the historical features of the site.

At present, raw materials are the main subject of many research programs (e.g., Horizon 2020). Within this context, quarrying should also be considered, as it is an activity that should evolve to meet the current standards for security, environment, health and other related issues. Many organisations and groups have the impression that quarrying activities, as well as mining activities, should be banned because they cannot guarantee complete adherence to all these standards, especially as they also undergo modification over time. However, it has been demonstrated that if extractive industries follow strict rules, they can contribute to the advancement of modern society with a minimal negative effect. This assumption is valid no matter the type of resource, metal or stone, the exploitation method, surface or underground, or the size of the affected area [5–9].

A conscientious effort has been made to maintain and preserve the historical buildings, though sometimes restoration involving replacement of the stone has not been the best option, largely due to the lack of information concerning the original materials. The International Union of Geological Sciences (IUGS) Subcommission on Heritage Stones (HSS) has been striving to encourage the dissemination of information concerning the original natural stones used in the construction of monuments and historic buildings and also regarding the need for preservation of historical quarries, so that restoration can be done with the original stone [10]. Such is the case of Salamanca, of which a large set of examples of good and bad restoration practices of its buildings has recently been published [4].

The most characteristic stone used in most of the foundations of the historic buildings in the centre of Salamanca is a leucogranite containing clusters of elongated tourmaline crystals, a fabric that inspired the local quarrymen to give it the name of Piedra Pajarilla (little bird stone) [4,10–12]. This granite has been proposed as a candidate for the designation of Global Heritage Stone Resource [10] and together with other stones that crop out in the province and are used in historic buildings of Salamanca, they have been proposed as a candidate for the designation of Global Heritage Stone Province [12]. All these papers make an effort to explain why this resource should be preserved and the quarries maintained, kept not only in reserve for restoration, but also for outreach and education. For these reasons, a detailed characterisation of the ancient quarries and geoconservation measures are needed.

The present paper describes the main characteristic of the Piedra Pajarilla granite and the systematic fracturing affecting the blocks. The intention is to show the local, province and regional governments the advantages of using the same original stone in the restoration of buildings and how the extraction of suitable blocks do not have to importantly alter the ecosystem of the former quarrying sites. It also highlights the potential interest of geoconservation activities, related to economic activities that could enhance the life of a society affected by aging and population exodus.

2. Materials and Methods

2.1. Martinamor Granite

Piedra Pajarilla is a leucogranite cropping out as thin sub-horizontal sheets (with less than 1 km width) within the Martinamor Complex, in the Spanish Central System [13]. These granitic sheets formed from the partial melting of a fertile meta-sedimentary protolith, which is part of the Schist–Greywacke complex [14,15], with references therein (Figure 1). Three different facies composed of light-colour minerals can be found—(1) a coarse-grained granite, with nodules of tourmaline, showing a luxullianitic texture; (2) a fine-grained granite with small crystals of tourmaline and (3) a two-mica granite, with the absence of tourmaline. The type most commonly used for construction purposes is the first one [16].

As explained above, luxullianitic textures sometimes resemble flying birds (Figure 2), leading to the local name 'Piedra Pajarilla', used to refer to the construction granite with such texture, and, unfortunately, to many others. 'Piedra Pajarilla' is also known as Martinamor granite, from the village where it was extracted. When different granites are given the same name (due to a lack of knowledge of the different materials by the construction professionals), aesthetic errors can occur, which is considered unacceptable in historic buildings of a World Heritage site [10]. The unique aesthetic characteristics of the 'Piedra Pajarilla' make it difficult to be replaced by other materials. No other material is known, resembling the same aesthetic of this granite. Several other granites originating from other local quarries were used for restoration/replacement of very damaged ashlars, with undesirable aesthetic results [4]. Such confusion can be avoided nowadays by using a certificate of origin (e.g., in Europe, UNE EN 12440 [17]). However, Martinamor granite was used long before the standardisation of stone names came into effect.

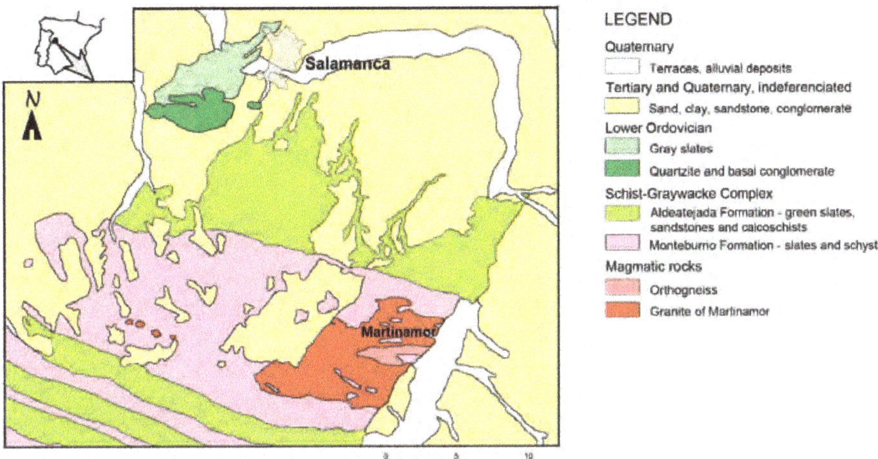

Figure 1. Geological setting of Martinamor granite (modified from [13]). The granite cuts the meta-sediments of Cambrian–Precambrian age (the Schist–Greywacke Complex).

Figure 2. Texture of the granite. The cumulates of tourmaline resemble flying birds.

This granite was stopped being used as a construction material in the early 20th century [10], before quality control was required for the commercial use of stones in construction. Therefore, data regarding its physical or mechanical behaviour are scarce. Previous published characterisations showed high variations in some of the physical and mechanical properties [5,11], but a work that is being carried out at present will show that these characteristics are closely related to the different states of weathering of the rock, the different outcrops that are found in the area and the direction of the testing, regarding the sub-horizontal foliation always present in this granite (Table 1). Weathered granites show high irregularity of the physical–mechanical properties, leading to a very different resistance to water action in ashlars from the same quarry [18]. The historic buildings in Salamanca do not show a high degree of alteration, and this behaviour is a sign that some selection was made by quarrymen at the time, in order to avoid the use of weathering prone granite.

Table 1. Abbreviated data of the physical and mechanical behaviour of Martinamor granite, depending on the weathering, the sampling area and the direction of the testing (Pereira in prep.).

Testing	Compressive Strength	Flexural Strength	Water Absorption
Martinamor Granite	161–252 MPa	12.7–25.3 MPa	0.9–0.24%

The restoration of monuments in Salamanca should be performed with appropriate ashlars to ensure good performance. Thus, the full physical–mechanical characterisation of this granite is highly recommended to ensure the utilisation of the material with the desired properties. All the old quarries and usable outcrops should be studied to ensure the best application according to their properties. A full characterisation of this granite will provide a clear picture of the appropriate use of the rock. Authors of this paper are already working on this aspect.

2.2. Methods

The first stage was the identification and demarcation of the granite outcrops, quarried or not, through field work and by using aerial photos. In these outcrops, the different spilling marks used in granite exploitation were identified.

In a portion of the outcrops, the strike and dip of fractures were measured with a compass. Rose and pole diagrams drawn with Richard Allmendinger's StereonetWin interface [19,20] allow for fracture assessment. For each fracture set, the spacing was measured perpendicular to the fractures and, afterwards, the histogram was drawn with all the joint spacing data. The mean joint spacing was calculated for the entire data. The volumetric index or volumetric joint count (Jv) [21,22] was calculated based on the mean joint spacing values. This general procedure (strike and joint spacing measurements) was repeated using aerial photos.

Finally, several proposals are presented to accomplish two main purposes. First, create a reserve area for the extraction of granite for the rehabilitation of the monuments in Salamanca. Second, keep the old quarries as an example of extraction techniques and make it possible to show the source of the granite used in many monuments.

3. Results and Discussion

3.1. Quarries and Extraction Techniques

The quarries are approximately 15 km south of Salamanca (Figure 3), in the outskirts of Martinamor village (Figure 4), which gives the formal name to the granite. At present, there are two main quarry areas outside Martinamor. The largest one (see Figure 3) is located on private land where cattle are currently kept. Others are located on public land and are composed mainly of intact outcrops (marked as 'Outcrops' in Figure 3). The quarry faces in these two areas are five meters high at their maximum and most of them are only one to two meters high. Other quarry areas are located nearby, but are much smaller in comparison, with no more than 10 m of historic quarrying, as shown by extraction marks.

So far, all these quarry areas are well preserved, i.e., the quarry fronts are accessible and without topographic modifications. It is possible to observe several quarry faces defined by joints. There are even some vestiges of earlier quarrying methods (Figure 5), which are similar to those observed in ancient Egypt and prehistoric sites in Europe and South America [23,24] (see also http://www.ancient-wisdom.com/quarrymarks.htm). These marks are due to tools used to separate the blocks, both in the vertical and in the horizontal plane. Several types of marks are observed, wedge marks, triangular-shaped drill holes and circular drill holes. Wedge marks are the oldest. Several researches mention the Roman period as the origin of this marks [25–29]. However, only a detailed archaeological investigation will permit accurate conclusions about the wedge marks observed in the area.

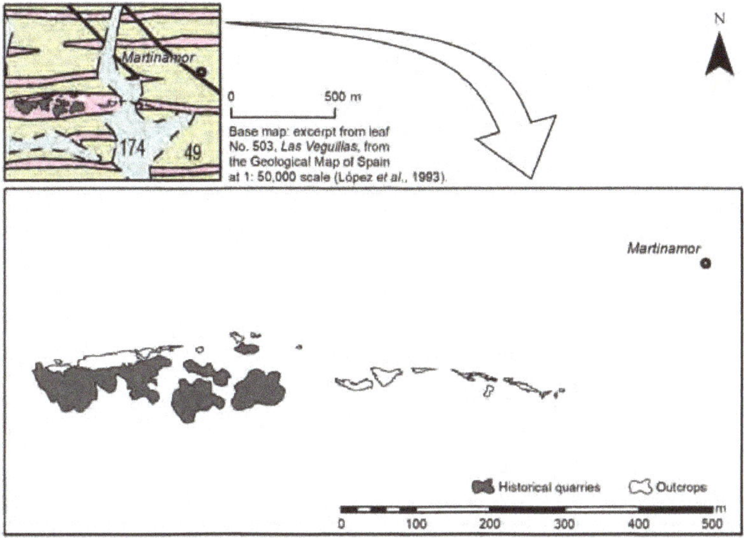

Figure 3. Location of the historical quarries near Martinamor (according to [13], 49, slates and limolites, Monterrubio Formation; 174, aluvial deposits; pink bodies, two-mica granites, leucogranites with tourmaline and sincinematics with Hercynian Phase II).

Figure 4. Photos of the ancient quarries. (**a**) aerial photomosaic of the largest quarry where it is possible to see a large amount of affected land due to the historic extractions; (**b,c**) details of the quarry front.

Figure 5. Splitting marks in ancient quarries of Martinamor granite: (**a**,**b**) wedge (rectangular) marks and (**c**,**d**) triangular-shaped drill holes.

The triangular-shaped drill holes are barely seen in quarries and not common in the literature. Hockensmith [30] mentions triangular holes with a maximum depth of 18 cm in a limestone quarry, probably produced by a pick. Gage [31] studied several triangular drill holes with a maximum depth of 1.5 m depth in a graphite mine. According to this author "there is a strong possibility that the points of the triangular shape holes were used to direct the line of fracture during the blast". Some of the triangular holes observed in the old quarries near Martinamor show one corner aligned with natural fractures, pointing out the practical purpose of such holes.

This topic is worth investigating further [32] in order to identify all the splitting marks in the quarried area, as well as to highlight the evolution of the drill techniques through time. This is an extra reason to work on a preservation figure for this area.

3.2. Fracturing and Resources Evaluation

The pattern and density of fracturing define the possibility of extracting granite blocks for further extraction and applications. The study of fractures has proven to be a useful method to discern whether a sufficient quantity of dimension stones can be extracted from a quarry [33] and whether the fracturing can be used in the extraction process, facilitating the quarrying and therefore lowering the cost. Our purpose is to investigate whether it would be feasible to use some of the blocks for restoration of monuments and historical buildings in Salamanca because it has been observed that in some cases other types of granite have been incorrectly used to replace deteriorated ashlars. This has led to undesirable aesthetic effects that should not be permitted in a historical site, especially for a World Heritage Site.

In a first stage, fracturing was evaluated in the outcrops near the Martinamor village (see Figure 3). The most important joint sets are N20°–40°W, dipping to NE, and N50°–60°E, dipping to NW. However, the span between these main directions is continuous (Figure 6). When the two main joint sets were present, the quarrymen took advantage of them (they are perpendicular) to facilitate the extraction of granite blocks. In this area, the joint spacing values are low (Figure 7), with most of them below one meter. The overall mean joint spacing of 78 values is 0.64 m, with slight differences between the N20°–40°W and N50°–60°E sets, with the mean values of 0.57 m (n = 40) and 0.69 m (n = 20), respectively. Neglecting the influence of the horizontal joints, the volumetric joint count (Jv) is 3.2. This value is far above the threshold usually considered for the financial revenue in quarries for ornamental purposes, Jv = 2.0 [21]; however, the joint spacing values are enough to obtain small blocks able to produce ashlars.

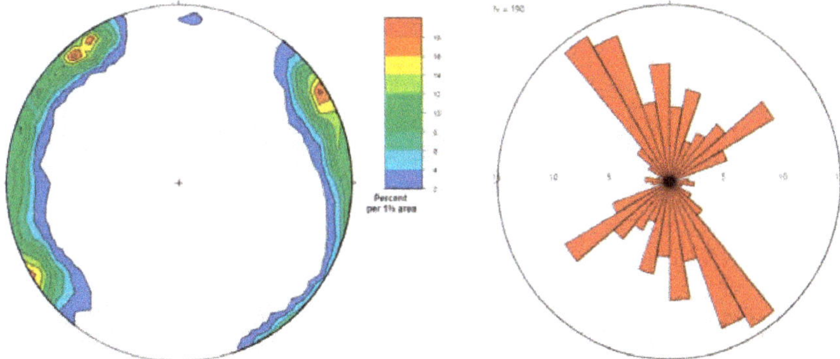

Figure 6. Distribution of the joints measured in the field work (n = 190; low hemisphere).

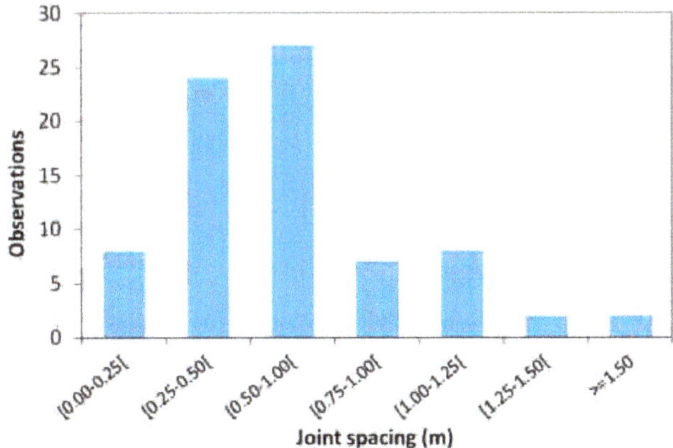

Figure 7. Distribution of the joint spacing data from field measurements.

Analysis of the fracture pattern can be easily carried out using photomosaics obtained by an unmanned aerial vehicle for aerial photographs. This technique was used to study the fracturing in the remaining outcrops (see Figure 3). An example of fracturing identification in outcrops is presented in Figures 8 and 9, regarding an outcrop located southwest of Martinamor. The N40°–50°W set fractures stand out, while the N70°–80°W, N30°–50°E and N70°–80°E sets are also important. The mean joint spacing (n = 104) is 0.87 m and volumetric joint count (Jv) is equal to 3.4. The same methodology was applied in the outcrops located near the ancient quarries (see Figure 3) and the results can be seen in Figure 10. In this area, the joints have a more random distribution with the predominance of the N50°–60°W, N20°–30°W, N60°–70°E and N80°–90°W sets. The overall mean joint spacing (n = 198) is equal to 1.3 m and the volumetric joint count (Jv) is 3.9.

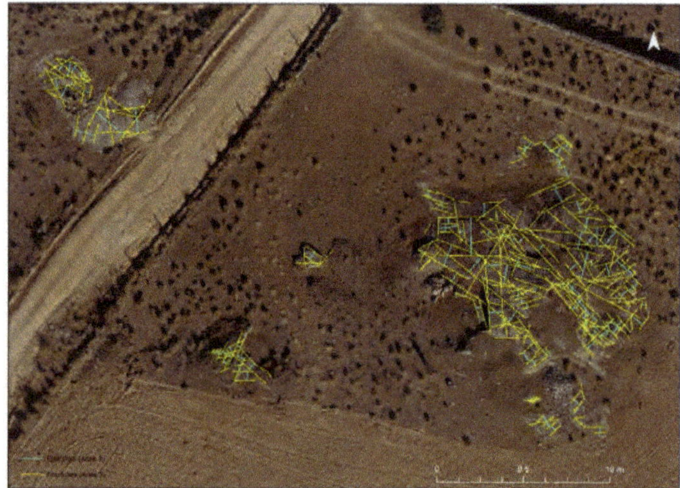

Figure 8. Example of fracturing identification in aerial photomosaic (yellow lines, joints; green lines, joint spacing).

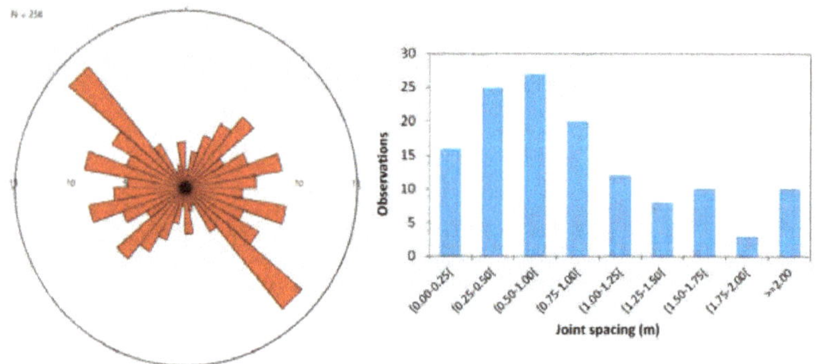

Figure 9. Fractures and joint spacing data from outcrops identified in Figure 8.

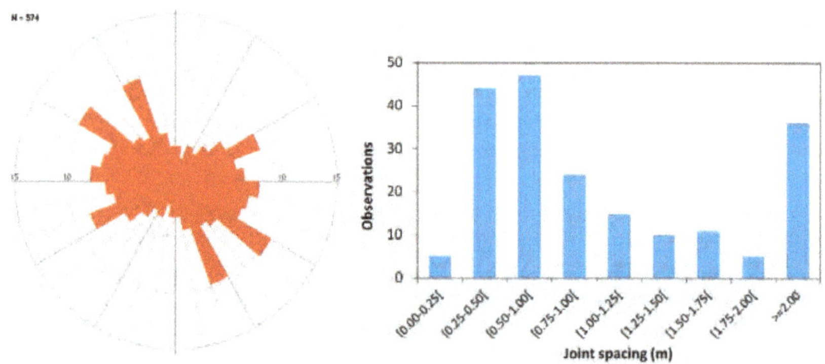

Figure 10. Rose diagram and joint space distribution relative to outcrops located near the ancient quarries (see Figure 3).

The fracturing pattern is slightly different within the studied areas, with the main joint set located in the N20°–60°W range, as well as other important joint sets. Consequently, fracturing patterns define natural blocks with a shape far away from the ideal cube or rectangular cuboid. The overall mean joint spacing ranges from 0.64 m to 1.3 m and volumetric joint count values are higher than 3.2 (3.2–3.9).

The results confirm the high fracturing density and the impossibility of extracting large blocks. Therefore, considering the available outcrops, a massive exploitation of this granite is unlikely. However, the possibility of extracting small blocks for restoration support the idea of protecting this area as a reserve for such purpose.

3.3. The Need for a Geoconservation Proposal

Our proposal considers different assets of the granite and its outcrops, the practical application of granite as source of restoration material, the scientific and educational potentiality of the quarries, the heritage values and recreational opportunities. The monuments located in Salamanca act as the trigger for further actions (Figure 11). In view of the historical importance of such monuments, all the restorations must be done taking into consideration the properties of the original stone. For that reason, the quarries should be studied and protected. In the scheme of Figure 11, the geoconservation measures at quarries are the consequence of monument heritage concern.

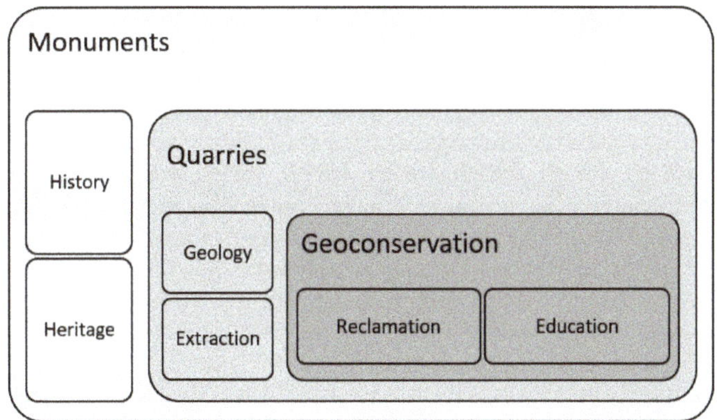

Figure 11. Geoconservation as an instrument for the conservation of monuments.

In order to avoid using stones that do not match the original ones, only designated materials, i.e., the same materials or materials with properties similar to those in the original fabrics should be used to avoid physical or aesthetic damage [34]. Therefore, regarding the outcrops of Martinamor granite, only the ones able to supply granite pieces for restoration purposes, need to be classified as a special reserve. Knowledge of the physical–mechanical properties of the granite is a key factor to avoid chaotic restorations that diminish the historical value of the monuments. Textural properties, e.g., texture, grain size and colour, should be considered when other stones are used. Such properties must resemble the original stone as much as possible. For this reason, initiatives such as the heritage stones designations by IUGS are important to provide the description and characterisation of stones that are not always commercialised or even extracted these days.

The location of the source of the material used in heritage monuments is important as a testimony of the progress in building techniques. It is also possible to make the connection between the monuments and the surrounding area, since the source of the stone is usually located not far away [35,36]. Old quarries allow us to understand the alteration process of the material, to establish their durability and provide valuable information in view of the restoration and conservation works. The Geomonumental Routes [37] are excellent examples of a holistic approach of how to promote the heritage interlocking history and science. The location of the village of Martinamor (a village dramatically affected by aging and population exodus) near the outcrops will permit the use of a rehabilitated house as a welcome centre, where visitors can have a first insight into the quarrying activities before the guided field trip. There are many examples of similar approaches in areas transformed by the mining activity that make it possible to outline the best options [38–40].

Other geological resources exploited in the neighbourhood can be added to this route, such as the tungsten mines and orthogneiss quarries. Salamanca province experienced active mining of tungsten, tin and cupper during and after the Second World War. The price of tungsten was very high, even higher than gold, because the Germans needed it for tungsten carbide tools and for armour-piercing munitions, among other things. The Allies tried to stop them from buying the tungsten, which was extracted from various mines in Spain and Portugal, so the price increased dramatically. There is a considerable amount of historical information regarding the efforts on both sides and even some fiction derived from the subject [41,42]. Tungsten mining in this area was closed in the 1970s because the price dropped. However, the landscape is full of abandoned mines which are part of the cultural heritage. The orthogneisses of San Pelayo, located nearby, are considered a Point of Geological Interest (POGI) [43]. The stone shows a typical compositional banded appearance defined by quartzfeldpsar and mica layers, biotite enriched and with lenticular eye-shaped feldspars. This orthogneiss is closely related to the enrichment of tungsten, and tailings with quite a lot of scheelite are witnesses of the mining that took place there as a main economic activity of the area.

In the case of Martinamor granite, the few small available outcrops are highly fractured for a regular exploitation; therefore, their present appearance will not be changed by quarrying activity. However, agroforest activities and cattle herding can affect the outcrops and their impact needs to be assessed. Protecting the outcrops is the priority and the municipality should create the legal bases for it. Another important issue is the attractiveness of the quarry landscape. Several factors can make the experience of a field trip more pleasant, such as a natural succession, quarry preservation, the presence of surface water, human activity, accessibility, among others [44]. Therefore, some improvements are necessary to enrich the area and increase its attractiveness. In some parts of the area, ecological restoration and rehabilitation actions are required to recover the degraded ecosystem. However, in the remaining portions of the outcrops/quarries, reclamation actions should not hide the quarry fronts, preventing access to the material and to features of scientific and educational interest, such as fracturing and evidence of the extraction techniques used in the past. This case can be an interesting example of post-mining land use as discussed by Kaźmierczak et al. [45] and Cross et al. [46].

In the above approach (Figure 11), monuments act as the starting point for a sequence of actions where geoconservation is the ultimate end. Another option, without changing the main actions to be undertaken, is the holistic point of view where monuments and the quarries are part of something wider and universal (Figure 12). In this case, monuments are an excellent example of the interaction between human beings and their geological environs, as with many other human actions such as the settlement location or territory dispute. The available stones influence construction techniques and the monumentality of the buildings; therefore, the quarry location is an important issue related to monument restoration. Monuments are examples of the interaction between human beings and their geological environs, particularly in terms of the available stones. Both options (Figures 11 and 12) have the same activities in the quarries/outcrops, since reclamation/ecological restoration/rehabilitation are required for environmental protection, as well as consensual extraction for restoration purposes. The only difference between the two approaches is the starting point, which can be the monument heritage (Figure 11) or the geological heritage (Figure 12).

The details of the geoconservation proposal will be presented in a subsequent publication, namely the reclamation and rehabilitation actions in a quarried area, with the delimitation of the outcrops available for granite extraction. Our aims are to design the overall arrangements in the outcrops area in order to make it easier to access and turn the field trip into an agreeable activity. Also, the visiting schema should be planned according to several options, such as the following: a guided or free visit; location of the starting point; educational level and age of the visitors. Our goal is to contribute to the preservation of the monumental heritage of Salamanca, aiming to promote awareness of the importance of the geological resources and to expand the initiative to other similar cases in

Europe and around the world. In the study area, as well as in many other places around the world, there is a need of further studies dedicated to understand ancient tools and ancient quarrying activities in order to facilitate the preservation of monuments and historic buildings for the mutual benefit of the local community, the environment, and science in general [47].

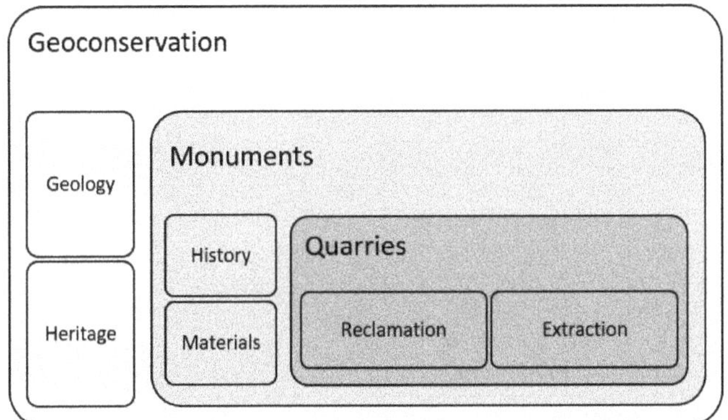

Figure 12. Holistic approach to the geoconservation actions.

4. Conclusions

Scientific research can show how past mining and quarrying activities have been carried out, but also how they can be part of much needed new endeavours. Traditional geological methods, such as mapping and petrology, contribute to the understanding and preservation of historical natural stone quarries that were used to build the architectural heritage of so many societies. Furthermore, the different extraction techniques and their evolution are recognisable in quarry faces.

A dense fracturing pattern cutting Martinamor granite doesn't allow the extraction of large blocks necessary for the modern stone industry. However, the extraction of small pieces for restoration, meeting the physical–mechanical and textural standards, is possible. This study will facilitate the potential re-use of an abandoned quarry, where stone to build a World Heritage city was quarried centuries ago, adapting to the very strict mining and environment laws regarding re-opening mines and quarries. An overall reclamation scheme is suitable for the quarried area in order to improve both its accessibility for educational/scientific visits and for extraction when restoration is necessary.

It is important to recognise that education and outreach programmes are very important activities to maintain the knowledge of past mining and quarrying traditions, as part of our cultural heritage, as well as to remind people that these are positive activities which are vital for the advance of modern societies. Therefore, a geoconservation safeguarding plan should be undertaken to ensure the protection and designation of this important quarry area.

This research is the first step for a sustainable use of the small outcrops of Martinamor granite. Considering the area available for potential quarrying and the fracturing density, only ashlars for restoration purposes can be extracted. The limited amount of granite resources highlights the inevitability of strict rules to access: what monuments can use these quarries for the potential restoration/conservation? What outcrops can be used? In fact, quarries are themselves an archaeological site. Therefore, a wider plan is necessary to manage the exploitation/rehabilitation of the area. Geoconservation proposals are tentative to put the quarries in a wider study perspective, make easier their protection and restricted utilisation. This means a comprehensive sustainability, including cultural heritage, scientific principles and environmental concerns.

Sustainability **2019**, *11*, 4352

Author Contributions: All authors contributed equally to this paper.

Funding: This research was part of a program funded by the following institutions: Fundación Universidad-Empresa, the Castilla y Leon regional government and the University of Salamanca. It is also part of the activities of the UNESCO IGCP-637. The FCT -Fundação para a Ciência e a Tecnologia, I.P., project UID/Multi/00073/2019 of the Geosciences Center, also supported this research.

Acknowledgments: The use of Stereonet software is acknowledged.

Conflicts of Interest: The authors declare no conflict of interest.

References

1. Luodes, N.; Panova, E.; Bellopede, R. Characterization of natural stone material used in the Nordic eastern urban and costal environment. *Environ. Earth Sci.* **2017**, *76*, 328. [CrossRef]
2. Freire-Lista, D.M.; Fort, R. Historical City Centres and Traditional Building Stones as Heritage: Barrio de las Letras, Madrid (Spain). *Geoheritage* **2019**, *11*, 71–85. [CrossRef]
3. Menningen, J.; Siegesmund, S.; Lopes, L.; Martins, R.; Sousa, L. The Estremoz marbles: An updated summary on the geological, mineralogical and rock physical characteristics. *Environ. Earth Sci.* **2018**, *77*, 191. [CrossRef]
4. Pereira, D. *Natural stone and World Heritage: Salamanca (Spain)*; CRC Press: Bota Raton, FL, USA, 2019.
5. Yang, Y.-Y.; Xu, Y.-S.; Shen, S.-L.; Yuan, Y.; Yin, Z.-Y. Mining-induced geo-hazards with environmental protection measures in Yunnan, China: An overview. *Bull. Eng. Geol. Environ.* **2015**, *74*, 141–150. [CrossRef]
6. Essalhi, A.; Essalhi, M.; Toummite, A. Environmental impact of mining exploitation: A case study of some mines of barite in the Eastern Anti-Atlas of Morocco. *J. Environ. Prot.* **2016**, *7*, 1473–1482. [CrossRef]
7. Ozcelik, M. Environmental pollution and its effect on water sources from marble quarries in western Turkey. *Environ. Earth. Sci.* **2016**, *75*, 796. [CrossRef]
8. OECD. *Managing environmental and health impacts of uranium mining*; NEA: Paris, France, 2014.
9. Mensah, A.K.; Mahiri, I.O.; Owusu, O.; Mireku, O.D.; Wireko, I.; Kissi, E.A. Environmental impacts of mining: A study of mining communities in Ghana. *Appl. Ecol. Env. Res.* **2015**, *3*, 81–94.
10. Pereira, D.; Gimeno, A.; del Barrio, S. Piedra Pajarilla: A candidacy as a Global Heritage Stone Resource for Martinamor granite. *Geol. Soc. London, Spéc. Publ.* **2014**, *407*, 93–100. [CrossRef]
11. Cooper, B.; Marker, B.; Pereira, D.; Schouenborg, B. Establishment of the "Heritage Stone Task Group" (HSTG). *Episodes* **2013**, *36*, 8–10.
12. Pereira, D.; Cooper, B.J. Building stone as a part of a World Heritage Site: 'Piedra Pajarilla' Granite and the city of Salamanca, Spain. *Geol. Soc. London, Spéc. Publ.* **2013**, *391*, 7–16. [CrossRef]
13. Díez Balda, M.A. *El Complejo Esquisto-Grauvaquico, las series Paleozoicas y la estructura Hercínica al Sur de Salamanca*; Ediciones Universidad de Salamanca: Salamasca, Spain, 1986; 162p.
14. Díez Balda, M.A.; Vegas, R.; González Lodeiro, F. Central Iberian Zone. In *Pre-Mesozoic Geology of Iberia*; Dallmeyer, R.D., Martinez-Garcia, E., Eds.; Springer: Berlín, Germany, 1990; pp. 172–188.
15. Pereira, M.D.; Rodríguez-Alonso, M.D. Duality of cordierite granites related to melt-restite segregation in the Peña Negra Anatectic Complex, central Spain. *Can. Mineral.* **2000**, *38*, 1329–1346.
16. López Plaza, M.; Sánchez, M.G.; Iñigo, A.C. La utilización del leucogranito turmalinífero de Martinamor en los monumentos de Salamanca y Alba de Tormes. *Stud. Geol. Salmant.* **2007**, *43*, 247–280.
17. *UNE EN 12440 Natural Stone: Denomination Criteria*; The National Standards Authority of Ireland: Dublin, Ireland, 2008.
18. Sousa, L.; Siegesmund, S.; Wedekind, W. Salt weathering in granitoids: An overview on the controlling factors. *J. Environ. Earth Sci.* **2018**, *77*, 502. [CrossRef]
19. Cardozo, N.; Allmendinger, R.W. Spherical projections with OSXStereonet. *Comput. Geosci.* **2013**, *51*, 193–205. [CrossRef]
20. Richard Allmendinger's Home Page. Available online: https://personalpages.manchester.ac.uk/staff/Richard.Allmendinger/ (accessed on 22 May 2019).
21. Sousa, L.M.O.; Oliveira, A.S.; Alves, I.M.C. Influence of fracture system on the exploitation of building stones: The case of the Mondim de Basto granite (north Portugal). *Environ. Earth Sci.* **2016**, *75*, 39. [CrossRef]
22. Palmström, A. The volumetric joint count—a useful and simple measure of the degree of jointing. In Proceedings of the 4th international congress of IAEG, New Delhi, India, 10–15 December 1982; pp. 182, 221–228.

23. Heldal, T. Constructing a quarry landscape from empirical data. General perspectives and a case study at the Aswan west bank, Egypt. In *QuarryScapes: Ancient stone quarry landscapes in the Eastern Mediterranean*; Abu-Jaber, N., Bloxam, E.G., Degryse, P., Heldal, T., Eds.; Geological Survey of Norway Special Publication: Trondheim, Norway, 2009; Volume 12, pp. 125–153.

24. Wootton, W.; Russell, B.; Rockwell, P. Stoneworking tools and toolmarks (version 1.0). The Art of Making in Antiquity: Stoneworking in the Roman World. 2013. Available online: http://www.artofmaking. ac.uk/content/essays/2-stoneworking-tools-and-toolmarks-w-wootton-b-russell-p-rockwell/ (accessed on 5 February 019).

25. Hori, Y. Ancient Quarry Techniques in the Ptolemaic Period in the Middle Egypt. In Proceedings of the XVII International Congress of Classical Archaeology, Roma, Italy, 22–26 September 2008.

26. Harrell, J.A.; Storemyr, P. Ancient Egyptian quarries—an illustrated overview. In *QuarryScapes: Ancient stone quarry landscapes in the Eastern Mediterranean*; Abu-Jaber, N., Bloxam, E.G., Degryse, P., Heldal, T., Eds.; Geological Survey of Norway Special Publication: Trondheim, Norway, 2009; Volume 12, pp. 7–50.

27. Coli, M.; Rosati, G.; Pini, G.; Baldi, M. The Roman quarries at Antinoopolis (Egypt): Development and techniques. *J. Archaeol. Sci.* **2011**, *38*, 2696–2707. [CrossRef]

28. Verhoeven, G.; Taelman, D.; Vermeulen, F. Computer vision-based orthophoto mapping of complex archaeological sites: The ancient quarry of Pitaranha (Portugal-Spain). *Archaeometry* **2012**, *54*, 1114–1129. [CrossRef]

29. Carfora, P.; Di Luzio, E. Archaeomorphological and Geological Studies on the Ancient Appian Way at the Aurunci Pass: Multidisciplinary Approaches for the Investigation of Ancient Quarries Siting and Exploitation. In Proceedings of the 3rd International Landscape Archaeology Conference, Rome, Italy, 17–20 September 2014.

30. Hockensmith, C.D. Archaeological Investigations at the Shrull Lime Kiln Near Russellville, Logan County, Kentucky. In *Current Archaeological Research in Kentucky*; Mills, E.N., Williamson, R.V., Davis, R.D., Eds.; Kentucky heritage council: Frankfort, KY, USA, 2007; Volume 9, pp. 1–26.

31. Gage, J. Osgood Graphite Mine, Nelson NH Site Report. 2002. Available online: http://www.stonestructures. org/html/graphite_mine.html (accessed on 6 February 2019).

32. Gage, M.; Gage, J. *The art of splitting stone. Early Rock Quarrying Methods in Pre-Industrial New England*; Powwow River Books: Amesbury, MA, USA, 2005.

33. Sousa, L.M.O. Granite fracture index to check suitability of granite outcrops for quarrying. *Eng. Geol.* **2007**, *92*, 146–159. [CrossRef]

34. Pereira, D.; Marker, B. The Value of Original Natural Stone in the Context of Architectural Heritage. *Geosciences* **2016**, *6*, 13. [CrossRef]

35. Freire-Lista, D.M.; Fort, R.; Varas-Muriel, M.-J. Alpedrete granite (Spain). A nomination for the "Global Heritage Stone Resource" designation. *Episodes* **2015**, *38*, 106–133.

36. Freire-Lista, D.M.; Fort, R.; Varas-Muriel, M.-J. San Pedro leucogranite from A Coruña, Northwest of Spain: Uses of a heritage stone. *Energy Procedia* **2016**, *97*, 554–561. [CrossRef]

37. Traditional building stone in Madrid monuments. Available online: https://www.madrimasd.org/English/ Science-Society/scientific-heritage/Geomonumental-Routes/madrid/default.asp (accessed on 8 May 2019).

38. Careddu, N.; Siotto, G.; Siotto, R.; Tilocca, C. From landfill to water, land and life: The creation of the Centre for stone materials aimed at secondary processing. *Resour. Policy* **2013**, *38*, 258–265. [CrossRef]

39. Carrión Mero, P.; Herrera Franco, G.; Briones, J.; Caldevilla, P.; Domínguez-Cuesta, M.; Berrezueta, E. Geotourism and Local Development Based on Geological and Mining Sites Utilization, Zaruma-Portovelo, Ecuador. *Geoscience* **2018**, *8*, 205. [CrossRef]

40. Prosser, C.D. Geoconservation, Quarrying and Mining: Opportunities and Challenges Illustrated Through Working in Partnership with the Mineral Extraction Industry in England. *Geoheritage* **2018**, *10*, 259–270. [CrossRef]

41. Guerra Garrido, R. *El año del wólfram*, 1st ed.; Catedra Ediciones: Madrid, Spain, 1984; 354p.

42. Wilson, R. *A small death in Lisbon*; Harper Collings Publishers: London, UK, 1999; 440p.

43. IGME Point of Geological Interest. Available online: http://info.igme.es/ielig/LIGInfo.aspx?codigo=CIs027 (accessed on 8 May 2019).

44. Baczyńska, E.; Lorenc, M.W.; Kaźmierczak, U. The Landscape Attractiveness of Abandoned Quarries. *Geoheritage* **2018**, *10*, 271. [CrossRef]

45. Kaźmierczak, U.; Lorenc, M.W.; Strzałkowski, P. The analysis of the existing terminology related to a post-mining land use: A proposal for new classification. *Environ. Earth Sci.* **2017**, *76*, 693. [CrossRef]

46. Cross, A.T.; Young, R.; Nevill, P.; McDonald, T.; Prach, K.; Aronson, J.; Wardell-Johnson, G.W.; Dixon, K.W. Appropriate aspirations for effective post-mining restoration and rehabilitation: A response to Kaźmierczak et al. *Environ. Earth Sci.* **2018**, *77*, 256. [CrossRef]

47. Goette, H.R.; Pajor, F. Eduard Schaubert's travel notes on Southern Euboea in May 1847. *Mediterr. Archaeol. Archaeom.* **2010**, *3*, 63–64.

Article

Historical and Contemporary Use of Natural Stones in the French West Indies. Conservation Aspects and Practices

Yves Mazabraud

Géosciences Montpellier, Université des Antilles, Université de Montpellier, CNRS, Morne Ferret, BP517, 97178 Les Abymes, France; yves.mazabraud@univ-antilles.fr; Tel.: +590-590-21-36-15

Received: 20 June 2019; Accepted: 19 August 2019; Published: 22 August 2019

Abstract: The French West Indies (F.W.I.), in the Eastern Caribbean, are part of a biodiversity hotspot and an archipelago of very rich geology. In this specific natural environment, the abundance or the lack of various natural resources has influenced society since the pre-Columbian era. The limited size of the islands and the growth of their economy demand a clear assessment of both the natural geoheritage and the historical heritage. This paper presents a brief review of the variety of the natural stone architectural heritage of the F.W.I. and of the available geomaterials. Some conservation issues and threats are evidenced, with particular emphasis on Guadeloupe. Some social practices are also evoked, with the long-term goal of studying the reciprocal influence of local geology and society on conservation aspects. Finally, this paper argues that unawareness is one of the main obstacles for the conservation of the geoheritage and the natural stone architectural heritage in the F.W.I.

Keywords: building stones; Guadeloupe; Martinique; French West Indies; eastern Caribbean; cultural heritage; geological heritage; historical and Archaeological sites

---

1. Introduction

In 2002, the law for a Democracy of Proximity [1] was voted by the French parliament. It stated that the State takes care of the conception, the animation, and the evaluation of the Natural Heritage. This heritage shall include all ecological, fauna, flora, geological, mineralogical, and paleontological items of interest. For the Geoheritage [2], in which human-related sites of interest are considered, Guadeloupe in the French West Indies (F.W.I.) was chosen as a test territory. Indeed, its geology is very rich and very peculiar [3,4], with a lot of variation. It includes, for example, a carbonate platform, coral reefs (both actives and fossils), mangrove-bearing mudflats, on-shore and submarine thermal springs, a great variety of volcanic deposits, tropical karsts, and an active volcano, La Soufrière. So, as recently as 2003, local stakeholders in Guadeloupe have been referencing several sites and created a collection of 33 outcrops' descriptions. In 2015, all these sites had been validated at the national level and entered into a national database (IGeotope) [5]. This inventory is meant to be permanent and a living collection. Since late 2018, the previous outcrops' descriptions are re-evaluated and new information (GIS mapping, petrological analysis) is being collected for later addition to sites descriptions. One of the goals of this second phase of inventory is also to add new sites, emphasizing on human-related sites. When finalized, a territorial geological and natural stone architectural heritage catalogue reveals itself a powerful tool for policy making and education. Despite the need to inventory the underground for resource purposes (water, building stones, mining) that has motivated geological research for centuries, such an inventory is meant to be used for protecting areas of natural interest, for teaching and for developing tourism.

The study presented in this article was conducted in order to further develop this inventory. One of the goals was to gather field observations and existing information with the long-term objective of evaluating the patrimonial value of the local geodiversity and the local natural stone structures This paper presents the first results of this study. A second, long-term, objective will be to document whether the local geology of different islands in the French West Indies has influenced building styles and society. Finally, another objective is to make a first assessment of the preservation of natural stones structures and to evidence potential threats. Indeed, in isolated environments with limited mineral resources, such as the Lesser Antilles archipelago, people have had to develop a specific knowledge of the local geology, for all purposes. In Martinique, clays were traditionally used for making roof tiles [6], but, in Guadeloupe, this activity has remained artisanal because of a lack of knowledge of the resource and the poor quality of clays [7]. They are still used only for medicinal and recreational purposes [8]. Paradoxically, the recreational use of clays in Guadeloupe tends to have a greater impact on the resource than the industrial use of clay in Martinique because of the limited resource and concentration of activity on a small area.

In buildings, the traditional creole architecture mainly uses wood, but some 19th and 20th century city center buildings and churches were made with volcanic stones (mostly andesite). Local limestones were processed in artisanal to semi-industrial limekilns. Stone-made houses and other buildings, limekilns and their extraction sites have a cultural value and represent an economic interest for tourism, but they suffer from a lack of protection, restoration and a poor awareness within the population. Indeed, although it is very present, most people remain unaware of this link between the local culture and the local geology. Nowadays, uncontrolled limestone quarries endanger many remarkable outcrops. Most of these quarries are small and designed solely to create a suitable flat area for building a concrete made house. The extracted limestone is used as a rather poor-quality road foundation material. With the densification of the population of the F.W.I. during the twentieth century, the anthropization of landscapes became important and limestones were excavated in many places, sometimes strongly impacting the landscape. Nowadays the population is slowly decreasing but the habit persists.

Finally, this article deals with the architectural heritage of the French West Indies, with emphasis on the use of dimension stone and other geomaterials, and exposes some related conservation issues. Then, with a main focus on Guadeloupe, it argues that unawareness is an important hindrance to the conservation of the local stone-built heritage in the French West Indies.

2. Context of the Study

Geological Context of the French West Indies

Located in the eastern Caribbean, the French West Indies are part of the Lesser Antilles arc [9]. They range south to north along the Caribbean plate boundary (Figure 1), where the North and the South American plates are subducting beneath the Caribbean plate. The geology of these islands is characteristic of subduction arcs, with a variety of volcanic formations along with scarce plutonic rocks and sedimentary deposits that form large carbonate platforms. The arc has been quasi-continuously active since the Eocene [10], with an inward migration of volcanism in its northern half. This has resulted in a variety of islands with either steep or shallow-shelf topography and an interbedding of volcanites and sediments. The Cenozoic migration of magmatism has also led to the formation of two distinctive sets of islands, the most recently active forming an internal arc (from Martinique to the Virgin Islands) and the others being associated with an inactive, older arc (Eocene to Miocene). From Dominica and further south, the two arcs, active and Eocene, are superimposed since very little migration occurred. Martinique island is formed by the two arcs. In the south and the east, Oligocene volcanic formations outcrop, the volcanic rocks are Miocene in the central part of the island and Plio-Pleistocene further west and north (Figure 2). The age variation of the volcanism, from east to west, from Oligocene to the Montagne Pelée active volcano, underlines the cenozoic westward migration of the magmatic activity. In Guadeloupe, the volcanic activity is occurring in the western part (Basse-Terre) and the eastern part (Grande-Terre) is covered with a carbonate platform (Figure 3).

The typology of rocks that outcrop is then separated, with massive andesites, suitable for producing dimension stones, in the West and limestones, suitable for lime production, in the East.

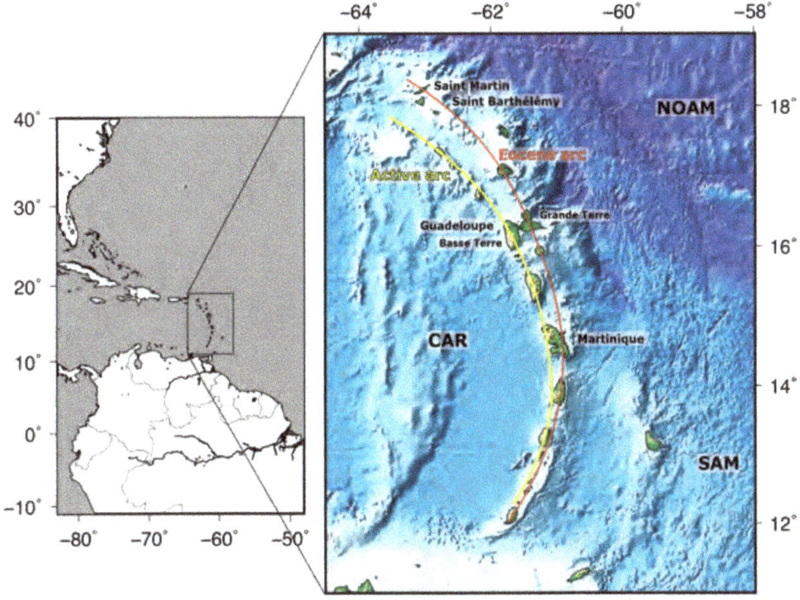

Figure 1. Map of the Eastern Caribbean. CAR: Caribbean plate, NOAM: North American plate, SAM: South American plate.

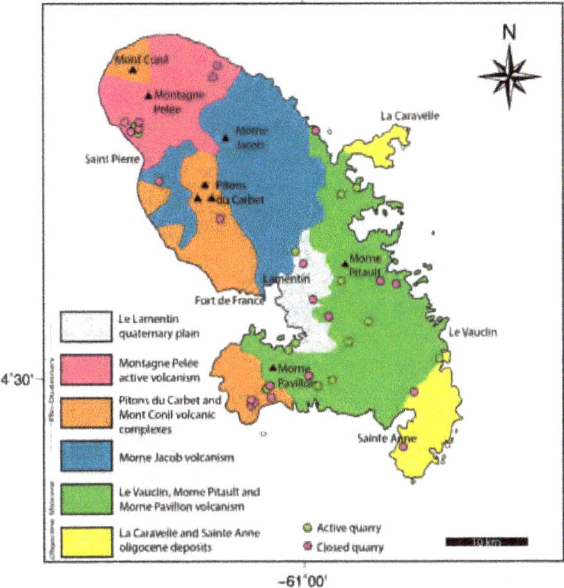

Figure 2. Simplified geological map of Martinique. Modified, after Westercamp et al., 1989 [11]. Active and closed quarries referenced in the national database InfoTerre [12].

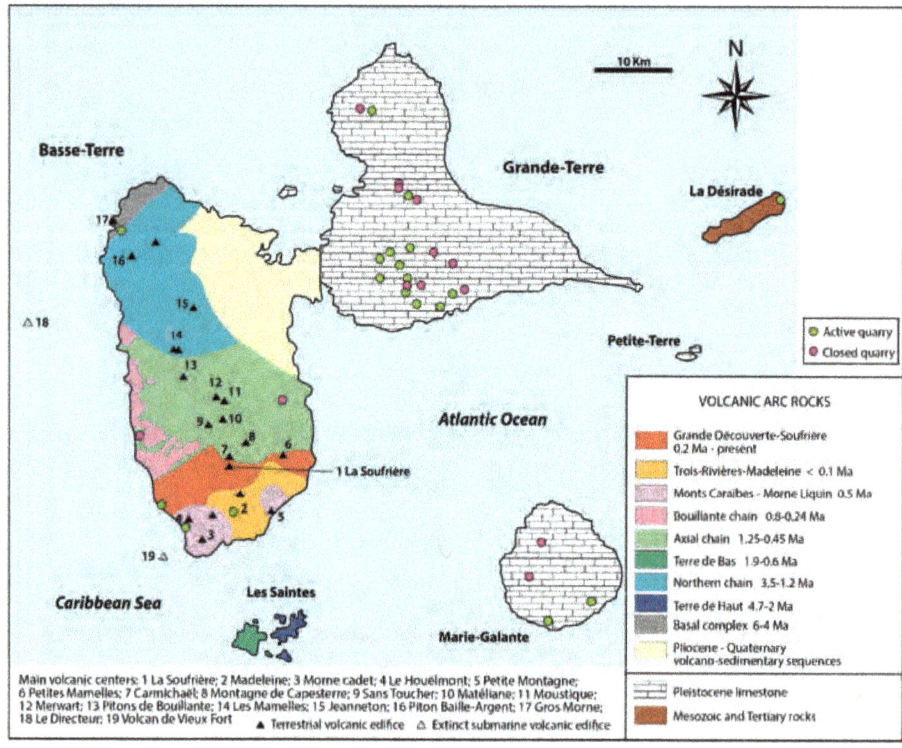

Figure 3. Simplified geological map of Guadeloupe, modified, after Komorowski et al., 2004 [13]. Active and closed quarries referenced in the national database InfoTerre [12] (for active quarries in 2019, see Figure 4).

3. Geoheritage Inventory

3.1. Geosites

In 2007, a National Geological Heritage Inventory [14] was launched in France. This state-funded program resulted in the description of several hundreds of geosites throughout the country, in both mainland and overseas territories. The majority of the referenced sites are natural (outcrops, landscapes, geological peculiarities, stratotypes, mineral deposits, various geological bodies), but some are also anthropogenic, especially sites located in quarries and mines, whether they are still in activity or not. Archeological sites are usually not included in this inventory, nor stone-built structures. The main purpose of this inventory is to gather information about the geological heritage and to provide it to the local authorities responsible for land-use policies. It is not an active protection measure, but it is designed to raise awareness about the geological heritage in order to avoid unintentional destruction of valuable outcrops. Each geosite is described in a technical note containing information such as a brief geological description, accessibility, any active protection measure, geological interest, educational interest, scarcity (local, regional, national, international), vulnerability and potential threats, relevant bibliographic references and the geographical perimeter of the site. All technical notes are included in public documents, but the information is not easily accessible to the general public. In Guadeloupe, Saint Martin and Saint Barthelemy the actual inventory has 33 sites, only one of which is located in Saint Martin and one in Saint Barthelemy.

One of the sites in Guadeloupe is inside the perimeter of an active quarry and one is a petroglyph archeological park, in Trois Rivières.

3.2. Present Day Geomaterials Extracted in the French West Indies

Today, a number of extraction sites are active in the French West Indies. Most of them are small limestone quarries and some extract andesite. None produces dimension stones on a regular basis. In the northern part of the arc, in Saint Martin, only one quarry is active (Hope Hill), it exploits andesite and volcanic tuffs [15]. In Martinique, 11 quarries exploit volcanic rocks, one limestone and two produce clay for the local roof-tile industry [16], a traditional specificity of this island in the Caribbean. In Guadeloupe, in 2019, 10 quarries produce stone from the limestone deposits, mostly in the Grand-Fonds area, one site exploits massive andesites (*Deshaies*) and two, in the south of Basse Terre, are extracting andesitic lapilli [15]. This section is mainly focused on Guadeloupe, where more information is available than in Martinique and Saint Martin. Figure 4 presents a map of the mineral resources of Guadeloupe, with the authorized quarries that are currently active.

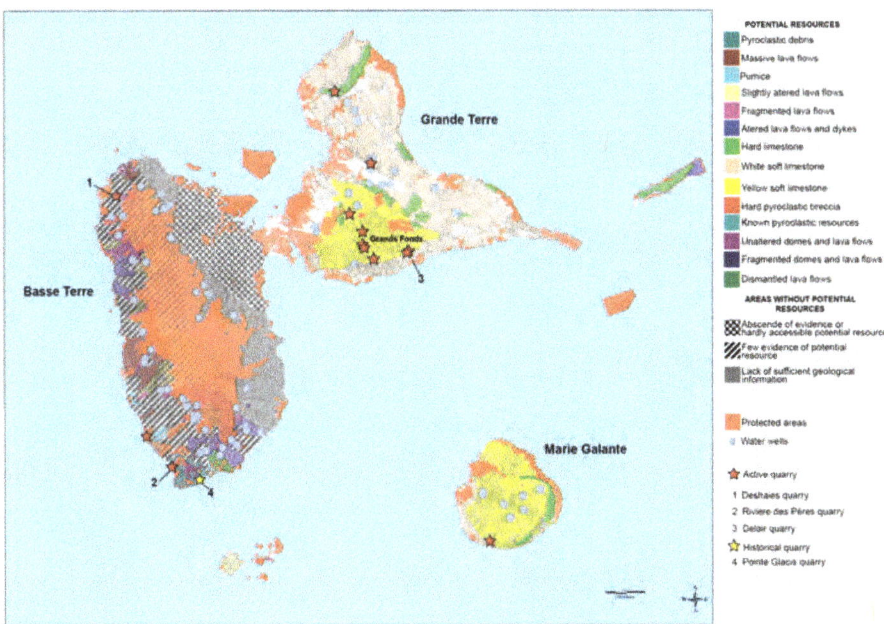

Figure 4. Map of the mineral resources of the Guadeloupe archipelago, modified after Bourdon and Chauvet, 2012 [15]. Red stars: quarries currently active in 2019.

The quarry in Deshaies is mining two different quality of rocks. Their average composition is given in Table 1. In the upper part of the pit, the volcanic rocks are more or less massive, with important weathering along the joints. Currently, this horizon is used for the manufacture of aggregates for hydraulic concretes. It is a massive basaltic andesite with a microlithic texture, plagioclases, pyroxenes and amphiboles phenocrysts and microcrystals of apatite and ilmenite (Figure 5a). There are abundant alteration zones that are rich in phyllosilicates (clay and chlorite). In the lower part of the exploitation the rocks are much less altered and fractured. They have a microlithic texture, with mostly pyroxenes and plagioclase unaltered phenocrysts (Figure 5b). This horizon is mined for producing road aggregates. Both types of aggregates produced from the Deshaies quarry require the use of explosives (the quarry uses 70 tons per year) and rock-crushing.

The Rivière des Pères quarry, at the extreme south of Basse-Terre, extracts hardened pyroclastic deposits, mostly lapilli, which require no explosive for the extraction. The aggregates are produced without rock-crushing and have a high vacuole content (Figure 5c,d).

Table 1. Chemical analysis of the Deshaies quarry rocks.

Content (%)	"Concrete" Rock	"Road Aggregate" Rock
SiO_2	58.33	59.00
Al_2O_3	17.76	17.91
Fe_2O_3	7.29	7.26
CaO	7.08	6.81
MgO	2.76	2.79
CO_2	1.41	0.77
Na_2O	3.22	3.35
K_2O	0.80	1.51

Data S.A.D.G. 2008 [17].

Figure 5. Basaltic andesite from Deshaies quarry, Guadeloupe (**a**) "Concrete" rock; (**b**) "Road aggregate" rock. Note the unaltered minerals on the fresh cut. Pictures a and b are from S.A.D.G. 2008 [17]. (**c**) and (**d**) volcanic deposits from Rivière des Pères, Monts Caraïbes, Guadeloupe. This outcrop is a geosite referenced in iGeotope; it is located one kilometer south of the Rivière des Pères quarry.

In the F.W.I., most quarries that are extracting limestones are in Guadeloupe (Marie Galante and Grande Terre), with only one in Martinique and none in Saint Martin and Saint Barthelemy. The carbonate platforms in Guadeloupe are well known and described [18–21]. They are constituted of an alternance of rodolitic algal limestones and reefal limestones, with intercalations of volcano-sedimentary detrital material. On Figure 6, the units 1 and 2 corresponds to the "Yellow" and "White soft limestones" of Figure 4, while units 3 and 4 are "hard limestones" (see also Figure 7).

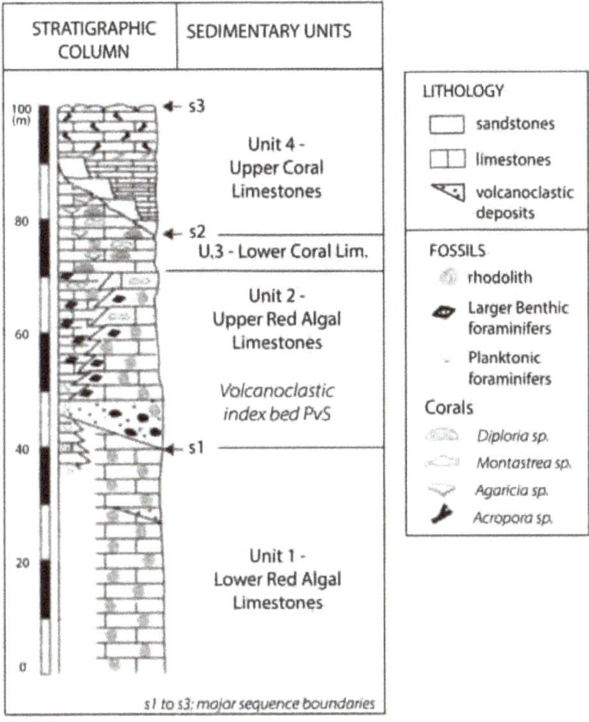

Figure 6. Synthetic stratigraphic log of the Grande-Terre and Marie Galante carbonate platform, Guadeloupe, from Conesa et al., 2012 [20].

(**a**) (**b**)

Figure 7. *Cont.*

(c)

Figure 7. (**a**) Reefal limestone in Delair quarry, Grande Terre. Note the large *Agaricia sp.* fossil colony above the hammer. Unit 3-Lower coral limestones; (**b**) *Acropora sp.* reefal limestone in Delair quarry. Unit 4-Upper coral limestones; (**c**) Rodolith from Unit 1-Lower red algal limestones, Grande Terre, Guadeloupe. Note the red algal carbonate concentric deposits around the volcanic debris.

3.3. Preservation Issues and Threats

Although the identification and the referencing in a national database of a heritage site does not automatically oblige the authorities to act for its conservation and protection, it is often seen as the first step towards active protection measures. In Guadeloupe, some geosites are protected, in fact, because they are located within regulated areas. Indeed, there are for example a National Park, a Natural Reserve and protected coastal areas. This is the case for the most spectacular sites, like the summit of the active volcano *La Soufrière*. On the contrary, some sites are neglected and some are endangered. The extraction activity appears to be often in tension with the conservation of the environment. In 2017, the quarry, south of Basse Terre (*Rivière des Pères*) was authorized to extend its activities in a natural area of high ecological value despite protests from the civil society [22]. In Grande Terre, almost all the quarries are extracting limestone in the Grand Fonds area. The Grand Fonds are a national heritage Geosite. It is a tropical karst that displays typical carbonate erosional features and remarkable sceneries. As the limestones have a low hardness, it is easy to open hill-side half-pits for the sole objective of providing a flat area for the construction of individual houses. This activity, sometimes illegally, strongly changes the landscape.

The first geosites in the F.W.I. were documented in Guadeloupe in 2003 [23] and in 2007 [24]. In early 2019, the author visited all sites from Guadeloupe (sites in Saint Martin and Saint Barthelemy were visited in 2017) in order to update their outcrop conditions and preservation status. Under tropical Caribbean conditions, the weathering can be very active, along with strong action of plant roots on rocks and soils. Indeed, 5 sites at least (over 33) were not easily visible anymore, due to vegetation growth or surface weathering. In particular, a site in southern Basse-Terre (*Pointe Glacis*—see Section 4.2.4). The sites that are the most vulnerable to weathering are small roadside outcrops, but even some landscapes associated with large geological formations are affected by anthropogenic activity. Two roadside sites have been covered with shotcrete for safety purposes. One of the sites, in Grand Fonds area, which is inside an active quarry (Delair), has been protected since its discovery. The quarry is not allowed to destroy the outcrop and has therefore concentrated its activity on other parts of the concession. The protection was then a success but, since then, the outcrop has been exposed to natural weathering and the sedimentary features that had motivated for its protection, although still present, are hard to read due to intense algae growth and carbonate surface dissolution (Figure 8a,b).

Apart from limestones, other georesources, that occur at locations that are not yet referenced within the geosites inventory, can also be faced with growing urbanization, or other activities, like for example, in Grande-Terre, the *Anse Babin* bay. In this area, local people come for mud bathing [8].

The mud is originally from the neighboring mangrove and is collected in shallow waters. This is done by the population only in Anse Babin, a small area with limited clay resource. As the bay is inside the lagoon, well protected from the ocean swell and any tidal current, clays accumulate on site, near the shore. However, the over-exploitation and collection for off-site personal use, possibly associated with the confusion between a natural resource and a renewable resource, has caused the rarefaction of the resource (Figure 8c,d).

Figure 8. (**a**) view from the south of the protected outcrop at Delair quarry, in Guadeloupe, in 2010. Several sedimentological facies were identifiable. The height of the escarpment is of 40 m; (**b**) view from the south-west of the same outcrop in 2019. Surface alteration has darkened the rocks, the structures are far less visible and a pile of materials from the quarry is blocking the access to the escarpment; (**c**) the bay of Babin, in Guadeloupe; (**d**) mud collected at the sea bottom for mud bathing and let onshore after use.

4. Stone-Built Heritage

4.1. Pre-Columbian Heritage

The first known use of geomaterial in the F.W.I. are pre-Columbian artefacts [25,26]. Carved rocks are also documented in most of the Eastern Caribbean islands [27] and protected, since 1981, in Guadeloupe, by an archeological Park. Besides protection and awareness, the Park also generates tourist attractivity. The best known and preserved petroglyphs are located in Trois-Rivières, south of Basse-Terre island, Guadeloupe. The carved rocks are boulders of andesite, from the Monts Caraïbes massif. These rocks are silicic and resistant to weathering, at least far more than the Grande-Terre limestones. It is likely that the pre-Columbian inhabitants have used the limestones as well for carving, but that the erosion has erased the petroglyphs. This assumption is supported by the attested pre-Columbian human occupation of some cavities of the Grande-Terre and Marie Galante karsts [28] where some altered engravings are still preserved [29].

4.2. Colonial Heritage

Later, during the Colonial period (16th to 20th century), most of the buildings were constructed from wood; few used rubble stones and fewer still were constructed from dimension stone. Most of the stone-built structures are military buildings, churches, houses in civic centers and windmills. Although some are preserved, many are known only by their archeological remains [30]. There is very few information about both the stones that were used and the extraction sites. The location of the historical quarries is almost never known. It is likely that they were small, that they were in activity over short periods of time and that stone was produced as close as possible to the needs.

4.2.1. Military Buildings

During the 18th century, military and religious natural stone buildings were constructed in the French West Indies. In Guadeloupe, the plans of the military forts followed the typical European engineering of that time. They were made using both soldiers' workforce and slave labor and set in the best-suited places for the defense of the harbors and main cities, usually on the top of hills near the sea. The fact that their locations was chosen for defense purposes implied a necessary adaptation to the ground and different quarrying strategies. In Grande Terre, the *Fort Fleur d'Epée* (built between 1750 and 1763) is on a limestone spur. There, the builders extracted the limestones directly on site and constructed on the rocks, following the topography with only little earthworks (Figure 9a). The excavation of tunnels and underground rooms for gunpowder storage provided material for the outer fortifications and small size buildings within the perimeter of the fort. The quality of the material makes it strong enough for construction and easy to carve, but it is coarse grained. For ornamentation, andesite was then chosen and imported from Basse Terre (Figure 9b).

In Basse Terre, the *Fort Delgrès* surmounts unconsolidated debris flow deposits (Figure 9c). Consequently, all materials have had to be brought to the site. The fortifications and barracks were made with cemented volcanic rock pebbles. No reliable historical recording of the origin of the materials could be found during this study, but it is likely that the pebbles have been collected on Basse Terre western beaches and that the cement used lime from Grande Terre.

(a)   (b)

(c)

Figure 9. (a) Foundations of the Fort Fleur d'Epée, Guadeloupe. Entrance of a tunnel, excavated in the limestones, and inner wall of the fort, directly on the rocks; (b) main gate of Fort Fleur d'Epée with dimension stone ornamentations made with Basse Terre andesite; (c) northern fortifications of Fort Delgrès, Guadeloupe, above the gorge of Galion river. Note the unconsolidated debris flow materials, from La Soufrière volcano, forming the escarpment.

4.2.2. Religious Buildings

In Guadeloupe, natural stone religious buildings from the colonial period are found only in Basse-Terre, where volcanic rocks were quarried. They principally consist in the church of the village of *Vieux-Habitants* and the cathedral of Basse Terre city (Figure 10a–e), as well as several small chapels. Vieux Habitants' church was built in 1703, after the original wooden construction was burned in an act of war. Most of the edifice is recent (rebuilt in the 20th century) and now covered with an impermeability coating. Nevertheless, its original ornamental dimension stone main gate is preserved. It is made of greenish and reddish andesite. The facade was carved by workers from central France (Massif Central), were similar rocks are traditionally used [31]. The Basse Terre cathedral was built shortly after, in 1730, and displays a more important use of natural stone [32], especially for the facade, the vault, and the ornamentation.

The preservation state of the materials of the buildings in the center of Basse Terre city, and especially the Basse Terre cathedral, shows highly variable deteriorations of the dimension stones. Although the origin of the blocks is uncertain, their coloration, the important variability of textures and the abundance of vacuoles suggest a provenance from the slightly weathered top of Basse Terre lava flows (Figure 10e). The deterioration is particularly important where the porosity of the material is high. Alteration also strongly follows the fabric of the material, especially the magmatic fluidity. This implies that for the eventual restoration of the cathedral, it will be likely that the use of andesite blocks from local lava flows or debris flow deposits will be required. Indeed, the andesitic rocks currently in exploitation at the quarry in northern Basse Terre are of different quality (color, fabric) and would presumably show a strong appearance difference with the stones used for the original construction of the building.

Figure 10. (**a**) facade of the Vieux Habitants church; (**b**) facade of the Basse Terre cathedral; (**c**) and (**d**) architectural elements of the Basse Terre cathedral; (**e**) rock alteration on the Basse Terre cathedral; (**f**) Basse Terre city center house, in front of the cathedral.

4.2.3. Houses

Natural stones houses in the French West Indies are mainly concentrated in the civic centers of Basse-Terre, in Guadeloupe (Figure 10f) and Saint Pierre, in Martinique (see Section 5.1 and Figure 13). There, some historical buildings are made with local dimension stones. Outside the main cities, only scarce natural stone houses are still preserved. They are usually made with boulders, but almost never with dimension stones. The house near the hamlet of Thomas, in Guadeloupe, is an example of local construction with pebbles collected directly on site, the house being only a few meters away from the beach (Figure 11). Unfortunately, this construction is not protected and in a very poor state of preservation. The detail view of the walls shows that the rocks used are of various volcanic origin and size, with little cement. As only a small quantity of lime was needed for these types of constructions, usually the builders were collecting limestones from beach debris or even extracting coral-bearing rocks in shallow waters.

Figure 11. (**a**) and (**b**) house at Thomas hamlet; (**c**) same house, detail view of the walls (**d**) beach, on site, where the pebbles were most likely collected.

4.2.4. Industrial Buildings

The industrial history of the West Indies is tightly linked to the culture of sugar cane. Thus, the main stone made industrial buildings are windmills and sugar factories. They are particularly abundant in Grande-Terre (eastern Guadeloupe) and Marie-Galante, where they are an important element of the landscapes [33]. Windmills are probably the natural stone architectural heritage that is the most identified as such within the population in the French West Indies. They are consequently rather well preserved and some are integrated in architectural projects, such as hotels and resorts (Figure 12a,b). For the production of lime, limekilns have been built in different places of the islands, near the resources. Some were even built on small islets where no rock outcrops above sea level (Ilet Fajou). They were using only materials extracted directly from the sea floor at shallow depth. Surprisingly, the most important in Guadeloupe is located not on the carbonated Grande Terre, but on the volcanic island of Basse Terre. It exploited the only limestone outcrop of this part of the island, in Trois Rivières, which consists in a lens of reefal limestones, interbedded with volcanic deposits. It had the advantage of being close to both the resource and the demand for lime of Basse Terre city. The limekiln is protected, but poorly maintained (Figure 12c–f) [34]. The former quarry (*Pointe Glacis*) is in the geosites inventory, but benefits from no protection measure and is currently covered with vegetation and not accessible easily.

Figure 12. (**a**) typical landscape in Grande Terre, with a former sugar cane windmill, in Saint François; (**b**) example of a restored windmill and adjacent building, used as the main entrance of a hotel, in Le Gosier; (**c**) photograph of the Trois Rivières limekiln, taken in 2012 [35] and in 2019 (**d**); (**e**) detail of the walls of the limekiln; (**f**) Pointe Glacis former quarry.

4.2.5. Dry-Stone Walls

In November 2018, the Intergovernmental Committee for the Safeguarding of the Intangible Cultural Heritage has inscribed the Art of Dry-Stone Walling, Knowledge and Techniques on the Representative List of the Intangible Cultural Heritage of Humanity [36]. In the French West Indies, dry-stone walls have been traditionally used only in the northernmost islands of Saint Martin and Saint Barthélemy. In Saint Martin, few of these walls are still preserved. In Saint Barthélemy, they have been widely used for the separation of fields and they are an important part of the islands' agricultural landscapes.

5. Discussion

5.1. Historical Cities and Volcanic Crisis

In the French West Indies, from the 17th to the early 20th century, the city of Saint Pierre in Martinique and the city of Basse Terre in Guadeloupe were the main economic centers. Consequently, this is where natural stone buildings were concentrated. These cities have benefited from abundant water resources and prosperous soils for agriculture due the proximity of volcanoes, as well as favorable conditions for sailing ships trade. In the 20th century however, the transformation of the economy has been accompanied with a translation of the activity toward the central zones of the islands. Furthermore, both sites were impacted by volcanic eruptions. In Martinique, the Montagne Pelée volcano erupted in 1902 and 1929 [37–39]. The eruption of 1902 resulted in the dramatic destruction of the city of Saint Pierre and the death of more than 28,000 inhabitants [40–42]. The city was hit by nuée ardentes and pyroclastic flows and the ruins are still preserved (Figure 13). This eruption is a milestone for the development of both volcanology and risk studies. Indeed, it is the first volcanic eruption that has been monitored with the tools of modern science and it is the first time a risk, known by the scientists, has led to a warning to the authorities that has been ignored due to political reasons (the local authorities refused to order the evacuation of the population as suggested by the scientists because some elections were about to take place) [43,44]. Thus, the patrimonial value of Saint Pierre dimension stone buildings is related not only to their architectural specificity within the French West Indies, but also to the history of the city. In Guadeloupe, the city of Basse Terre was never struck by a catastrophic volcanic event, since the last magmatic eruption date from 1530, before the city was built [45]. However, several phreatomagmatic events have occurred on La Soufrière volcano since. The last one, in 1976, triggered intense debate among the scientists and the authorities. It resulted in the evacuation of the south of Basse-Terre for several months [46,47]. This accelerated the process of translation of the economic activity toward the geographical central part of Guadeloupe. However, unlike in Saint Pierre, the main territorial administrations are still located in Basse Terre city, allowing the preservation of its dynamism and its population density. Nowadays, although the historical dimension stone buildings are still in use, they suffer from the lack of sufficient preservation efforts.

In Martinique, the destruction of Saint Pierre resulted in a drastic change with regard to the use of natural stones as a construction material. Posterior constructions don't show a use of this type of material at such a scale, preferring wood and concrete materials. In Guadeloupe also, no city center has been built with intensive use natural stone after the 19th century.

(**a**) (**b**)

Figure 13. *Cont.*

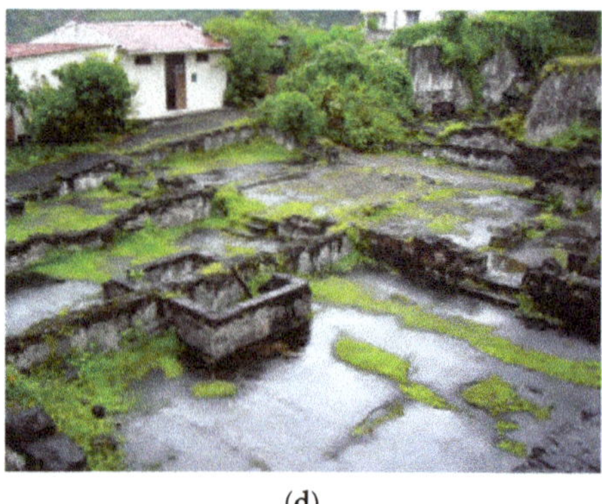

(c)   (d)

Figure 13. (**a**) Montagne Pelée volcano overlooking Saint Pierre; (**b**), (**c**) and (**d**) ruins of Saint Pierre, after the eruption of 1902.

5.2. Geomaterials Exploitation and Landscape Preservation

With the transformation of the construction style, the need for dimension stones decreased in the 19th and 20th centuries, while the need for raw building materials increased. The quarrying activity changed consequently, which led to the abandon of small extraction sites, like the Pointe Glacis quarry, and the reduction of uncontrolled rock collection for building stones or lime fabrication. Although the construction of traditional buildings needs small amounts of rock, modern constructions and roads require a much higher supply. In the early 21st century, the quarries in the French West Indies had an overall production of more than 4.5 millions of tons per year (details in Table 2), and the production has been growing ever since.

Table 2. Volumes of production in the French West Indies.

Unit: (10^3.T/y)	Volcanic Rocks	Carbonated Rocks	Clay
Guadeloupe [1]	1500	300	-
Saint Martin [1]	250	-	-
Martinique [2]	2450	45	40

[1] Data from 2007 [15], [2] data from 2004 [16].

In order to follow the increasing demand, the production needs to rise and the impacted surface widens. On small territories, and the West Indies being a hotspot for biodiversity, this is a source of debate and conflicting interests within the society [48]. Recently, some quarries have been authorized to mine adjacent areas, providing they take biodiversity loss compensation measures [14]. Although the extraction of andesites, which requires mining techniques, is restricted to only a few sites, the softness of the carbonate platform limestones in Guadeloupe allows an easy mechanical extraction. In the Grands Fonds area, south of Grande Terre, Guadeloupe, seven authorized quarries are active and three others are located around the Grand Fonds and in Marie-Galante. However, in many places, the limestones are extracted without mining authorization, for earthwork purposes. The landscape appears to be strongly impacted (Figure 14).

(a)   (b)

Figure 14. (**a**) Delair active quarry, in Les Grands Fonds area, Guadeloupe; (**b**) example of landscape transformation for individual houses building in Les Grands Fonds.

5.3. *Transmission of Knowledge and Natural Stone Heritage Preservation*

The transformation of the society in small island territories and the intense weathering of geomaterials under tropical climate are putting the preservation of the natural stone architectural heritage, as well as the geological natural heritage, under growing tension. In Martinique, the ruins of Saint Pierre are remarkable because of both the history of the destruction of the city and their rare dimension stone architecture. The volcanic crisis of 1902 and 1929, accelerated to the abandon of the architectural style of the 19th century and, with the use of modern materials, the know-how has not been transmitted. In Guadeloupe as well, the natural stone architectural heritage is poorly preserved and not all historical buildings benefit from a protection. However, many constructions demonstrating the preservation of the traditional know-how can be seen in the French West Indies. For example, some natural stones walls are still made with the same techniques as in the 17th to 19th century (Figure 15a). In Saint Barthélémy, preservation efforts are being made and stone walls are still being built and keep structuring the landscapes of the island. However, they are no longer used for field separation but rather along the roads. According to modern standards, these walls use natural stone, but not dry stone (Figure 15c,d). This solution, that doesn't follow the traditional building modes, has been favored for esthetical considerations as well as for preventing earth, dirt, or any objects to be wind-pushed on the road. Traditionally, dry stones walls were preferred in Saint Barthelemy, which allowed the rain water to flow downhill during the rainy season. Today, during the heavy rain events in Saint Barthelemy, the water occasionally accumulates along the road-side stone walls, sometimes damaging the soils and infrastructures. In Europe, in areas submitted to strong wind episodes, dry stones were also favored for their resistance, as the space between the stones allows the wind to go through. In the Burren, Ireland, the stones are oriented in a manner that minimizes the resistance to the wind and that leave significant empty spaces in the wall, in order to let the airflow easily (Figure 15b).

(a)   (b)

Figure 15. *Cont.*

(c)   (d)

Figure 15. (**a**) 1970's natural stones wall in the University campus, Guadeloupe. To be compared with Figure 11c; (**b**) dry stone wall in the Burren, County Clare, Ireland. The stones are oriented in a manner that minimizes the resistance to the wind; (**c**) dry stone wall at the bottom of a hill in Saint Barthelemy; (**d**) modern road-side stone wall, Saint Barthelemy.

6. Conclusions

In the French West Indies, the natural stones architectural heritage and the natural geological heritage are notions that are still in development in the society. For example, in Guadeloupe, some traditional buildings and walls are made with boulders. Surprisingly this specific local know-how appears to be preserved and still used despite the fact that it is not an identified cultural heritage and that very few modern constructions use natural stones. In other places, like Saint Barthélémy, efforts are being made for the preservation of the natural stone architectural heritage. Paradoxically, some efforts forget to consider the environmental reasons at the origin of the traditional dry stones walling techniques. The limited abundance of natural resources in small island territories is also an important factor to consider. Indeed, unlike other natural resources, like wind, solar or geothermal energies or wood and halieutic resources, geological resources are not renewable at a rapid rate. Georesources, which are local and natural resources, are sometimes confused with renewable resources [49]. This can lead to unsustainable exploitation of georesources, associated with significant landscape changes and permanent loss of geodiversity. The brief assessment of the natural stone architectural heritage in the French West Indies presented in this paper also evidenced a lack of sufficient data for its conservation. Most historical extraction sites remain unknown. In order to better preserve this heritage for future generations, efforts are required for both raising geological awareness and science education within the population and better characterizing the geological heritage. In that sense, further studies are needed in the French West Indies. In particular, the geoheritage inventory should be extended to new geosites and its links with the architectural heritage emphasized. Petrological and geochemical analysis of the stones used to build the most significant architectural heritage buildings should be the subject of a specific study in order to identify historical quarries that could be re-used for their future restoration.

Author Contributions: Y.M. rose funding, designed the study and wrote the paper. All pictures are from the author, except Figure 5a,b [17] and Figure 12c [35] and the pictures of the dry-stone walls in Saint Barth (Figure 15c,d), that are courtesy of Hélène Bernier.

Funding: This research was funded by the French Ministry of Ecological and Social Transition (DEAL de la Guadeloupe), grant number 971-2019-04-09-003 (Caribbean Academy of Sciences).

Acknowledgments: The author wishes to thank Hélène Bernier for providing the pictures of the dry-stone walls in St Barth. Yaëll Conchis and Laure Bouriat participated in some field work as part of their engineer degree.

Conflicts of Interest: The author declares no conflict of interest. The funders had no role in the design of the study; in the collection, analyses, or interpretation of data; in the writing of the manuscript, or in the decision to publish the results.

References

1. Loi n° 2002-276 du 27 février 2002 relative à la démocratie de proximité. Available online: https://www.legifrance.gouv.fr/affichTexte.do?cidTexte=JORFTEXT000000593100 (accessed on 6 May 2019).
2. DeWever, P.; Egoroff, G.; Cornée, A.; Lalanne, A. *National Geoheritage in France*; Mémoire Hors-Séries de la Société Géologique de France: Paris, France, 2014; Volume 14, p. 180.
3. Garrabé, F. *Geological Map of Grande-Terre, 1/50 000*; Bureau des Recherches Géologiques et Minières, BRGM: Paris, France, 1988.
4. De Reynald De Saint Michel, A. *Geological Map of Basse-Terre, 1/50 000*; Bureau des Recherches Géologiques et Minières, BRGM: Paris, France, 1966.
5. Igeotope. Available online: http://igeotope.brgm.fr (accessed on 6 May 2019).
6. Javey, C.; Baudet, G. *Inventaire des Ressources en Argiles de la Guadeloupe, Service Géologique des Antilles*; Rapport BRGM n76, ANT, BRGM: Paris, France, 1976; p. 34.
7. Baudet, G. *Etude D'orientation Réalisée sur des Matières Premières Pour Céramique en Provenance de la GUADELOUPE, Service Géologique National*; Rapport BRGM 76 SGN 347 MIN; BRGM: Paris, France, 1976.
8. Gros, O. Analyses chimique et bactériologique des boues marines de Babin. In *Rapport Commune de Morne à l'Eau*; Université des Antilles: Pointe à Pitre, France, 2012.
9. Maury, R.C.; Westbrook, G.K.; Baker, P.E.; Bouysse, P.; Westercamp, D. Geology of the Lesser Antilles. In *The Geology of North America*; Geological Society of America: Boulder, CO, USA, 1990; pp. 141–166.
10. Legendre, L.; Philippon, M.; Léticée, J.L.; Münch, P.; Maincent, G.; Cornée, J.J.; Thomas, L.; Lebrun, J.-F.; Mazabraud, Y. Upper plate strain pattern and volcanism accommodating trench curvature. Insights for the Lesser Antilles arc, Saint Barthélemy Island, FWI. *Tectonics* **2018**, *37*, 2777–2797. [CrossRef]
11. Westercamp, D.; Andrieff, P.; Bouysse, P.; Cottez, S.; Battistini, R. *Geological Map of Martinique, 1/50 000*; BRGM: Paris, France, 1989.
12. National Geological Database InfoTerre, BRGM. Available online: http://infoterre.brgm.fr/ (accessed on 27 July 2019).
13. Komorowski, J.C.; Boudon, G.; Semet, M.; Beauducel, F.; Anthénor-Habazac, C.; Bazin, S.; Hammouya, G. *Volcanic Atlas of the Lesser Antilles, Guadeloupe*; University of the West Indies: Mona, Jamaica, 2004; pp. 65–102.
14. Avoine, J.; Baillet, L.; Charbonnier, S.; de Kermadec, C.; Duranthon, F.; Egoroff, G.; Graviou, P.; Guiomar, M.; Guyetan, G.; Hoblea, F.; et al. Inventory of the Geological Heritage. Available online: https://inpn.mnhn.fr/programme/patrimoine-geologique/presentation?lg=en (accessed on 17 May 2019).
15. Bourdon, E.; Chauvet, M. *Schéma des Carrières de la Guadeloupe*; Rapport BRGM/RP-57157-FR; BRGM: Paris, France, 2012.
16. Comte, J.P.; Le Berre, P.; Maurin, C. *Schéma des Carrières de la Martinique*; Rapport BRGM/RP-53465-FR; BRGM: Paris, France, 2005.
17. SADG, Carrière de Deshaies, Guadeloupe. *Investigations Visant à Rechercher L'origine de la Dérive des Valeurs au Bleu Mesurées en Fin de Production*; CNI7.8.425-Rapport 01, BRGM: Paris, France, 2008.
18. Garrabé, F. Evolution sédimentaire et structurale de la Grande-Terre de Guadeloupe. Ph.D. Thesis, Paris-Sud University, Paris, France, 1983; p. 171.
19. Léticée, J.L. Architecture d'une Plate-Forme Carbonatée Insulaire Plio-Pléistocène en Domaine de Marge Active (Avant-Arc des PETITES Antilles, Guadeloupe): Chronostratigraphie, Sédimentologie et Paléoenvironnements. Ph.D. Thesis, Université des Antilles, Guadeloupe, France, 2008; p. 229.
20. Conesa, G.; Vernhet, E.; Baden, D.; Guglielmi, Y.; Viseur, S.; Marié, L. New petrophysical approach in the study of Cenozoic coral carbonate rocks as reservoirs: Example of Pleistocene platforms, Guadeloupe, French West indies. In Proceedings of the AAPG/SPE/SEG Hedberg Research Conference Fundamental Controls on Flow in Carbonates, Saint-Cyr-sur-Mer, France, 8–13 July 2012.
21. Vernhet, E.; Conesa, G.; Bruna, P.O. Reworking processes and deposits in coral reefs during (very) high-energy events: Example from a Pleistocene coral formation (125 ka), La Désirade Island, Lesser Antilles. *Paleogeogr. Paleoclimatol. Paleoecol.* **2018**, *490*, 293–304. [CrossRef]

22. Exploitation de la carrière de Rivière Sens. *Avis du Conseil National de la Protection de la Nature*; Ref. Onagre 2018-04-14a-00629; Ministère de la Transition Ecologique et Solidaire: Paris, France, 2018.

23. Des Garets, E. *Inventaires des Sites Géologiques Remarquables de la Guadeloupe des Geosites de Guadeloupe, Phase 1*; Rapport BRGM/RP-52728-FR; BRGM: Paris, France, 2003.

24. Bès de Berc, S.; Chauvet, M.; Lebrun, J.F.; Léticée, J.L.; Randrianasolo, A.; Traineau, A. *Inventaires des Sites Géologiques Remarquables de la Guadeloupe des Geosites de Guadeloupe, Phase 2*; Rapport BRGM/RP-55737-FR; BRGM: Paris, France, 2007.

25. Lammers-Keijers, Y. *Tracing Traces from Present to Past—A Functional Analysis of Pre-Columbian Shell and Stone Artefacts from Anse à la Gourde and Morel, Guadeloupe, FWI*; Leiden University Press (Archeological Studies Leiden University): Leiden, The Netherlands, 2007.

26. Reynaud, G. Evaluation du potentiel scientifique du matériel de mouture, de broyage et de polissage des sites amérindiens. In *Bilan Scientifique Régional de la Région Guadeloupe 2010*; Direction Régionale des Affaires Culturelles, Service Régional de l'Archéologie: Guadeloupe, France, 2010.

27. Dubelaar, C.N. *The Petroglyphs of the Lesser Antilles, the Virgin Islands and Trinidad*; Foundation for Scientific Research in the Caribbean Region: Amsterdam, The Netherlands, 1995; p. 492.

28. Grouard, S.; Bonnissent, D.; Courtaud, P.; Fouéré, P.; Lenoble, A.; Richard, G.; Romon, T.; Serrand, N.; Stouvenot, C. Fréquentation Amérindienne des Cavités des Petites Antilles. In Proceedings of the International Association of Caribbean Archeology 24th Congress, Pre-Columbian Archeology General Session, Martinique, France, 25–30 July 2011.

29. Lenoble, A.; Queffelec, A.; Bonnissent, D.; Stouvenot, C. Rock art taphonomy in Lesser Antilles: Study of wall weathering and engravings preservation in two pre-Columbian caves on Marie-Galante Island. In Proceedings of the 25th International Congress for Caribbean Archaeology, San Juan, Puerto Rico, 15–21 July 2013; pp. 634–657.

30. Laborie, S.; Fourteau, A.-M.; Stouvenot, C.; Yvon, T. *Bilan Scientifique Régional de la Région Guadeloupe 2006–2007–2008*; Direction Régionale des Affaires Culturelles, Service Régional de l'Archéologie: Guadeloupe, France, 2010; ISSN 1262-887.

31. Joseph, E.S. *Patrimoine Architectural Mérimée*; PA00105877; Ministère de la Culture: Paris, France, 1993.

32. Cathédrale Notre-Dame de Guadeloupe. *Patrimoine Architectural Mérimée*; PA00105849; Ministère de la Culture: Paris, France, 1993.

33. Atlas des Paysages de la Guadeloupe, Région Guadeloupe, DEAL de la Guadeloupe. 2013. Available online: http://www.guadeloupe.developpement-durable.gouv.fr/atlas-des-paysages-de-l-archipel-guadeloupe-r651.html (accessed on 1 June 2019).

34. Ancien four à chaux. *Patrimoine Architectural Mérimée*; PA97100023; Ministère de la Culture: Paris, France, 2018.

35. Aristoi. 2012. Available online: https://commons.wikimedia.org/wiki/File:Ancien_four_à_chaux_01.JPG (accessed on 3 June 2019).

36. Convention for the Safeguarding of the Intangible Cultural Heritage. In Proceedings of the 13th Session of Intergovernmental Committee for the Safeguarding of the Intangible Cultural Heritage, Port Louis, Republic of Mauritius, 26 November–1 December 2018.

37. Tanguy, J.C. The 1902–1905 eruptions of Montagne Pelée, Martinique: Anatomy and retrospection. *J. Volcanol. Geotherm. Res.* **1994**, *60*, 87–107. [CrossRef]

38. Gourgaud, A.; Fichaut, M.; Joron, J.L. Magmatology of Mt. Pelée (Martinique, F.W.I.). I: Magma mixing and triggering of the 1902 and 1929 Pelean nuées ardentes. *J. Volcanol. Geotherm. Res.* **1989**, *38*, 143–169. [CrossRef]

39. Martel, C.; Bourdier, J.L.; Pichavant, M.; Traineau, H. Textures, water content and degassing of silicic andesites from recent Plinian and dome-forming eruptions at Mount Pelée volcano (Martinique, Lesser Antilles arc). *J. Volcanol. Geotherm. Res.* **2000**, *96*, 191–206. [CrossRef]

40. Fisher, R.V.; Smith, A.L.; Roobol, M.J. Destruction of St. Pierre, Martinique, by ash-cloud surges, May 8 and 20, 1902. *Geology* **1980**, *8*, 472–476. [CrossRef]

41. Bourdier, J.L.; Boudon, G.; Gourgaud, A. Stratigraphy of the 1902 and 1929 nuée-ardente deposits, Mt. Pelée, Martinique. *J. Volcanol. Geotherm. Res.* **1989**, *38*, 77–96. [CrossRef]

42. D'Ercole, R.; Rançon, J.P. La future éruption de la Montagne Pelée: Risque et représentations. *MappeMonde* **1994**, *4*, 31–36.

43. Smith, A.L.; Roobol, M.J. Mt Pelée, Martinique: A study of an island arc volcano. In *GSA Memoirs*; Geological Society of America: Boulder, CO, USA, 1990; Volume 175.

44. Heilprin, A. *Mont Pelée and the Tragedy of Martinique: A Study of the Great Catastrophes of 1902, with Observations and Experiences in the Field*; The Geographical Society of Philadelphia, JB Lippincott Company: Philadelphia, PA, USA; London, UK, 1903.

45. Boudon, G.; Komorowski, J.C.; Villemant, B.; Semet, M.P. A new scenario for the last magmatic eruption of La Soufrière de Guadeloupe (Lesser Antilles) in 1530 A.D.: Evidence from stratigraphy, radiocarbon dating and magmatic evolution of erupted products. *J. Volcanol. Geotherm. Res.* **2008**, *178*, 474–490. [CrossRef]

46. Tazieff, A. La Soufrière, volcanology and forecasting. *Nature* **1977**, *269*, 96–97.

47. Le Guern, F.; Bernard, A.; Chevrier, R.M. Soufrière of Guadeloupe, 1976-1977 eruption—Mass and energy transfer and volcanic health hazards. *Bull. Volcanol.* **1980**, *43*, 577–593. [CrossRef]

48. Myers, N.; Mittermeier, R.A.; Mittermeier, C.G.; da Fonseca, G.A.B.; Kent, J. Biodiversity hotspots for conservation priorities. *Nature* **2000**, *403*, 853–858. [CrossRef] [PubMed]

49. Anjou, C.; Fournier, F.; Mazabraud, Y.; Forissier, T. Learners' Conceptions about Geothermal Energy in Three Caribbean Islands. In Proceedings of the World Geothermal Congress 2020, Reykjavik, Iceland, 26 April–1 May 2020; in press.

 *sustainability*

Article

The Candoglia Marble and the "Veneranda Fabbrica del Duomo di Milano": A Renowned Georesource to Be Potentially Designed as Global Heritage Stone

Giovanna Antonella Dino [1,*], Alessandro Borghi [1], Daniele Castelli [1], Francesco Canali [2], Elio Corbetta [2] and Barry Cooper [3]

[1] Department of Earth Sciences, University of Torino, Via Valperga Caluso, 35, 10125 Torino, Italy
[2] Veneranda Fabbrica del Duomo di Milano, Via Carlo Maria Martini, 1, 20121 Milano, Italy
[3] School of Natural and Built Environments, University of South Australia, Adelaide SA 5001, Australia
* Correspondence: giovanna.dino@unito.it; Tel.: +39-011-670-5150

Received: 20 June 2019; Accepted: 24 July 2019; Published: 29 August 2019

Abstract: Marbles from Alpine area have been widely employed to build and decorate masterpieces and buildings which often represent the cultural heritage of an area (statuary, historic buildings and sculptures). Candoglia marble, object of the present research, is one of the most famous and appreciated marbles from Alpine area; it has been quarried since Roman times in the Verbano-Cusio-Ossola (VCO; Piemonte—NW Italy) extractive area. Candoglia Marble outcrops are present as lenses within the high-grade paragneisses of the Ivrea Zone, a visible section of deep continental crust characterised by amphibolite- to granulite-facies metamorphism (Palaeozoic period). Candoglia calcitic marble (80–85% $CaCO_3$ and the 15–20% other minerals) shows a characteristic pink to gray colour and a coarse-grained texture (>3 mm): frequent centimetre-thick dark-greenish silicate layers (mainly represented by diopside and tremolite) characterize the texture of the marble. It has been largely used in local rural constructions and historical buildings, but its most famous application has been (and still is) for the "Duomo di Milano" construction (fourteenth century). The Veneranda Fabbrica del Duomo di Milano carried out the anthropogenic activities dealing with the Candoglia marble exploitation; it has to be highlighted that the company have managed the Marble exploitation during the last seven centuries and that the quarry itself is a tangible sign of the development of extraction and heritage in the VCO area. Candoglia marble can be recognized as a significant example of a "Global Heritage Stone Resource": its exploitation from quarry to building (the Duomo di Milano) well represents the close correlation between stone and cultural heritage, between georesources and humankind development.

Keywords: Candoglia marble; Duomo di Milano; petrographic analysis; geoheritage; cultural stone

---

1. Introduction

Italy has been the cradle of several cultures and architectonic styles, from the Roman period to the present. This has left tangible traces, such as historical buildings, sculptures, etc., many of which are made of a multitude of ornamental stones, frequently sourced from the quarries present in the area. This cultural heritage represents the tangible signs of the historical and cultural wealth of an era in specific territories. A clear example of this "stone culture" is represented by the wide utilization of white marble (e.g., the Luni marble, now recognized as Carrara marble) in both *statuaria* and architecture since the Greek and Roman times [1].

Marble varieties from the Italian Western Alps, even though they are not as famous as the renowned Carrara varieties, have been largely employed for sculptures and buildings construction.

The Piemonte region has been and still is strongly interested by intensive quarrying activity: exploited stones were and are largely employed in cultural heritage, from the Roman times (e.i. Arc of Augustus in Susa (9th BC.) [2]) to the late Baroque period (e.i. Savoy architecture in Turin [3]). Nevertheless, few publications report on the petrographic and geochemical characteristics from an archaeometric point of view; moreover, often, the Piemonte varieties are not included in the provenance databases of classical marble [4–7].

Generally, the stones used for ornamental purposes are documented only via historical documents or macroscopic data [8]. For a few cases, for example for the Arc of Augustus or the churches of St. Cristina and St. Filippo Neri in Turin, a deeper scientific analysis was carried out on marble from the Western Alps for stone provenance definition [9,10]. More recently, a minero-petrographic and isotopic overview of the main Piemonte marble varieties has been reported [11]. Thus, knowledge on stone resources (mineral-petrographic characteristics, past and present quarrying and working techniques, applications in local, national and international buildings and/or architectures, etc.) is fundamental to emphasize the historical and cultural relevance of such georesources, highlighting the importance of an economic activity, which is fundamental for the cultures and customs of the different heritages that have characterized and strengthened the Mediterranean area through the centuries.

It is fundamental to share knowledge on these historical and cultural stones among a wider audience rather than storing this knowledge and disseminating it only among researchers and experts, so as not only to improve general knowledge on natural resources but also to enhance the self-consciousness of the deep connection between environment and exploitation, balancing both sustainability and cultural heritage. The aim of this paper is to disseminate knowledge on Candoglia marble, a pink, coarse-grained marble, locally known from the late Middle Ages (with renowned applications for the Madonna di Campagna Church at Verbania and the San Giovanni Battista Church at Montorfano, VCO Province, Northern Piemonte), and made famous thanks to its use in the Veneranda Fabrica del Duomo di Milano in the fourteenth century. It was also used for the facades of different monuments such as the Certosa di Pavia (Carthusian monastery), the Cappella Colleoni (Bergamo) and the San Petronio Cathedral (Bologna); moreover, in the past, it was used for altars and columns in Roman buildings (e.g., two columns in the Chiostro Arcivescovile at Novara, Piemonte) [12].

The study of the investigated material was carried out using different analytical techniques, both traditional (such as optical microscopy and stable isotope analysis) and more modern (Scanning Electron Microscope (SEM), equipped with an Energy Dispersion Spectrometry (EDS) analytical facilities) such approach rarely has been applied, to date, on Candoglia marble.

Furthermore, the development of quarrying and working techniques, together with some information about the Veneranda Fabbrica del Duomo di Milano, are presented in this paper.

2. Geological Setting

Palaeozoic marbles include the Ornavasso and Candoglia ones which pertain to the Verbano–Cusio Ossola (VCO) quarrying basin (Southalpine geologic domain).

In Piedmont territory, Southalpine domain is represented by deep and intermediate continental crust units, denominated Ivrea-Verbano zone and "Serie dei Laghi". They are separated by tectonic lineaments of Cossato-Mergozzo-Brissago. Ivrea-Verbano zone is mainly composed of Permian age igneous bodies ([13], with refs.). One of the igneous body is represented by the renowned Ivrea gabbro batholith, which was put in place at the base of a thinner gneissic crust (Kinzigitic complex; [14]). The "Serie dei Laghi", which crops out in S-E Ivrea-Verbano, consists of upper-intermediate tectonic unit which suffered an earlier metamorphic event in the Ordovician age.

Lenses of marble (up to 30m thick) are inserted within the amphibolite-granulite paragneisses characterising the Ivrea Zone. The Ornavasso and Candoglia marbles are quarried in the two described areas in which the lens of marble outcrops (on the right and left sides respectively of the Toce River–Ossola Valley), few kilometres west of the mouth of the river in the Lake Maggiore (Figure 1b).

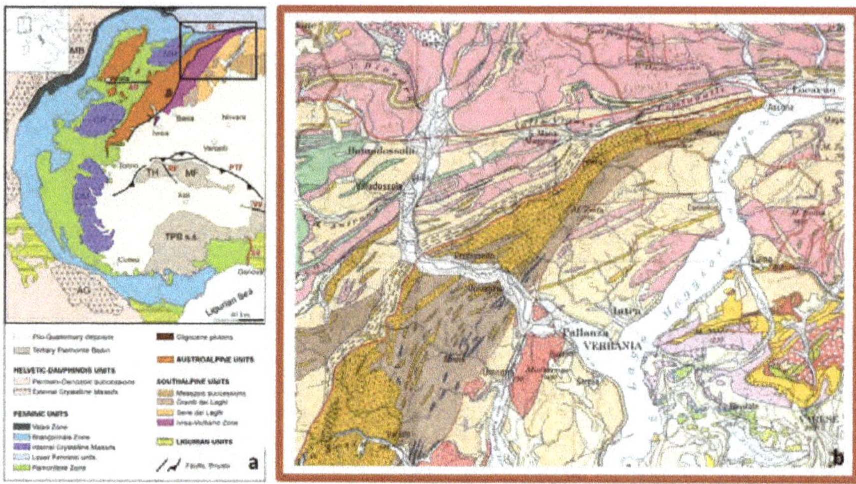

Figure 1. (**a**) Tectonic sketch-map of the Western Alps. Tectonic lines: IL = Insubric Line; SL = Simplon Line; SV = Sestri Voltaggio Line, RF = Rio Freddo Line, VV = Villalvernia-Varzi Line; AR: Aosta Ranzola Fault; PTF: Apennine Front Thrust. MB = Monte Bianco; MR = Monte Rosa; GP = Gran Paradiso; DM = Dora Maira; AG = Argentera; TH = Torino Hill; MF = Monferrato; TPB = Tertiary Piedmont Basin. The square correspond to the geologic map reported in Figure 1b. (**b**) Geological map of the lower Ossola Valley (from Domodossola to Verbania, on the western branch of Lago Maggiore) [15]. Marble of the Ivrea Zone are reported as North-North East (NNE) South-South West (SSW) trending, thin lenses (given in blue colour) within the high-grade paragneiss (given in brown).

3. Candoglia Marble Characterisation: Materials and Methods

3.1. Petrographic and Geochemical Characterisation

Petrographic analysis by optical microscope and SEM, together with the minero-chemical analysis of the main and accessory minerals by EDS microanalysis and the micro X-ray florescence determination of trace elements were used to characterise six samples representative of two historical quarries (Table 1).

Table 1. Candoglia marble samples localisation.

Sample	Quarry Name
canAB1	Cava Madre underground pit
canAB2	Cava Madre underground pit
canAB3	Cava Madre underground pit
canAB4	Cava Corte Nuova open pit
canAB5	Cava Corte Nuova open pit
canAB6	Cava Corte Nuova oper pit

Petrographic analyses were carried out on polished thin sections, using a Cambridge S360 scanning electron microscope, according to the analytical conditions reported in [16]. In particular, the working distance was 25 mm; the probe current was 200 pA; the accelerating potential was 15 kV; the counting time resulted of 60 s. All the analyses were recalculated with MINSORT program by Petrakakis and Dietrich [17].

Micro-XRF Eagle III-XPL (Röntgen analytic Messtechnik GmbH, Butzbach, Germany) was used for trace elements analysis of calcite and dolomite in rock samples. Analytical condition are reported in [18]. The Candoglia marble archaeometric characterization has been carried out using stable isotopic data reported by Antonelli and Lazzarini [19].

3.2. Physico-Mechanical Characterisation

To characterise a stone, it is fundamental, together with the minero-petrographic characterisation, to know its physico-mechanical behaviour. Data about physico-mechanical characterization (such as apparent density, water absorption, uniaxial compressive strength, compressive strength after freeze and thaw, indirect tensile strength and impact strength test) have been collected from published papers and official web sites. The legislation followed to carry out the different tests is reported in the consulted publications [20–22].

4. Candoglia Marble Characterization: Results

4.1. Petrographic and Geochemical Characterisation

Candoglia marble shows a characteristic pink to gray colour and a coarse-grained texture (>3 mm); furthermore it is characterised by the 80–85% $CaCO_3$ and the 15–20% other minerals, concentrated along discrete layers. The greyish aspect is due to diopside and tremolite (centimetre-thick) (Figure 2a); minor minerals include epidote, quartz, barite, Ba-feldspar, sulphides and, rarely, phlogopite. It shows a heteroblastic texture with a sutured grain boundary shape of the carbonate crystals (Figure 2b). The structure is isotropic as no preferred crystallographic orientation of calcite grains has been evidenced. The average grain size is 0.28 ± 0.1 mm and the high standard deviation is linked to the heteroblastic microstructure, which results in a maximum grain size (MGS) showing a median value of 3.3 mm (from 2.6 to 4.9 mm). Such a coarse grain characteristic, exceeded by varieties of Mediterranean marble only from Naxos marble and Brossasco marble [11,19], did not prevent this marble from being widely used in architecture and statuary. The marbles coming from VCO quarrying basin are also characterized by values of oxygen more negative than other marble of the Western Alps. For Candoglia marble, a range between −7 and −13 $^{18}\delta O$ is reported by [19]. This feature is probably due to the high metamorphic grade reached by the Ivrea Zone, the geological unit to which Candoglia marble pertains.

Figure 2. (**a**) Macroscopic appearance of marble: the presence of dark-greenish silicate layers (centimetre-thick) which highlight the suffered ductile deformation is clearly visible. (**b**) Photomicrograph of marble (can AB2), showing the heteroblastic texture, with the sutured grain boundary shape of the carbonatic crystals. The structure is isotropic as no preferred crystallographic orientation of calcite grains has been evidenced. Optical microscope crossed nicols.

Candoglia marble also shows distinctive minor and trace element contents, which help in characterizing and distinguishing it from all the other Mediterranean marbles. The manganese and iron contents are respectively 119 ± 3 and 186 ± 4 ppm, reflecting the chromatic features of the marble. Despite remarkable macroscopic heterogeneities, the Fe and Mn concentrations do not show significant variations within all the analysed areas. The Sr concentration results in a mean of 212 ± 30 ppm, which is significantly higher than most of the investigated Mediterranean varieties. Quantitative results, referring to micro-X-Ray Fluorescence (XRF) calcite trace element concentration, are reported in Table 2.

Table 2. Geochemical and minero-petrographic variables characterizing Candoglia marble reference samples. Micro-XRF trace element composition of calcite has been recalculated based on 43.97% CO_2. AGS and MGS refer to average and maximum grain size, respectively, while the microstructural and textural features are expressed as categorical variables. Ho/He refers to homeoblastic vs. heteroblastic microstructure; Iso/Aniso to isotropic vs. anisotropic microstructure. The grain boundary shape (GBS) is, in this, case sutured (Sut).

Sample	Mg (%)	Ti (ppm)	Mn (ppm)	Fe (ppm)	Zn (ppm)	Sr (ppm)	AGS	MGS	Ho/He	Fabric	GBS
canAB1	0.418	2	119	179	4	228	0.29	3.51	He/Ho	Iso	Sut
canAB2	0.296	2	119	189	5	216	0.32	3.12	He/Ho	Iso	Sut
canAB3	0.378	2	117	189	7	241	0.20	3.09	He/Ho	Iso	Sut
canAB4	0.393	2	116	187	3	186	0.39	3.39	He/Ho	Iso	Sut
canAB5	0.417	2	120	187	2	202	0.46	3.47	He/Ho	Iso	Sut
canAB6	0.489	2	122	187	2	202	0.34	3.16	He/Ho	Iso	Sut

The peculiar pink colour, the coarse grain size and the isotopic value of 18δO are all archaeometric characteristics that make it easy to identify this marble from any other Mediterranean marble used in local and international cultural heritage.

4.2. Physico-Mechanical Characterization

Table 3 reports the more common physico-mechanical characteristics tested for Candoglia marble, compared to the famous Carrara white marble. The characteristics of both are comparable except for three values: the water absorption (higher in Candoglia marble), compressive strength after freeze and thaw (lower in Candoglia marble) and impact strength (lower in Candoglia marble). The first two characteristics are linked to the feature of the rocks: coarse grain size in Candoglia marble and finer grain size in Carrara marble.

Table 3. Physico-mechanical characterisation of Candoglia marble.

Characteristics	Unit	Candoglia Marble	Carrara Marble ***
Apparent density	kg/m^3	2620–2830 *	2688
Water absorption	%	0.8 **	0.16
Uniaxial compressive strength	MPa	44–155 *	118.6
Compressive strength after freeze and thaw	MPa	95 **	115.8
Indirect tensile strength	MPa	1.8–12.3 *	17
Impact strength test	cm	38 **	73.8

* data from [20]; ** data from [21]; *** data from [22].

5. Candoglia Marble Exploitation and Processing

The first traces of the exploitation of marble outcrops in the Ornavasso territory belongs to the Roman period; the marble presents in Ornavasso pertains to the same marble lenses of Candoglia marble (Figure 1) and is locally recognized as the Candoglia "bastard brother". Both the Candoglia and Ornavasso marbles have been widely employed in local buildings and infrastructures; the more noble variety, Candoglia marble, is renowned for its applications in the Milan Cathedral: in the 1387 Gian Galeazzo Visconti obtained the authorisation to exploit the Candoglia Marble. From that time, Candoglia marble quarries have been quarried by the Veneranda Fabbrica del Duomo of Milano and the exploited material, nearly 1.000 t/year, has been and still is employed uniquely for building and maintenance of the Cathedral [23,24].

The quarrying activity evolved over the centuries: from small open pits, near Candoglia village, to the large underground quarry (Cava Madre) (Figure 3; Figure 4). Cava Madre is characterized by homogeneous materials, exploited in subvertical benches (Figure 5), first using sledgehammers, *punciotti* (plug and feather), chisels and gunpowder (Figure 6a) and later using heilcoidal wires (Figure 6b) and diamond wires [12]. The bench is interested by side forces, that need to be constantly monitored and contained [20,25–27]. The activities in the quarry yard were characterized by initial selection of the best blocks, chosen on the basis of the aesthetical, physical, and lithological characteristics.

Figure 3. (**a**) View (and sketch) of the Candoglia quarries (from San Nicola church in Ornavasso): 1. Cava Madre underground pit; 2–4. Cava Corte Nuova open pits, active up to the early XX cent. (**b**) View of the Candoglia open pit quarries and harbour (from Ornavasso side). Late XIX cent. (Veneranda Fabrica Historical Archive).

Figure 4. (**a**) Cava Madre quarry profile (1814) (Veneranda Fabrica Historical Archive). (**b**) Cava Madre entrance: view from San Nicola Church bell tower (Ornavasso).

Figure 5. (**a**) Entrance of Cava Madre underground quarry. (**b**) Cava Madre underground pit. XX century. (Veneranda Fabrica Historical Archive).

Figure 6. (**a**) Quarry worker with a sledgehammer, *punciotti* (plug and feather), and chisels. (**b**) Candoglia open pit yard: helicoidal wire. (Veneranda Fabrica Historical Archive).

The selected blocks were squared directly in the quarry yard: a first rough sketch of the sculpture to be produced was arranged in the yard, based on sketch drawings of blocks and sculptures (Figure 7), before being transported to the working plants to be chiselled and refined. The less valuable blocks (not aesthetically good) were sold as a secondary category material, which was crushed and used to produce lime (traces of kilns are still visible in region Calmatta, near Ornavasso Village) [28].

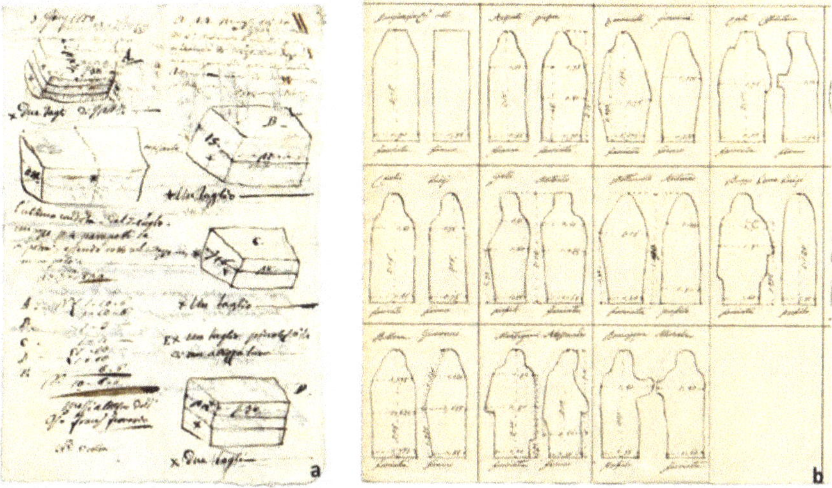

Figure 7. (**a**) Sketch drawings of blocks, with defects. XVIII Cent. (**b**) Sketch of sculptures, each assigned to a single stone worker. (Veneranda Fabrica Historical Archive).

After being quarried and squared, the blocks were transported from the quarry to the manufacturing plants on running routes, thanks to the employment of large wooden sledges—"struse" (Figure 8). More recently (from the 1920), the "binda" (jack) or the derrick were used to move the blocks. The blocks were then transported to the river docks by means of wagons towed by pairs of oxen (Figure 9a). From the River Toce docks, the marble was transported to small manufacturing plants (Figure 9b), in Suna and Baveno (along Lake Maggiore); in these laboratories, all the locally exploited blocks were squared, cut and worked. It has to be highlighted that the first mechanized saw for blocks was built for the processing of Candoglia marble.

Figure 8. Running routes made of quarry waste (left) and big wooden sledges—"struse"—to transport the blocks from the quarry to the valley (right) [29].

Figure 9. (**a**) Wagons towed by pairs of oxen which transported the blocks from the valley to the river pier. (**b**) Veneranda Fabbrica working yard in Milan (XIX–XX cent.) [29].

After the working phase, the blocks were transported, using big barges, from the Candoglia pier (Figure 10a) to the Navigli in Milan (Figure 10b), up to the Duomo di Milano working yard: the path was through the River Toce, the Lake Maggiore and the River Ticino. The presence of rivers (Toce and Ticino) and the construction of canals (Navigli) guaranteed the direct and easy transport from the quarry area to the Cathedral yard. Thus, Candoglia marble was preferred to other more famous Italian marble (e.g. Carrara marble) even if its minero-petrographic features (coarse grain-size) were not the most suitable to use for statuary applications. It has to be highlighted that until mid-XV century, approximately 400 people were employed as stone workers for the Duomo construction, with another 400 people as transport workers [29].

Figure 10. (**a**) Candoglia pier (late XIX–XX century). (**b**) Navigli di Milano during the Duomo di Milano construction activities (portrait). (Veneranda Fabrica Historical Archive).

Nowadays, all the quarried material is exploited and used in different applications: from the best quality (used to produce sculptures and masterpieces for the maintenance of the Cathedral) to irregular or not aesthetically suitable blocks (employed as armour stones).

The working activities are run out in Candoglia and Milan. In particular, the first block squaring by means of a diamond-wire block cutter is set in the working area close to the Cava Madre quarry.

The squared blocks are sent to the main working plant near Milan to be mechanically worked for the production of a first rough sketch of the sculpture. After this rough working phase, the blocks are sent back to the Candoglia mason stone cutter laboratory (near the offices of the Venaranda Fabrica del Duomo di Milano) to be chiselled by expert stone workers, to obtain the copy of the pieces to substitute (Figure 11).

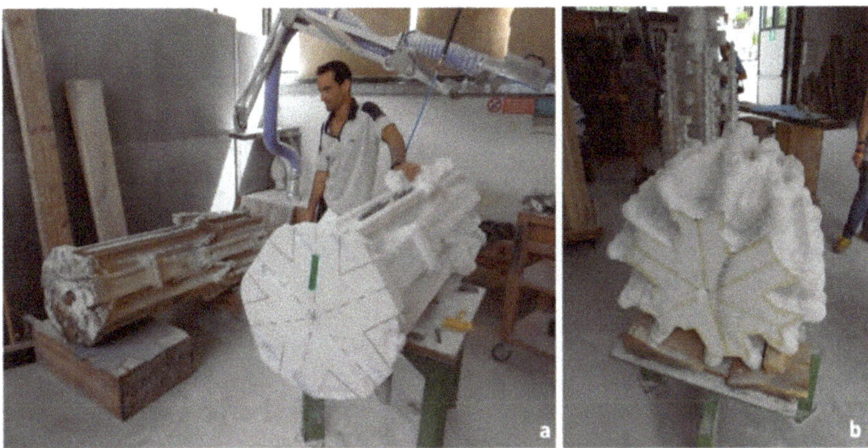

Figure 11. (**a**) Rough working sketch of the sculpture. (**b**) Chiselled sculpture to substitute the original one.

6. The Duomo di Milano and the Museum of the Venaranda Fabbrica del Duomo di Milano

The Milan Cathedral (Figure 12) is the seat of the Archbishop of Milan and is named to St Mary of the Nativity (Santa Maria Nascente). It could be considered as a "still alive" yard; indeed, Duomo di Milano has been completely built but its maintenance continues (the overlapping of maintenance with construction started in the mid-XIX century, when a large piece of the main spire fell onto the terraces below). The Cathedral is one of the most important and famous worldwide known Catholic church (5th place in term of size) and in Italy it comes after the Basilica of San Pietro, in Rome.

Figure 12. Duomo di Milano: general view from the main square and sketch of sculptures and steeples (different colours due to the different maintenance phases have to be highlighted).

The plan consists of a central nave with four side aisles, crossed by a trampset; opposite to the entrance it is possible to admire the choir and the apse. The main nave shows a 45 m height; it represents the he highest Gothic vaults of a completed church. Tourists can reach the roof, from which it is possible to appreciate some spectacular sculptures that otherwise would not be enjoyed.

The Museum of the Veneranda Fabbrica (Figure 13a) hosts several precious sculptures made of Candoglia marble, such as a Saint Ambrogio statue (Figure 13b).

Figure 13. (**a**) Museum of the Veneranda Fabbrica del Duomo di Milano: located at Sforza palace, South of the Duomo. (**b**) Sant Ambrogio—Milan patron saint—sculpture. (**c**) sculptures stored in the museum.

7. Conclusions

Candoglia marble represents an example of heritage stone valorisation for the next generation: a unique and priceless material, a witness of the culture change of a wide territory in Piemonte and Lombardia (the quarrying area in Candoglia and the Milan cathedral). Indeed, thanks to its unique characteristics (physical, petrological and mineralogical) and to the peculiar history of its exploitation during the centuries for the construction and maintenance of an unique heritage building, Candoglia marble is worldwide well-known and appreciated.

Candoglia marble exploitation has influenced the culture and industrial development of an entire area over the centuries. Candoglia has to be deemed as a place where humankind and nature interact. Indeed, visiting the Candoglia area, the audience is fascinated by the immeasurable twist of natural landscape and human activities, which are tangible in:

- the need to discover and exploit a unique local resource, Candoglia marble, aesthetically suitable for the realisation and the maintenance of a worldwide known cultural and religious building(the Duomo di Milano);
- the presence of working and historical quarries which developed over the centuries from open pits to underground yard that reach into the heart of the mountain and exploit the most prospective part of the formation. Furthermore, the need to preserve the resource has brought about an upgrade of the quarrying technologies: from open pits exploited with explosives to underground pit yards, exploited with more modern techniques(the helicoidal and then the diamond wire machines);
- the continuing upgrade of working techniques, from local working activities (in Candoglia, Suna and Baveno) to the more recent activities managed both in Candoglia (in the quarry yard and the mason stone cutter laboratory) and Milan;
- the presence of historical means of transport, such as the "vie di lizza", the wagons being towed by pairs of oxen, the big barges, developed utilizing more modern transport by truck.

All these natural and human features of the area are fundamental to indicate the Candoglia marble area as geoheritage, with the consequence that Candoglia marble will be indicated for its designation as "Global Heritage Stone Resource".

A geotourist path "from quarry to building" can be organised, starting from the Candoglia quarry area (including both the still active Cava Madre and some yet accessible historic open pits) to the laboratory, and finally visiting the Duomo di Milano. It is also possible to consider a geotourist path "from building back to quarry", starting from the more famous Duomo di Milano, directing people to discover where everything started: an example of this approach is the path from the famous Arena of Arles back to the ancient quarries, exploited for the production of the employed stones.

Author Contributions: Conceptualization, G.A.D., A.B., D.C. and B.C.; Data curation, G.A.D., A.B., D.C., F.C. and E.C.; Funding acquisition, A.B. and D.C.; Investigation, G.A.D., A.B., D.C., F.C. and E.C.; Methodology, G.A.D., A.B. and B.C.; Supervision, G.A.D., A.B. and D.C.; Validation, G.A.D., D.C. and F.C.; Writing—original draft, G.A.D. and A.B.; Writing—review and editing, G.A.D., A.B., D.C. and F.C.

Funding: This research was funded by Compagnia San Paolo and University of Torino in the frame of the GeoDIVE Project, by MIUR (Italy) and CNR-IGG U.O of Torino (Italy).

Acknowledgments: Vittorio Barella and Enrico Allais (ISO4—Torino) are thanked for isotope analysis.

Conflicts of Interest: The authors declare no conflict of interest.

References

1. Primavori, P. Carrara Marble: A nomination for 'Global Heritage Stone Resource' from Italy. *Geol. Soc. Lond. Spéc. Publ.* **2015**, *407*, 137–154. [CrossRef]

2. Fiora, L.; Audagnotti, S. *Il Marmo di Chianocco e Foresto: Caratterizzazione minero–Petrografica ed Utilizzi*; Atti 1° Congresso nazionale di Archeometria Patron Editore: Bologna, Italy, 2001; pp. 235–246.

3. Borghi, A.; D'atri, A.; Martire, L.; Castelli, D.; Costa, E.; Dino, G.; Favero Longo, S.E.; Ferrando, S.; Gallo, L.M.; Giardino, M.; et al. Fragments of the Western Alpine chain as historic ornamental stones in Turin (Italy): A new geotouristic approach for the enhancement of urban geological heritage. *Geoheritage* **2014**, *6*, 41–55. [CrossRef]

4. Attanasio, D.; Brilli, M.; Ogle, N. *The Isotopic Signature of Classical Marbles*; "L'Erma" di Bretschneider: Roma, Italy, 2006; p. 297.

5. Gorgoni, C.; Lazzarini, L.; Pallante, P.; Turi, B. An Updated and Detailed Mineropetrographic and C-O Stable Isotopic Reference Database for the Main Mediterranean Marbles Used in Antiquity. In *ASMOSIA V: Interdisciplinary Studies on Ancient Stone*; Hermann, J.J., Herz, N., Newton, R., Eds.; Archetype Publications: London, UK, 2002; pp. 115–131.

6. Moens, L.; Roos, P.; De Rudder, J.; De Paepe, P.; Van Hende, J.; Marechal, R.; Waelkens, M. A multi-method approach to the identification of white marbles used in antique artifacts. In *ASMOSIA I: Classical Marble: Geochemistry, Technology, Trade*; Herz, N., Waelkensb, M., Eds.; Kluwer Publishing Company: Dordrecht, The Netherlands, 1988; pp. 243–250.

7. Belfiore, C.M.; Ricca, M.; La Russa, M.F.; Ruffolo, S.A.; Galli, G.; Barca, D.; Malagodi, M.; Vallefuoco, M.; Sprovieri, M.; Pezzino, A. Provenance study of building and statuary marbles from the Roman archaeological site of Villa dei Quintili (Rome, Italy). *Ital. J. Geosci.* **2016**, *135*, 236–249. [CrossRef]

8. Frisa Morandini, A.; Gomez Serito, M. Indagini Sulla Provenienza dei Materiali Lapidei Usati Nell'architettura e Nella Scultura di Epoca Romana. In *Archeologia in Piemonte, II: L'età Romana*; Mercando, L., Ed.; Allemandi: Torino, Italy, 1998; pp. 223–233.

9. Chiari, G.; Colombo, F.; Compagnoni, R.; Fiora, L.; Rava, A.; Di Macco, M.; Ormezzano, F. Santa Cristina: Restoration of a Façade by Filippo Juvarr. In Proceedings of the International Congress on Deterioration and Conservation of Stone, Lisbon, Portugal, 15–18 June 1994; pp. 1393–1402.

10. Chiari, G.; Fiora, L.; Compagnoni, R. Studio dei materiali di facciata della Chiesa di S. Filippo in Torino. In *Atti Terzo Simposio Internazionale 'La conservazione dei monumenti nel bacino del Mediterraneo'*; Grafo Edizioni, a cura di Fulvio Zezza: Venezia, Italy, 1994; pp. 773–778.

11. Borghi, A.; Vaggelli, G.; Marcon, C.; Fiora, L. The Piedmont White Marbles Used in Antiquity: An Archaeometric Distinction Inferred by a Minero-Petrographic and C–O stable isotope study. *Archaeometry* **2009**, *51*, 913–931. [CrossRef]

12. Cavallo, A.; Bigioggero, B.; Colombo, A.; Tunesi, A. The Verbano Cusio Ossola province: A land of quarries in northern Italy (Piedmont). SPECIAL ISSUE 3: A showcase of the Italian research in applied petrology. *Per. Mineral.* **2004**, *73*, 197–210.

13. Quick, J.; Sinigoi, S.; Mayer, S. Emplacement dynamics of a large mafic intrusion in the lower crust, Ivrea–Verbano Zone. *J. Geophys. Res.* **1994**, *99*, 21559–21573. [CrossRef]

14. Peressini, G.; Quick, J.E.; Sinigoi, S.; Hofmann, A.W.; Fanning, C.M. Duration of a Large Mafic Intrusion and Heat Transfer in the Lower Crust: A SHRIMP U-Pb Zircon Study in the Ivrea-Verbano Zone (Western Alps, Italy). *J. Pet.* **2007**, *48*, 1185–1218. [CrossRef]

15. Carte géologique de la Suisse; scale 1:500.000. Schweizerische Geologischen Kommission, 1980. Available online: https://shop.swisstopo.admin.ch/fr/products/maps/geology/GK500/GK500_DIGITAL (accessed on 18 July 2019).

16. Vaggelli, G.; Sebar, L.E.; Borghi, A.; Cossio, R.; Re, A.; Lo Giudice, A. Improvements to the analytical protocol of lapis lazuli provenance: first study on Myanmar samples. *Eur. Phys. J. Plus* **2019**, *134*, 104. [CrossRef]

17. Petrakakis, K.; Dietrich, H. *MINSORT: A Program for the Processing and Archivation of Microprobe Analyses of Silicate and Oxide Minerals*; Neues Jahrbuchfür Mineralogie—Monatshefte: Stuttgart, Germany, 1985; Volume 8, pp. 379–384.

18. Vaggelli, G.; Cossio, R. μ-XRF analysis of glasses: A non-destructive utility for cultural heritage applications. *Analyst* **2012**, *137*, 662–667. [CrossRef] [PubMed]

19. Antonelli, F.; Lazzarini, L. An updated petrographic and isotopic reference database for white marbles used in antiquity. *RENDICONTI Lince* **2015**, *26*, 399–413. [CrossRef]

20. Morlin Visconti Castiglione, B.; Oggeri, C.; Oreste, P. Tecnica estrattiva e storia nella cava madre di Candoglia. In *Geologia e Turismo: Dall'affioramento al Costruit*; ISPRA—Istituto Superiore per la Protezione e la Ricerca Ambientale: Bologna, Italy, 2013; pp. 273–279.

21. Selecting Candoglia Marble. Available online: http://www.italithos.uniroma3.it/welcome.php (accessed on 14 February 2019).

22. Dino, G.A.; Chappino, C.; Rossetti, P.; Franchi, A.; Baccioli, G. Extractive waste management: A new territorial and industrial approach in Carrara quarry basin. *Ital. J. Eng. Geol. Environ. Spec. Issue 2* **2017**, 47–55.

23. Ferrari Da Passano, C. La Veneranda Fabbrica del Duomo di Milano. *Quarry-Laboratory-Monument. Int. Congr. Pavia Proceeding* **2000**, *1*, 23–34.

24. Canali, F. Il Marmo di Candoglia. In *Bollettino Italia Nostra 487*; Gangemi Editore: Rome, Italy, 2015; pp. 17–18.

25. Pelizza, S.; Oreste, P.; Peila, D.; Oggeri, C.; Oreste, P. Stability analysis of a large cavern in Italy for quarrying exploitation of a pink marble. *Tunn. Undergr. Space Technol.* **2000**, *15*, 421–435. [CrossRef]

26. Pelizza, S.; Oggeri, C.; Oreste, P.; Peila, D. The Candoglia quarry: The stability of the large room and the enlargement. In Proceedings of the 15th International Symposium on Mine and Planning Equipment Selection MPES 2006, Turin, Italy, 20–22 September 2006; Volume 1.

27. Pelizza, S.; Oggeri, C.; Oreste, P.; Peila, D.; Morlin Visconti Castiglione, B.; Corbetta, E. Una Grande Caverna Per L'estrazione di Pietra Ornamentale: Il Caso Della Cava Madre del Duomo di Milano a Candoglia. In *Gallerie e Grandi Opere Sotterranee*; Patron Editore: Granarolo dell'Emilia, Italy, 2010; Volume 95, pp. 23–30.

28. Dino, G.A.; Cavallo, A. Ornamental stones of the Verbano Cusio Ossola quarry district: Characterization of materials, quarrying techniques and history and relevance to local and national heritage. In *Global Heritage Stone: Towards International Recognition of Building and Ornamental Stones*; Pereira, D., Marker, B.R., Kramar, S., Cooper, B.J., Schouenborg, B.E., Eds.; Geological Society of London, Special Publications: London, UK, 2015; Volume 407, pp. 187–200.

29. Moschini, C. *Il Percorso dei Marmi: Dalle Cave di Candoglia ed Ornavasso al Duomo di Milano*; Skira Editore: Milan, Italy, 2005; p. 158.

MDPI

St. Alban-Anlage 66

4052 Basel

Switzerland

Tel. +41 61 683 77 34

Fax +41 61 302 89 18

www.mdpi.com

Sustainability Editorial Office

E-mail: sustainability@mdpi.com

www.mdpi.com/journal/sustainability